Environmental

D. Srinivasan

Former Professor
Department of Chemical Engineering
Algappa College of Technology
Anna University
Chennai

PHI Learning Private Limited
Delhi-110092
2015

₹ 350.00

ENVIRONMENTAL ENGINEERING
D. Srinivasan

ISBN-978-81-203-3600-1

The export rights of this book are vested solely with the publisher.

Fourth Printing ... ... **May, 2015**

Published by Asoke K. Ghosh, PHI Learning Private Limited, Rimjhim House, 111, Patparganj Industrial Estate, Delhi-110092 and Printed by Syndicate Binders, A-20, Hosiery Complex, Noida, Phase-II Extension, Noida-201305 (N.C.R. Delhi).

To

My late loving parents

Contents

Preface

During the last two decades, the environmental pollution regulations have undergone a vast change. Attempts have been made to refine the conventional technologies and to develop new technologies to meet increasingly more stringent environmental quality criteria. The challenge that one faces today is to meet these stringent requirements in both an environmentally acceptable and cost-effective manner.

While reviewing the existing methods of treatment of effluents, the present book addresses the application of the state of art technology to the solutions to today's problems in industrial effluent pollution control and environmental protection.

Industries are striving to minimize waste generation at the source, to reuse more of waste materials that are generated and to design processes which are environment-friendly. The overall objective is to minimize end-of-pipe treatment although some effluent treatments will always be needed.

Industry has started accepting the concept of pollution prevention because management has seen the economic benefits resulting from it. A knowledge of pollution prevention principles should allow the engineer to include environmental consequences in decision processes in the same way that economic and safety factors are considered.

The objective of this book is to present the principles of pollution prevention of environmentally important products coming out of processes and manufacturing systems.

Organization of the Book

In Chapter 1, a brief introduction to the environment, the interaction between humans and environment and the role of an environmental engineer are presented. It also includes the information on the contribution of the environmental engineer in the methodology for making pollution control decisions. Chapter 2 deals with the biogeochemical cycles. The chemistry of waste waters and the waste water treatment systems is dealt with in Chapter 3. Chapter 4 is devoted to the presentation of water quality which consists of the physical water quality parameters and the chemical water quality parameters including the chemicals and metals found in water. A discussion on the engineered systems for liquid effluent treatment and its disposal is given in Chapter 5. Chapter 6 deals with air quality, air quality management concepts, sources of air pollution and the pollutants present in air. Chapter 7 covers the engineered systems for air pollution control (both control devices for particulate contaminants and treatment of gaseous effluents). It also includes the economics of particulate emissions, control summary of particulate matter, the factors to be taken into account in the process of selection of particulate control equipment and a brief discussion on the atmospheric sampling and analysis and the measurement methods for ambient air quality parameters. In Chapter 8, pollution problems of a few chemical and processing industries are presented. A brief discussion on the location of industries based on environmental consideration is also discussed. Chapter 9 deals with the sources and disposal of solid waste and the engineered systems for solid waste management. In Chapters 10 through 12, the salient features of process design, process modification and other important and effective methods and techniques for waste minimization, and the strategies for pollution prevention are explained. In Chapter 13, the material on hazardous waste management is included which encompasses some details about all relevant aspects in the field. Finally, case studies on environmental pollution problems and their prevention are provided at the end of the chapters.

D. Srinivasan

Acknowledgements

I am greatly indebted to my wife S. Saroja for her love, patience and unstinted support during my work on this book.

A major portion of this book was written during my stay in the US last year with my children. I greatly appreciate the steadfastness of my son, Srinath; daughters, Vasumathi and Priya; daughter-in-law, Shyamala; sons-in-law, Narayanan and Raghav in extending support at various stages of this work. I acknowledge their readiness in providing me considerable computer logistics support.

I am delighted to include the names of my grandchildren Varun, Varshini, Aarthi, Rohan, Archana and Sanjay for their co-operation which was conducive for me to focus my attention and efforts on this task.

Finally, I would like to thank all my well-wishers and former colleagues who have been expressing their appreciation of my endeavours of this kind.

I invite the readers of this book for their valuable feedback which may help in updating this book to the great extent.

D. Srinivasan

1 Introduction

Our environment comprises atmosphere, earth, water and space. The interaction between the various strata of the atmosphere has been continuing for years. The composition and nature of environment have been undergoing continuous changes owing to activities like industrialization, transportation, construction, etc. The natural environment is clean, but owing to the aforementioned activities of man, it becomes polluted giving rise to what is termed as environmental pollution.

1.1 THE ENVIRONMENT

The environment can be defined as one's surroundings. From the environmental engineer's perspective, the word environment may assume global dimensions, may refer or connote to a very localized area in which a specific or particular problem must be addressed or may, in the case of contained environments, refer to a small volume of liquid, gaseous or solid materials within a treatment plant sector.

The global environment comprises the atmosphere, the hydrosphere and the lithosphere in which life-sustaining resources of the earth are contained or accommodated. The atmosphere is a mixture of gases extending outward from the surface of the earth, evolved from the elements of the earth that were gasified during the formation and metamorphosis with the latter implying change of form or structure. The hydrosphere comprises the oceans, the lakes and streams and the shallow groundwater bodies that interflow within the surface water. The lithosphere is the soil mantle that wraps the core of the earth.

The biosphere, a thin shell that surrounds the earth, is made up of the atmosphere and the lithosphere adjacent to the surface of the earth, together with the hydrosphere. It is within the biosphere, that the life forms of earth including humans are existing. Life–sustaining materials in gaseous, liquid and solid forms are cycled through the biosphere, providing sustenance to all living organisms.

Life-sustaining resources—air, food and water are withdrawn from the biosphere. Waste products in gaseous, liquid and solid forms are discharged into the biosphere. From the beginning of time, the biosphere has got and assimilated the wastes generated by plant and animal life. Natural systems have been ever functional and carrying out processes such as dispersing smoke from forest fires, diluting animal wastes washed into streams and rivers and converting debris of the past generations of plant and animal life into soil, rich enough to support increase in population.

For every natural act of contamination, for every undesirable alteration on the physical, chemical or biological characteristics of the environment, for every incident that caused deterioration in the quality of the immediate or local environment, there were natural actions that restored that quality. Only in recent years, it has become evident or apparent that the sustaining and assimilative capacity of the biosphere, though tremendous, is not after all infinite. Though the system has operated for millions of years, it has started showing signs of stress, primarily owing to the impact of human activities upon the environment.

1.2 THE INTERACTION BETWEEN HUMANS AND ENVIRONMENT

Since early history, the earth was essentially a closed system and materials were recycled and reused in this closed system. The basic links in nature's closed materials systems are (1) the energy source—sunlight (2) the earth's non-living resources—oxygen, water, etc. (3) the plants, from planktons to fully grown trees that generate carbohydrates via photosynthesis (4) the consumers that feed on plant products or other consumers, and (5) the decomposers—bacteria, fungi and insects that close the cycle by breaking down the dead producers and consumers and send back the chemical compounds to the earth's pool of resources.

Over a very long period of time man has been on earth, he has both been changed by and has brought about changes in the earth's environment. Until very recently, however, these changes have represented only a small perturbation in a large system and as a matter of fact, have not threatened the self-healing balance of nature's closed system. At the root of our growing environmental crisis is the fact that man has now the capability to make large enough perturbations in the total earth's system which will upset the present balance of nature.

The reason why man is capable of creating such large perturbations is his control of energy. Man alone, on earth, is capable of harnessing energy other than that got through the conversion of food and air inside its body. This event dates back to 300,000 to 400,000 years ago when man first used fire. Until about 800 years ago, the sole source of man's auxiliary energy was his daily sunlight, in other words, solar energy. This was stored in wood, which was burned for heat or in the food, eaten by the beasts of burden which helped man in his work.

Then man discovered coal, a reserve of concentrated energy created over the last half billion of years of so when a small trickle of the sun's energy had been trapped and stored in the form of fossil fuels. The successful harnessing of sizeable amounts of this concentrated energy only a few hundred years ago, first as inputs to agriculture and then to industrial products, is the real source of man's wealth and also his present-day problems.

1.3 THE ROLE OF THE ENVIRONMENTAL ENGINEER

The definition of *environmental engineering* must be attached or tied to the definition of engineering itself. Engineering may be defined as 'the application under constraints of scientific principles to the planning, designing, construction and operation of structures, equipments and systems for the benefit of society'. If the tasks carried out by environmental engineers were analyzed, it would be observed that the engineers deal with structures, equipments and systems that are designed to protect, safeguard and enhance public health and well-being. For instance, the environmental engineer plans, designs, constructs and operates sewage treatment and industrial effluent treatment plants to prevent the pollution of receiving streams and rivers. In other words, these structures are constructed to protect and enhance water quality. Environmental engineers plan, design, construct and operate air pollution control equipments. The resulting cleaner air is conducive to people's good health and prevents the deterioration of materials through the harmful effects of air contamination. Such an equipment thus safeguards and also enhances public health and welfare.

It has to be indicated in this context that all these activities are carried out by environmental engineers under constraints which may involve shortage of funds, political pressure, inadequate spaces in which the structures are to be built, social and community considerations and similar factors that bring about a limitation on the freedom of design. Environmental engineering will now be defined as the application of engineering principles and concepts, under constraint, to the protection and enhancement of the quality of the environment and to enhancement and protection of public health and well-being.

Developing countries have been envisaging a rapid rate of industrialization to achieve a comfortable level of economic growth. However, uncontrolled growth of industries is likely to result in a severe deterioration in the quality of environment if precaution towards pollution from industrial wastes and chemicals is not taken in terms of pollution control and abatement measures. As pollutants enter air, water or soil, natural processes such as dilution, biological conversions and some of the atmospheric chemical reactions convert waste material to more acceptable forms and disperse them through a larger volume. Yet these natural processes can no longer perform the *cleanup* by themselves. The treatment techniques or facilities designed and developed by the environmental engineer are based on the principles of self-cleansing observed in nature, but the engineered processes amplify and optimize the operations observed in nature to handle larger volumes of pollutants and to treat them more rapidly. Engineers adopt the concept of natural mechanisms to engineered systems for the purposes of pollution control when they construct tall stacks to disperse and dilute air pollutants, design biological effluent treatment facilities for the removal of organics from waste water, employ chemicals to oxidize and precipitate out the iron and manganese in drinking water supplies or bury solid wastes in controlled landfill operations.

Occasionally, the environmental engineer may have to design with an idea of reversing or counteracting natural process. For instance, the containers used for the disposal of hazardous waste materials, such as toxic chemicals and radioactive substances must isolate those materials from the environment so as to prevent the onset of the natural, but highly desirable processes of dilution and dispersion.

A good understanding of natural and engineered purification processes calls for an understanding of the biological and chemical reactions involved in these processes. Thus, in addition to being knowledgeable in mathematical, physical and engineering sciences, the environmental engineer should have an expertise in the subject areas of chemistry, microbiology and a few other subject areas not usually emphasized in engineering curricula. As a matter of fact, an understanding of biological and chemical principles is as essential to the environmental engineer as the understanding of statics and strength of materials is to the structural engineer.

The environmental engineer's unique role is to establish a connection between biology and technology by applying all the techniques made available by modern engineering technology to the job of cleaning up the debris left in the wake of an indiscriminate and unplanned use of that technology. The delicate balance of our biosphere has been tampered with and the state in which we now find ourselves is a direct consequence of our having ignored the limits of the earth's ability to overcome heavy pollution loads and of our having been ignorant of the constraints imposed by the limited activity of the self-cleansing mechanisms of our biosphere.

Needless to mention a keen awareness of these natural constraints plays an important role in the work of environmental engineers. For instance, the laws of conservation of mass and energy prevent the destruction of pollutants and the engineer is bound or constrained by these limits. The principles of effluent treatment must therefore be to convert the objectionable materials to other less objectionable forms to disperse the pollutants so that their concentrations are minimal or to concentrate them for isolation from the environment.

In all situations the end products of the treatment of polluted air or water or of disposal of solid wastes must be compatible with the existing environmental resources and must never overburden the assimilative powers of hydrosphere, atmosphere or lithosphere. In structural engineering, the engineers can simply specify a larger or stronger beam to carry a heavier load. The environmental engineer, on the other hand, must accept the carrying capability of a stream, an air shed or a landmass because they cannot be altered.

The environmental engineer functioning within these constraints, employs all available technological tools to design efficient and effective control and treatment devices that are modelled after natural processes that have, so long, preserved our biosphere. Only by bringing technology into harmony with natural environment, the engineer harbours hope to achieve the goals and objectives of the profession—the protection of the environment from the potentially deleterious effects of human activity, the safeguarding of human populations from the effects of adverse environmental factors and the enhancement of environmental quality for human health and welfare.

1.4 CONTROL TECHNIQUES—GENERAL CONSIDERATIONS

The general considerations in respect of pollution control are given hereinafter.

1.4.1 Control by Dilution in the Atmosphere by Dispersion

The most positive way to combat air pollution is to prevent pollution from coming into being. Smoke stacks are employed to keep reduction of ground-level concentration of pollutants by providing natural atmospheric turbulence, an opportunity to dilute the pollutant before it gets to ground-level receptors in harmful concentrations. This technique controls emission to some extent and may help to get the desired air quality since the atmosphere has tremendous powers to dilute, disperse and even destroy a wide variety of substances that are discharged into it.

The effective approximate stack height is a minimum two to one-half times the height of the tallest building in the neighbourhood. This height should allow dispersion of the plume (the path and extent in the atmosphere of the gaseous effluent released from a source usually a stack) downward 5 to 10 times the height of the building. This effective height also depends on the wind factors, dilution, diffusion and prevailing temperature.

1.4.2 Control at Source

This could be achieved by keeping the pollutant from coming into existence or by destroying, altering or trapping it before it reaches the atmosphere.

Source relocation

This is one method of control at source. In the study of meteorological effects in community air zoning, it is, in certain situations, possible to determine a more satisfactory location for an industry that is causing unacceptable air pollution in its present location. By relocating away from heavily populated area and taking advantage of prevailing winds, an acceptable level of air pollution may be attained.

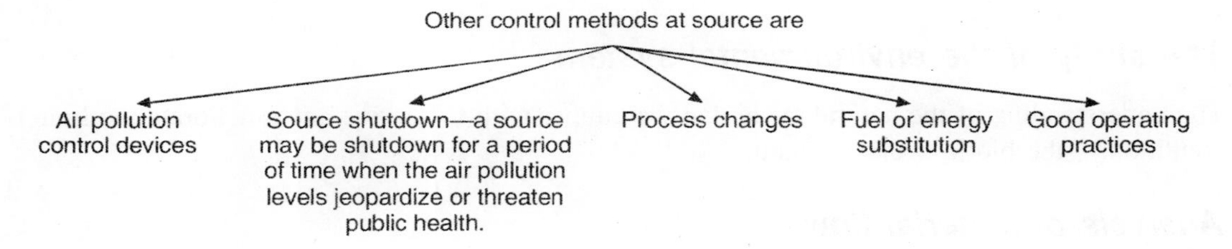

Fuel or energy substitution

This may be accomplished by replacing soft coal with hard coal, residual oil, distillate oil or natural gas. An even more drastic improvement would be obtained by replacing fossil fuels with hydraulic, electric, nuclear, solar or geothermal energy. Fuel can be treated before combustion by desulphurizing coal and fuel oil or by refining coal or natural gas to Liquefied Natural Gas (LNG) or LPG which is sulphur-free.

Process changes

A good example in this method is in steel industry. Open-hearth furnaces have been replaced with controlled basic oxygen furnaces or electric furnaces that emit less pollution into the atmosphere by reduction of smoke, carbon monoxide and metal fumes. Such alterations coupled with various combinations of gas-cleaning devices could be very effective in combating pollution.

1.4.3 Good Operating Practices

Irrespective of the type of equipment that is installed, the fuel burned or the raw material used, the operator can play a key role in abating air pollution from a given source. The possible sources of errors are:

1. Introduction of excess liquid sulphur in the burner at an H_2SO_4 plant without enough excess air—this may result in excessive emission of SO_2.
2. Excessive emissions of fly ash from a power plant due to the oversight of an operator by introducing too much excess air into the boiler furnace.

It is to be borne in mind that, the fuel industry, equipment manufacturers and government agencies have prepared guidelines for good operating practices.

1.5 STRATEGY OF POLLUTION CONTROL

A comprehensive strategy is more desirable for effective control of pollution menace rather than thinking of solutions (then and there) to problems in pollution control.

An effective pollution control strategy would include

The study of the environmental system

The understanding of the natural cycle that we tamper with to survive and ameliorate the living conditions—the biotic cycles of nature.

Analysis of material flow

To come to grips with any pollution problem, facts have to be gathered among the most important of which are estimates of the quantities and compositions of the offending streams. The material balance is perhaps the only most valuable tool we have at our disposal in assessing the magnitude of a pollution problem and the effects of proposed pollution abatement policies.

Examples Computation of say how much SO_2 pollutant must leave in the stack gas of a power plant and the second example is the assessment of the leakage of Hg from a chlor-alkali plant from operating data.

Analysis of energy flow

Without a good or sound understanding of certain basic principles of energy management, it may be impossible to predict the effects of our actions as our economic and environmental systems are interconnected by the flow of materials and energy.

Example Electricity is generated by the conversion of thermal energy (as one of the methods) which is accomplished by the combustion of coal. Principles of energy management dictate that only 1/3 of the useful energy in coal can appear as electricity and the remaining 2/3 must be dissipated into the environment as thermal pollution.

Strategic use of chemistry

A technique for transforming the pollutants into other less harmful materials by chemical reaction or by changing the waste into the secondary source of raw material.

Processing with living organisms

May be without our knowledge quietly and out of sight similar chemical reactions take place in association with the metabolism bacteria, fungi and other microorganisms. These organisms thrive in soil and water in the natural environment and our idea is to accelerate these natural reactions.

Materials' separations—technologists' domain

Possibility of separating materials is what is implied in many of the proposed pollution control systems.

The principle involved is that materials can be separated from each other only if they differ in some chemical or physical way.

The example is ammonia removal from water by stripping by the injection of a non-condensable gas, separation based on the differences in volatility or vapour pressure of the ammonia and water.

Systems integration

If the individual operations involved in the pollution control measures are integrated into a complete system, greater efficiency can be reached which leads to the development of an integrated pollution abatement system. The savings in operating cost and energy consumption will turn out to be significant and the improvement in system performance astonishing.

Example An extraction process effects transfer of a solute from the original solvent to a wash solvent and a distillation process is adopted to separate the wash solvent from the solute to be removed. Such an extraction-distillation system must be considered as an integrated rather than as separate operation.

Policy studies

Political and socio-economic considerations will also come in besides the technology of pollution control while handling environmental degradation problems.

- Here in comes the establishment of pollution control policies.
- Setting of legal standards on emissions.
- The societal decision is also a contributory factor while analyzing the pros and cons of any contemplated pollution abatement and prevention proposals.

1.6 METHODOLOGY FOR POLLUTION CONTROL DECISIONS

Analysis is an organized approach to decision-making, whereby alternate methods and courses of action are reviewed, prior to the decision-making. Analysis of potential from future facilities or equipments is normally investigated by the engineering department in an industrial organization. By adopting a flexible approach the pollution engineer can deal effectively with the complexities of environmental control. Then an analysis can be presented in an organized manner, examining the alternatives so that a decision can be made in respect of the type of pollution control system.

A plan to build a new facility or modify an existing one constitutes need for action. This need is communicated to the pollution engineering department for analysis. This problem is then divided into several work areas.

In analyzing the problem, the pollution engineer should include the three items given here, in his work.

Acquisition of information

Site selection surveys, searches of existing and proposed regulations, analysis of processes and study of techniques of controlling air, water and land pollution.

Processing of information

Data organized into a form that will enable management of the organization concerned to make a decision.

Display of information

Organized presentation of alternatives for a pollution abatement programme is desirable.

Information is acquired by thorough investigation of all sources which could in any way affect the use of the land, building, production processes or equipments. A pollution site survey should be made for a new plant and existing plants, prior to initiating a construction programme. Sites should be evaluated in terms of

Meteorology: Wind speed and direction, temperature variations, stability conditions and precipitation.

Air or water quality: Background levels.

Effects: Observable adverse effects, proximity to sensitive receptors.

Topography: The surface features of the region is its topography. A search of existing and proposed regulations should be conducted to find all pollution and land use restrictions which would affect the operation.

Also an analysis of the process should be made to determine number, location and physical characteristics of discharge points, character and rate of emissions. Then processes should be analyzed in terms of the pollution control systems and equipment necessary to comply with regulations.

Information affecting pollution abatement plans should be identified and evaluated. All information must be considered in the light of specific policy criteria established by the industrial organization under consideration. These requirements include the following:

Goals of the organization

No new facility will be planned without adopting pollution control measures.

Economics

Each pollution control device or abatement plan must be considered in relation to the profit-making capacity of the plant, including tax relief and possible by-product recovery.

Personnel required

The number and technical training of staff required to operate the environmental control systems must be taken into consideration.

Abatement planning must be a balance of the equities between control economics and desired environmental quality. It is during this process the alternative solutions and possible outcomes must be determined by the pollution engineer. Following the processing of information, data should be displayed to facilitate decision-making by management. The key to such a display is a visual presentation of the maximum number of alternatives available from which a final decision may be arrived at. By tabulating data in an orderly manner obvious conclusions can be made by simply comparing alternatives.

1.7 POLLUTION PREVENTION AROUND THE WORLD

The perennially increasing globalization of the chemical process industries is entailing an environmental awareness to various places of the world at a rate never anticipated in the past. Innumerable ecological calamities or disasters have made people realize that pollution prevention is necessary for humanity's survival. Trading nations have begun to understand that pollution problems cross international and geographical borders, and that they cannot be mortgaged through national legislation alone. International corporations are realizing that environmental stewardship everywhere they operate is necessary, for their long-term survival and prosperity.

It is absolutely necessary that an international corporation adopts progressive environmental practices aimed at pollution prevention in every facility it owns or operates throughout the world. The advantages or benefits include reduced environmental risks and costs, higher manufacturing efficiency and productivity and lower manufacturing cost, a positive public image and last but not the least, progress towards sustainable development. In the following section, some general strategies that a Chemical and Processing Industry (CPI) can adopt when it starts up, expands and diversifies or acquires overseas manufacturing operations.

1.7.1 Waste Water Discharge

An obvious and fundamental pollution problem at any location is waste water discharge. Most effluents contain sanitary waste generated by the employees at the industrial site and other streams like kitchen wastes as well as process waste waters. If the industry generates appreciable quantities of process waste water in excess of the sanitary waste volume, then it necessitates separate treatment of the process waste water. In such a circumstance, it is assumed that the sanitary waste water can be treated satisfactorily by the municipal waste water treatment system and that it is segregated from the process contaminants. In some instances partial treatments of the sanitary waste may be necessary at the site, especially if the local municipality has an inadequate treatment system.

The contaminants in industrial effluents that are commonly regulated include Total Suspended Solids (TSS), free oil and grease, pH, Biochemical Oxygen Demand (BOD) and certain heavy metals. End of pipe treatment facility shall be installed to control these basic pollutants to the same extent as normally required by the relevant national standards, even if the local regulations do not mandate such performance. The treatment should adequately address not only industrial discharges, but also the human waste and waste from any kitchen facilities on the premises.

1.7.2 Indoor Air Quality

Indoor air quality is hard to measure, mainly because human activities affect indoors' air quality issues to a much greater extent than those of outdoors particularly in small work areas. Therefore, facilities must plan, at the initial design stages, to install proper manufacturing process and equipment, ventilation devices and control measures to assure a safe indoor work environment that meets applicable national or international standards.

Once operations commence in an overseas plant, an annual industrial hygiene survey should be conducted if the facility uses any Volatile Organic Compounds (VOCs) or toxic substances that become airborne, if dust is visually apparent, or if there is a persistent odour. The permissible exposure limits and Threshold Limit Values (TLVs) applicable, should be used as a general guide to set the minimum standards for the most chemicals.

1.7.3 Outdoor Air Quality

Emissions from a manufacturing site not only contaminate the air in the immediate vicinity, but such pollutants can migrate without regard to territorial boundaries. Thus, any facility known to emit such airborne pollutants should go in for the installation of control devices to achieve a level of air emissions that complies with the standards applicable to a similar established and smooth-running plant.

For example, particulate control typically involves the use of baghouses with high efficiency particulate arrestor filter media. To control air pollutants emitted at high temperatures, after coolers and other design considerations to reduce discharge temperatures, should be employed. And volatile organic compounds may be controlled by such technologies as thermal or catalytic incineration, activated carbon adsorption, scrubbing, biofiltration and membrane separation.

1.7.4 Contaminants Entering with Raw Material

Companies, sometimes establish overseas operations to take advantages of low-cost raw materials as is often the situation with commodity chemicals and fertilizers manufactured in bulk and the processing of ores. The inclusion of very low or minute concentrations of toxic and/or hazardous pollutants in such large-scale operations may appear to be of little significance in relation to the bulk handling, storage and processing problems faced by a chemical processing industry/manufacturing facility. However, these contaminants may become concentrated or may turn out to be a significant factor in pollution as they pass through subsequent manufacturing steps.

A systematic inspection programme of the raw materials should be established. This programme should include a means of evaluating the dangers from contaminants that are of minimal concern to product quality, but environmentally harmful.

1.7.5 Onsite Analytical Services

It is advantageous for a chemical plant to have comprehensive data, including safety information on the chemicals entering the facility in order to manage both short-term operations and long-term risks. Onsite analytical capabilities, if available, enhance the knowledge of the operating personnel and also enable the development of cost-effective, industry-specific analyses to address quality, health and safety risks, environmental compliance and other related issues. Plants should compile record and make readily available data on chemicals entering the facility as would be required on material safety data sheets.

1.7.6 Spill Prevention

Chemical processing industry plants need to be concerned about both indoor spills and outdoor spills into bodies of water or the soil. Spills are often associated with poor quality storage facilities for the chemicals involved. The primary factors that go along with the storage problems include corrosion of the container walls due to material incompatibility and/or inadequate maintenance, the absence of a level or weight measuring device and high/low alarms and/or faulty control logic leading to vessel over filling; or lack of proper procedures and/or training of the operating personnel. Properly designed and instrumented storage systems and employee training can minimize release of harmful chemicals into the environment.

In case an outdoor storage vessel leaks or fails, the release of stored liquid is likely to affect the environment instantaneously. Unfortunately, outdoor releases are seldom noticed immediately. Thus, it is absolutely necessary that the every outdoor liquid storage tank be designed with containment provisions. Besides this, a spill prevention control and countermeasures plan should be developed for every outdoor liquid storage facility. This plan will keep to the minimum, the chances of a spill and outline damage control measures (in case the inevitable takes place).

1.7.7 Pollution Prevention Pays

This point is to be borne by every international organization. When the waste is low, manufacturing efficiency is high, so since pollution prevention reduces waste, it also increases manufacturing efficiency.

Owing to the fact that profits and pollution prevention success have direct correlation, a pollution prevention team comprising a group of plant personnel from diverse areas of responsibility (production, maintenance, engineering, management) should be chartered. This team should be assigned the job of developing a written pollution prevention plan, generating and helping in the implementation of the ideas that will bring about a reduction in wastes and manufacturing costs and eventually monitoring the progress of the process of pollution prevention.

1.7.8 A Competent and Dedicated Environmental Professional

Any industrial organization that takes up on overseas manufacturing venture with employees exceeding fifty, should have at least one full-time environmental staff person. It is the responsibility of this person to develop and implement the plant's environmental management system and for overseeing the daily as well as periodic activities necessary to carry out the facility's environmental policies and procedures. This should include regularly scheduled training for all levels of employees (including management) with good record-keeping of employee attendance. Such an environmental coordinator must be a person who has credibility in the organization and must be given adequate freedom to do his or her job.

1.7.9 The Need for an Environmental Advisory Group

If the organization's activities involve routine environmental issues and interfacing with customers, then an advisory group is desirable. This should be a formal group consisting of the environmental coordinator, design engineers, operators, maintenance personnel, and marketing and other administrative staff. It is expected of this group to review the environmental issues involving the organization's products and services, then brainstorm and provide suggestions to improve the weaknesses that are identified.

1.7.10 Environmental Accounting

Even among the most environmentally progressive corporations there is a lack of accounting systems that track the generation of wastes and the costs of waste disposal and environmental compliance to provide data on such costs per unit of product. Dividing total cost for all matters relating to environmental compliance by the total quantity of product manufactured is inadequate and does not encourage pollution prevention.

Tracking the costs of chemicals wasted at each stage of manufacturing, the capital and operating costs associated with pollution control equipment, and the costs of intangibles such as public relations and the corporation's environmental reputation can shed some light on pollution problems. Data of this kind should be included in budgets as line items.

1.7.11 Providing Incentives for Progress

It is well known that even the best plan without incentives to encourage achieving higher goals rarely provides sustainable progress. A financial compensation plan based on meeting the environmental management system's objectives and exceeding specific goals is recommended. The plan should involve all key technical and administrative personnel (including design, manufacturing, marketing and upper-level management) and goals should be reviewed periodically and updated as appropriate.

1.7.12 Environmental Audits

No system can survive in the global markets in the long run without a meaningful third party review from time to time. Pollution prevention systems are no exceptions. Organizations should have a provision for conducting comprehensive third party reviews and inspections without much advance notice. Such audits must be conducted with a high degree of transparency and professional freedom.

1.7.13 Management of Chemical Purchases

Facilities can benefit from implementing a system for tracking all chemical purchases including even routine maintenance supplies. Purchasing authority should be limited to a centralized group of trained professionals and purchases limited to a list of pre-approved chemicals in specific quantities supplied by screened and pre-approved vendors.

Such a system is of paramount importance in (and even crucial to) tracking consumption as well as waste generation. It minimizes the proliferation of unwanted chemicals being bought randomly or used without management's knowledge and it is not only a powerful pollution prevention tool, but also an excellent cost control measure.

1.7.14 Preparation of an Annual Environmental Report

Just as every public corporation publishes an annual report for its shareholders, it should also issue an annual summary of the wastes generated by each of its facilities. This involves instituting a formal administrative procedure for preparing annual environmental progress reports that provide an accounting of pollutants at their point of entry and exit. The information in the report should include compliance data, costs involved and historical trends, as well as written plans to correct known problem areas with target dates for completion.

1.7.15 Recognition of Success and Reward for the Employees' Contribution

The organization should perform a formal and written appraisal of the facility at least once a year and assign an overall rating or *grade*. Significant contributors to the plant's pollution prevention efforts should be recognized. Such a recognition programme can serve as stepping-stones to further progress and can encourage employees to work towards continuous improvement.

The aforementioned strategies are applicable to any large multinational organization in the world. Needless to mention that they must be customized to meet the specific requirement of each organization. For example, corporate styles for environmental management systems and pollution prevention efforts will depend on such organization's leadership and the leader's vision. The successful implementation must, therefore, be dynamic in order to meet the changing challenges in the field of pollution prevention and environmental protection.

EXERCISES

1.1 Describe the methodology to be followed by an environmental engineer in making pollution control decisions.

1.2 Give an account of the general considerations in respect of pollution control.

2

Biogeochemical Cycles

A biogeochemical cycle is a summary of the different chemical centres where a particular element resides, coupled with the pathways that convert and transport the element from one centre to another. The chemical elements which include all the essential elements for life, follow some form of biogeochemical cycles. Many such cycles exist in the globe. Of them; carbon, oxygen, nitrogen, phosphorus and sulphur cycles are more important than the others.

Global biogeochemical cycles distribute nutrients throughout the earth's atmosphere and upper geologic layers. The cycles of interest to environmental engineers include carbon, nitrogen, oxygen and water. Most of these cycles have at least one gaseous or atmospheric phase, the only exception being phosphorus cycle which has no significant atmospheric presence.

All these cycles are composed of a reservoir pool and a cycling pool, with the reservoir containing the majority of the elements. This non-biological portion of the elements is relatively slow to transform. The reservoir pool can be either solid, contained in sedimentary geologic forms or gaseous phase contained in the atmosphere. The cycling pool is more chemically or biochemically active. Elements in it move fast between organisms and their immediate environment.

Most of the natural systems are cybernetic. A cybernetic system is one in which feedback causes self-regulation and stability to some extent. Many natural processes, including the carbon cycle, are to some extent cybernetic. As the carbon dioxide level increases, photosynthetic organisms increase their rate of conversion of carbon dioxide. This results in the consumption of additional amounts of atmospheric and aqueous carbon dioxide. So the carbon enhancement is dampened by the increased consumption. The problem for humans at present is that we are destroying forests that carry out most of world's photosynthesis, thus, negating positive effects of the cybernetic systems.

2.1 CARBON CYCLE

Carbon, as it is well known, is the building block of life on earth. The carbon-carbon bond is relatively stable and it helps in the construction of complex macromolecules. Compounds containing carbon, excluding the carbonates and cyanides are considered organic. Organic chemistry is the chemistry of life.

Carbon is present in the air primarily as CO_2. It is present in both fresh water and sea water as dissolved carbon dioxide, carbonic acid, bicarbonate and carbonate. In minerals it exists in the form of carbonates such as limestone or as petroleum or coal. In all life forms it exists as proteins, fat and carbohydrates. Around eighty-three per cent of all carbon exists in the form of inorganic chemicals. The remainder is contained in organic sedimentary minerals. This segment is basically kerogen, the remains of tissues of ancient plants and animals. Only a very small fraction exists as petroleum and coal. It has also to be borne in mind that very little of the carbon on earth is contained in the atmosphere, in surface water or groundwater or in living things. Over 100,000 times as much carbon is contained in sedimentary rocks as in all living things.

Carbon can be converted between the various living and non-living forms by the processes of photosynthesis and degradation. A key to carbon's diversity and distribution is its gaseous phase which makes it possible for carbon to be transported all over the earth. A simplified schematic representation of the carbon cycle is shown in Figure 2.1. Photosynthetic organisms convert inorganic carbon into plant biomass. On land, plants perform the photosynthetic process, making use of the carbon dioxide present in the atmosphere. In the oceans and bodies of freshwater photosynthetic organisms such as planktonic algae perform synonymous conversion processes using dissolved inorganic carbon.

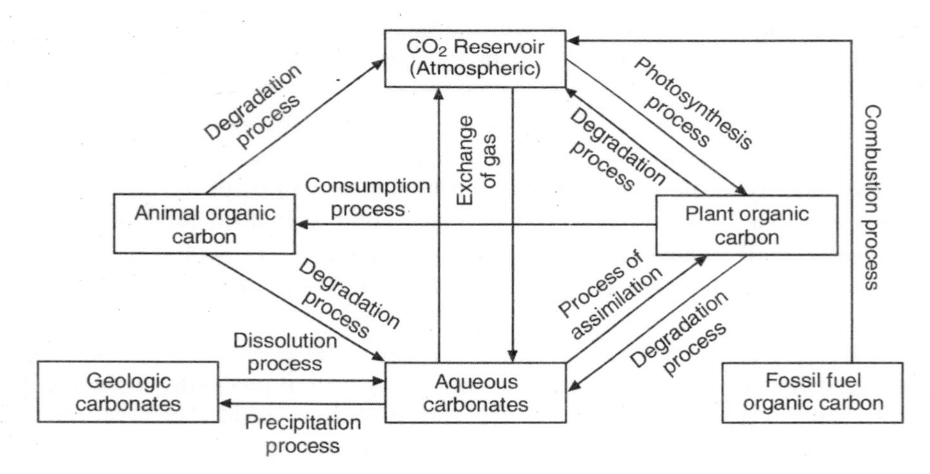

Figure 2.1 The carbon cycle.

Much of this organic carbon is either converted into animal biomass by consumers or back to inorganic carbon by decomposers. However, a fraction of it remains as organic decay products on the bottom of oceans or lake floors. These photosynthetic processes are primary means by which carbon dioxide is removed from the earth's atmosphere. In addition, aquatic organisms utilize carbon in building shell and skeletal structures. This inorganic carbon, primarily carbonate, is finally deposited on the floor of oceans and lakes. This process also makes the removal of a substantial amount of inorganic carbon possible.

Animals that exist at the second tropic level consume plant matter for energy production and biomass syntheses. These life processes facilitate the transformation of the plant organic carbon into other forms. The result of this transformation is the production of carbon dioxide and organic carbon. At death, organic matter is converted into simpler organic compounds and into inorganic forms by decomposers. The decomposition process returns the carbon dioxide to the atmosphere and to water.

The various inorganic forms of carbon can be interconverted from one reservoir to another by chemical processes like precipitation, dissolution and gas transfer. The mineral carbonates can be dissolved by undersaturated or acidic waters seeping through the formations, conversely, additional carbonates can be deposited though chemical precipitations. However, most geologic carbonates have been formed by aquatic invertebrates using dissolved inorganic carbon to form carbonate shells. These shells are eventually deposited on the ocean or lake floors.

Carbon dioxide in the atmosphere dissolves into both ocean and fresh water and is released back into the atmosphere. This takes place in accordance with Henry's law, where carbonate minerals reach great depths, the heat within the earth makes them revert to carbon dioxide. This results in the release of carbon dioxide from volcanoes.

During the past two centuries human activity has been causing a change in the global carbon balance. The largest changes are occurring as a consequence of the introduction of carbon dioxide into the atmosphere from fossil fuel combustion in developed countries. In addition, large amounts of forest cover have been destroyed in the past 100 years both in developed and developing nations. This combination has brought about a marked increase in atmospheric carbon dioxide. Although atmospheric carbon dioxide levels have been recorded for only the past five decades or so, coring in the polar ice caps has established atmospheric carbon dioxide levels for the past several hundred years. The world has known reserves of oil, natural gas and coal adequate to last for the next 200 to 300 years at the current consumption rates but there are presently no practical methods of removing carbon dioxide from combustion processes. So we can anticipate global carbon dioxide levels to continue to rise over that same period. There are strong indications that increased global carbon dioxide is entailing warming of the earth which is known as the **greenhouse effect**. This greenhouse effect has got a great significance in air pollution.

2.2 NITROGEN CYCLE

Nitrogen is the major constituent of the atmosphere, a major dissolved gas in water and a component of amino acids which are the building blocks for proteins comprising all life forms. Amino acids form bonds with other amino acids resulting in protein structures. Nitrogen is also a chemical of life like carbon. It is a major nutrient for both plant and animal forms. In addition, nitrogenous compounds are major components of all municipal water and some of the industrial waste water. In view of their playing such a critical role in the environment, the nitrogenous compounds turn out to be of lot of significance to environmental engineers.

Figure 2.2 illustrates the nitrogen cycle. As with carbon, nitrogen undergoes conversion into different forms by several processes which are both biotic and abiotic. Nitrogen gas present in the atmosphere is oxidized and deposited on earth by lightening. Also some organisms can convert atmospheric nitrogen to nitrate by a process known as nitrogen fixation. Nitrogen present in the soil or water can be used by plants to produce organic nitrogen, amino acids and proteins. Animals get nitrogen by consuming plants or other animals. Plant and animal life is eventually decomposed and sent back to the soil or water as inorganic forms of nitrogen. Organic nitrogen

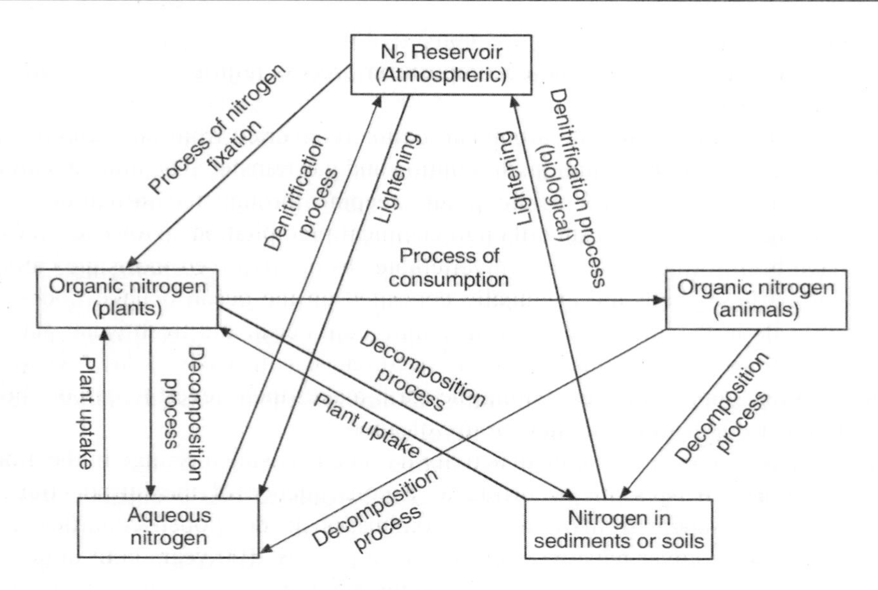

Figure 2.2 The nitrogen cycle.

is nitrogen contained in carbon compounds such as amino acids and urea. Inorganic nitrogen is primarily nitrogen gas, ammonia, nitrite or nitrate. Animals expel excess nitrogen as urea. This is subsequently broken down into ammonia and carbon dioxide. The biological processes of nitrification and denitrification can bring about the conversion of the different inorganic forms of nitrogen to nitrogen gas and back again. Nitrification is the conversion of ammonia (or ammonium) to nitrate.

$$NH_3 + 1\frac{1}{2}O_2 \longrightarrow H^+ + NO_2^- + H_2O$$

and then

$$NO_2^- + \frac{1}{2}O_2 \longrightarrow NO_3^-$$

In the presence of organic matter, nitrate can be converted to atmospheric nitrogen by the denitrification process.

$$6H^+ + 6NO_3 + 5CH_3OH \longrightarrow 3N_2 + 5CO_2 + 13H_2O$$

Human activities effect changes in the nitrogen cycle. Nitrates and ammonia are often added to agricultural crop to augment yields. Combustion processes particularly those in automobiles, oxidize atmospheric nitrogen into nitrate, producing nitric acid, a component of acid precipitation. Domestic waste water, treated and then discharged into lakes, streams or oceans, often contain large amounts of nitrogen, usually as nitrate or ammonia. The excess nitrogen that is contained in surface waters often causes increased photosynthetic activity.

2.3 HYDROLOGIC CYCLE

The hydrologic cycle is the movement of water in the earth's atmosphere on the surface and underneath—a process powered by the sun's energy. Water is of paramount importance in view of its requirement for all life processes on earth. Also it receives much of the contaminations and pollution that humans generate.

Figure 2.3 gives a simplified schematic representation of the hydrologic cycle.

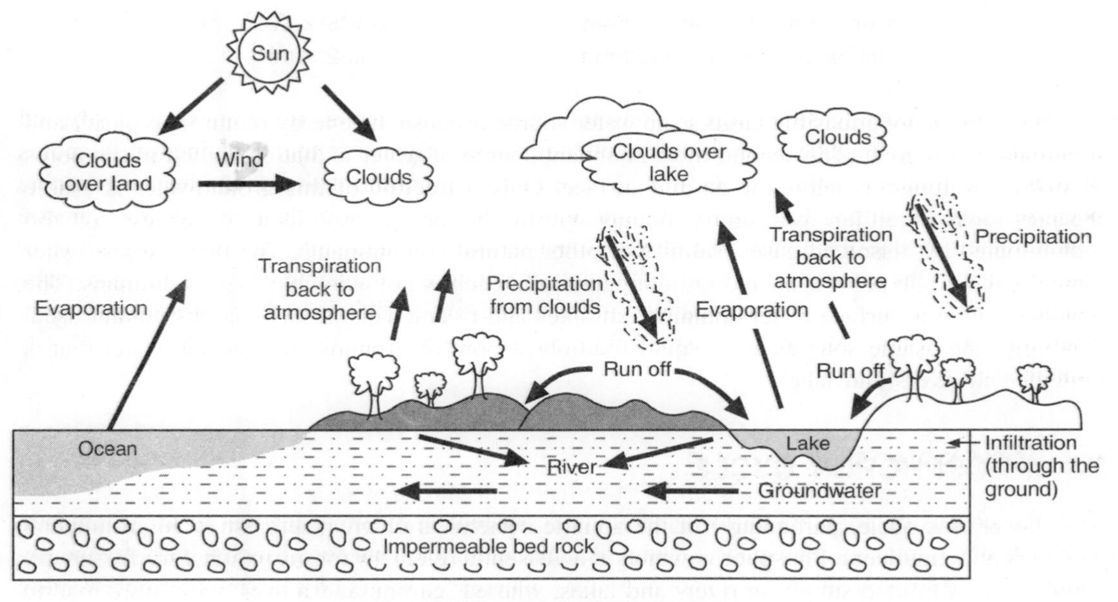

Figure 2.3 The hydrologic cycle.

In this cycle the sun's radiant energy supplies the power to evaporate water from lakes, rivers and plants. In view of the fact that the sun also supplies the energy to drive the winds, it is responsible for the transport of moisture in the atmosphere. Much water is sent back to the ocean by precipitation before ever reaching land. However, winds move part of the water vapour to land, where it gets deposited as precipitation. Of the water that falls over land, a portion infiltrates through the ground to the water-table recharging the groundwater. A portion of it is intercepted by vegetation and directly sent back to the atmosphere as evaporation. The portion of water absorbed by vegetation is sent back to the atmosphere through transpiration. Part of the precipitation flows over land to lakes and rivers. And part falls directly onto lakes and rivers. Ultimately when fresh water is sent back to the sea, the cycle is completed.

The wastes generated due to our activities are discharged into water. In some cases, this water is then treated to remove part of the waste material before being discharged into streams, rivers, lakes or oceans. Many a time untreated water is discharged into the environment and nature is left with the task of dealing with the pollution.

Water covers two-third of the earth's surface. This makes it appear to be an inexhaustible resource, but in reality amount of water available for human use is a very small fraction of the total quantity. As is indicated in Table 2.1, most water on earth is sea water.

Table 2.1 The Hydrosphere

Water sources	*Mass,* kg
Oceans	$13{,}700 \times 10^{17}$
Groundwater	$3{,}200 \times 10^{17}$
Water locked in ice	165×10^{17}
Water available in the polar caps and mountain tops, lakes, rivers	0.34×10^{17}
Water present in atmosphere	0.105×10^{17}
Total yearly stream discharge	0.32×10^{17}

Sea water is not a feasible large-scale water source because the energy required to purify and desalinate is too great. The second most abundant source of water is that contained in the pores of rocks and minerals below the earth's surface. Only a fraction of this groundwater is usable because most is either too tightly bound within the pores, too deep to retrieve or too contaminated by dissolved gases, salinity or other natural contaminants. The third largest water source is ice in the polar caps and on mountaintops which is not available to most humans. The smallest source is surface water contained in lakes and rivers. This is early harnessed and used. Therefore, the usable sources are a small fraction of pore or groundwater and the water that is contained in rivers and lakes.

2.4 PHOSPHORUS CYCLE

The phosphorus cycle, particularly in the aquatic system, is of great interest to environmental scientists and engineers. Phosphorus which is an essential element for growth is very frequently found to be in limited supply in rivers and lakes, whereas carbon and nitrogen are more readily available. Therefore, excessive growth of algae and aquatic weeds in rivers and lakes can often be reduced or prevented by limiting the supply of phosphorus alone. Phosphorus is thus a limiting factor.

Phosphorus is found in soils and rocks as calcium phosphate ($Ca_3(PO_4)_2$) and as hydroxyapatite ($Ca_5(PO_4)_3OH$). Owing to the limited solubility of phosphate rock, very small amount of phosphorus is leached into solution resulting in concentrations as low as 1 ppb. Since phosphorus is needed for all life processes, its concentration in natural waters is further reduced by the biological system, owing to the seasonal changes in plant and animal production and because of increased phosphorus input to natural waters from spring run-off, the concentration of phosphorus in water varies markedly over the year.

The input of phosphorus from human activity can be much higher than that from natural sources. Domestic sewage contains phosphorus in faeces and from commercial detergents in which phosphates are utilized as wetting agents, although the latter contribution has been greatly reduced in a number of places following legislation. Run-off from agricultural areas that have received fertilizers (which normally contain nitrogen, phosphorus and potassium) can be another significant source of phosphorus. Therefore, soluble phosphorus can reach much higher concentrations in many polluted waters than it can in non-contaminated waters. This readily available phosphorus can often lead to the growth of nuisance organisms such as filamentous algae which can cause taste and odour problems in water supplies and distributions and also bring about clogging of filters in water treatment plants.

Phosphorus is a constituent of nucleic acids, phospholipids and numerous phosphorylated compounds. It has been observed that the ratio of phosphorus to other elements in organisms, tends to be considerably greater than the ratio of phosphorus to other elements in external sources such as, soil or water indicating that the supply of phosphorus is very important to biological growth in lakes. For their nutrition, plants (and bacteria) need phosphorus in the phosphate (dissolved) form generally as orthophosphate (PO_4). They assimilate it by directly converting the PO_4 to the organic (insoluble) form in their protoplasm. Decay of these organisms dissolves and releases (mineralizes) the phosphorus for reuse. However, in lakes much of the phosphate is removed from the water by the sediment which eliminates it from the seasonal water circulation.

A simplified schematic representation of the phosphorus cycle is shown in Figure 2.4 in which the solid arrows represent major pathways of the flow and dashed arrows designate the flows that are of much less importance.

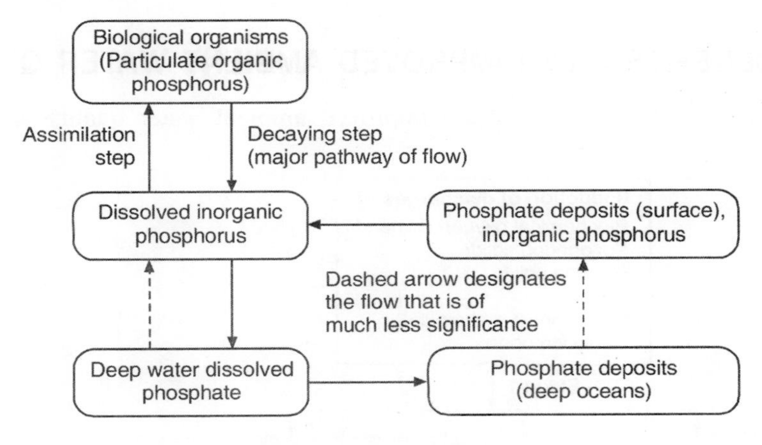

Figure 2.4 Phosphorus cycle.

Source: Henry, J. Glynn, and Gary, W. Heinke, *Environmental Science and Engineering*, Prentice Hall Inc., Upper Saddle River, NJ, 1996.

Particulate organic phosphorus is contained within dead and living cells and part of the dissolved inorganic phosphorus in the water is derived from the organic material by excretion and decomposition.

2.5 THE BENEFITS FROM IMPROVED AMBIENT AIR QUALITY

The benefits that could be derived from improved ambient air quality are indicated in Figure 2.5.

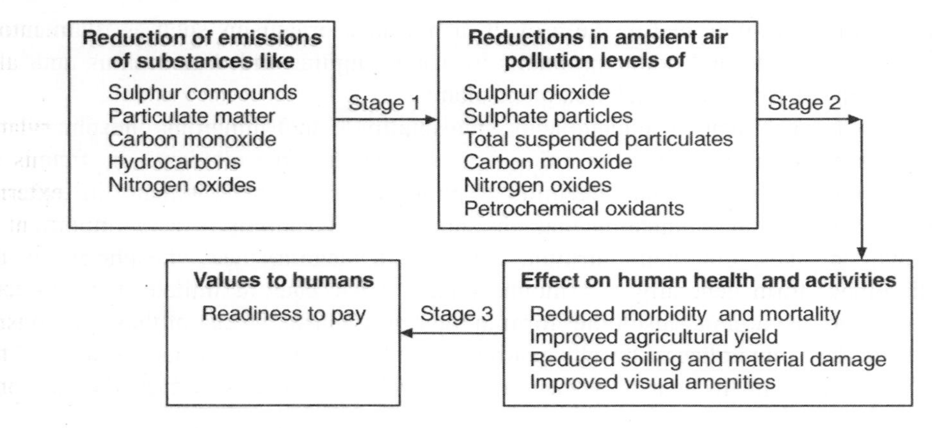

Figure 2.5 The benefits from improved ambient air quality.

Adapted from A. Myrick Freeman, Jr., *Air and Water Pollution Control—A benefit cost assessment*, Wiley Interscience, New York, (1982).

2.6 THE BENEFITS FROM IMPROVED AMBIENT WATER QUALITY

The benefits that could be derived from improved ambient water quality are indicated in Figure 2.6.

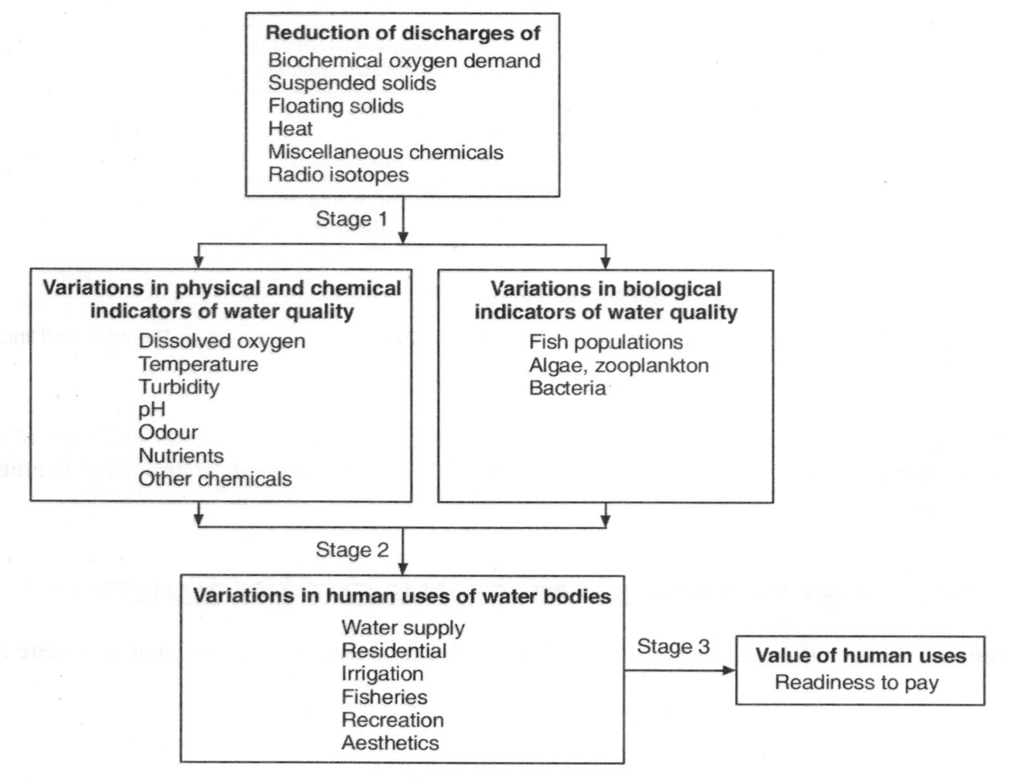

Figure 2.6 The benefits from improved ambient water quality.

2.7 EXCHANGES AND INTERACTIONS OF POLLUTANTS BETWEEN AIR AND WATER, AIR AND LAND, AND WATER AND LAND

The exchanges and interactions of pollutants between air and water, air and land, and water and land are presented in Figures 2.7, 2.8 and 2.9, respectively.

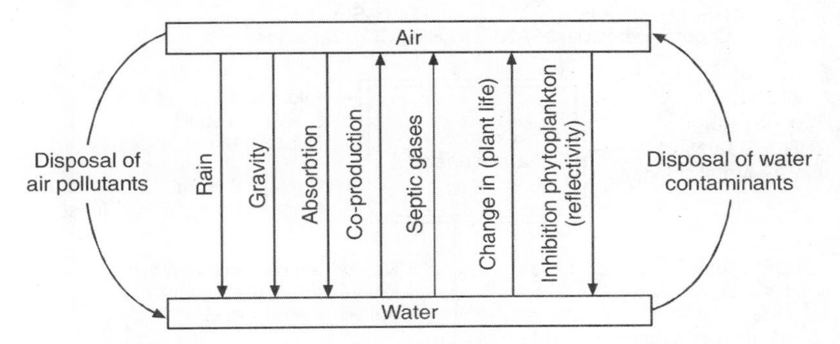

Figure 2.7 Some exchanges and interactions between air and water.

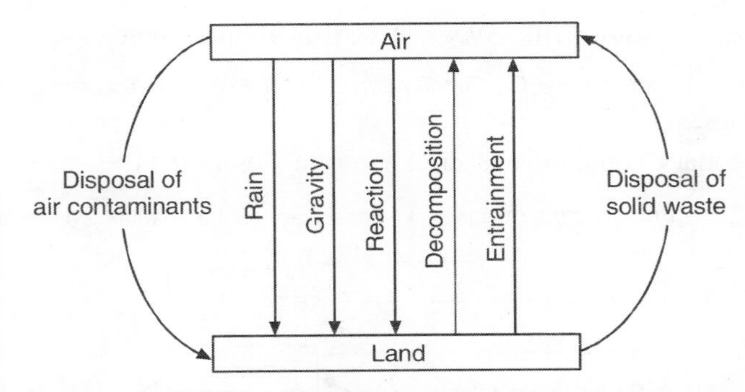

Figure 2.8 Some exchanges and interactions of pollutants between air and land.

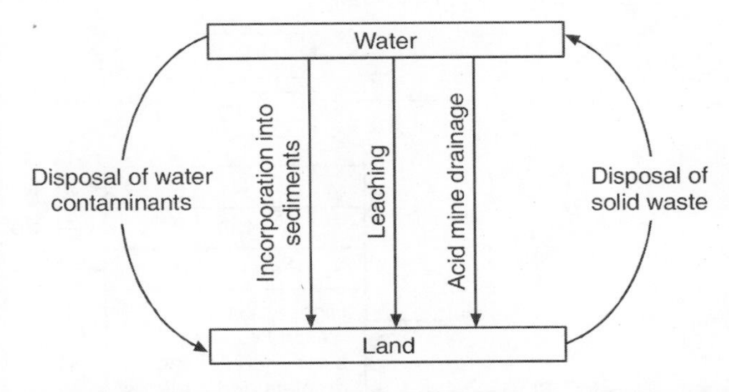

Figure 2.9 Some exchanges and interactions of pollutants between water and land.

2.8 TYPES OF CORROSIVE ATMOSPHERES, PRIMARY AND SECONDARY POLLUTANTS

The types of corrosive atmospheres are shown in Figure 2.10.

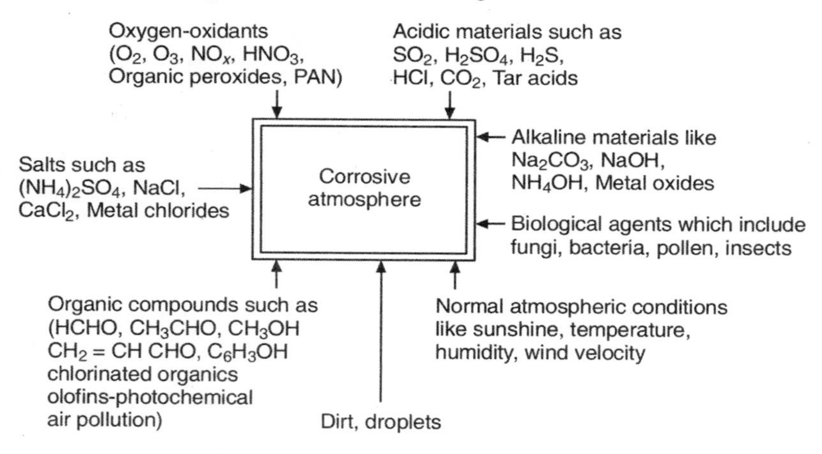

Figure 2.10 Types of corrosive atmospheres.

Adapted from Stern, Arthur C. and Henry, C., *Fundamentals of Air Pollution*, Academic Press, New York, 1973.

The primary and secodary contaminants are shown in Figure 2.11.

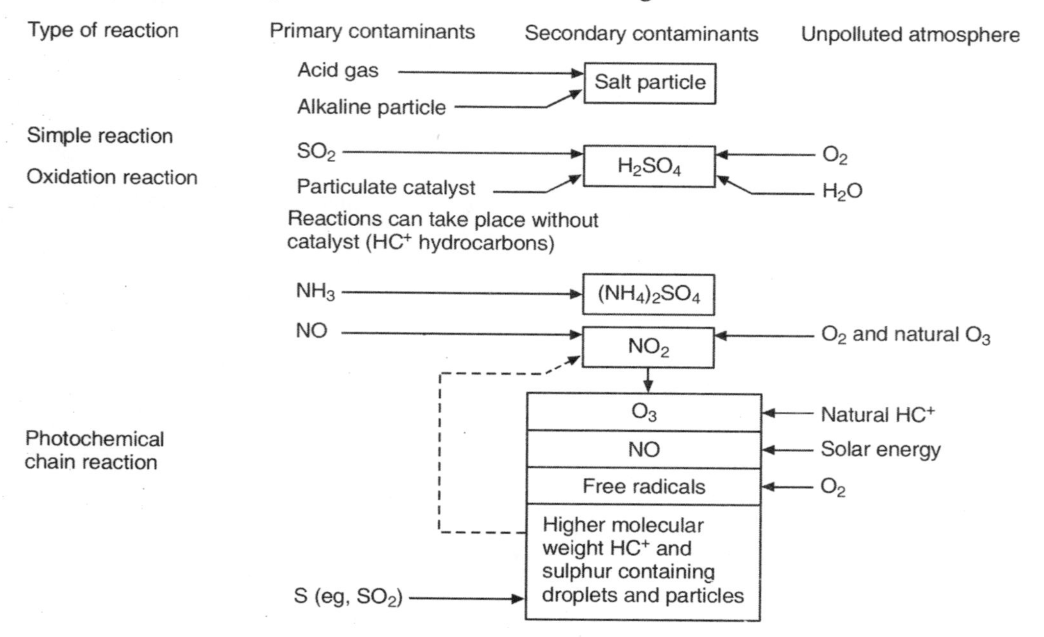

Figure 2.11 Primary and secondary pollutants.

Adapted from Stern, Arthur C. and Henry, C., *Fundamentals of Air Pollution*, Academic Press, New York, 1973.

2.9 POLLUTION CONTROL METHODS

2.9.1 Applicable to All Emissions

Decrease or eliminate production of emissions

- Change specification of product
- Change the design of product
- Change process temperature, pressure or cycle
- Change specification of materials
- Change the product

Confine the emissions

- Enclose the sources of emissions
- Capture the emissions in an industrial exhaust system
- Prevent drafts

Separate the contaminant from effluent gas

- Scrub with liquid

2.9.2 Applicable Specifically to Particulate Emissions

Decrease or eliminate particulate emissions

Change to process that does not need blasting, blending, buffing, calcining, chipping, crushing, drilling, drying, grinding, milling, polishing, pulverizing, sanding, sawing, spraying, tumbling, etc.

- Change from solid to liquid or gaseous material
- Change from dry to wet solid material
- Change particle size of solid material
- Change to process that does not require particulate material

Separate the contaminant from effluent gas stream

- Gravity separator
- Centrifugal separator
- Filter
- Electrostatic precipitators

2.9.3 Applicable Only to Gaseous Emissions

Decrease or eliminate gas or vapour production

- Change to process that does not require annealing, baking, boiling, burning, casting, cooking, dehydrating, dipping, distilling, expelling, galvanizing, melting, pickling, plating, quenching, reducing, rendering, roasting, smelting, etc.
- Change from liquid or gaseous to solid material
- Change to process that does not require gaseous or liquid material

Burn the contaminant to CO_2 and H_2O

- Incinerator
- Catalytic burner

Adsorb the contaminant

- Activated carbon

EXERCISES

2.1 Explain with the help of a diagram any three biogeochemical cycles which distribute nutrients throughout the earth's atmosphere and upper geologic layers.

2.2 What are the benefits derived from improved ambient air quality and water quality?

2.3 Write a short note on

(a) The types of corrosive atmosphere

(b) Primary and secondary pollutants.

3

The Chemistry of Waste Waters

The design of treatment system for an industrial liquid effluent will have a bearing on both the quantity and the quality of the waste to be treated. In this section, the common analyses carried out on industrial wastes together with methods of sampling are presented. Analyses of industrial waste waters are carried out for a variety of reasons which include

1. The need to ascertain their strength and nature (and to detect the presence of any industrial contaminant) so that the type of treatment plant can be decided when an extension of a new project is envisaged.
2. To evaluate the quantity of sludge to be anticipated.
3. To determine the overall effectiveness of an existing work or the effectiveness of any section of a work.

The commonly practised analyses are presented here. These include tests for oxygen demand to indicate the quantity of organic matter; solid contents to provide information on settling tank and sludge treatment system design; nitrogen compounds (which are quite important if the standard set by the river authority limits the quantity of unoxidized ammonia in the effluent), and a pH measurement—as this has a bearing on the type of treatment to be employed. Other tests may be required depending on the origin of the waste and the type of treatment proposed, but it is not usually necessary to analyze domestic sewage in any more detail. Analyses of industrial effluents must cover both the organic and inorganic impurities besides their temperature, any toxic metals and any other difficult constituents like cyanide, phenols, oil and grease, and colour.

3.1 BIOCHEMICAL OXYGEN DEMAND (BOD)

A variety of tests can be employed to determine the organic matter in effluent streams and their affinity for oxygen. The result of these tests indicates the quantity of biodegradable organic matter present and the tendency of the organic matter to oxidize.

The test for biochemical oxygen demand is a measurement of the weight of dissolved oxygen consumed by chemical and microbiological action during incubation of a sample for a definite period at a defined temperature. The sample is diluted with a known amount of standard dilution water (tap water is mostly employed for this purpose) to provide the needed dissolved oxygen; the actual rate of dilution will be governed by various factors, but usually this will be either 1:49 or 1:99. The result of the test based on a comparison of dissolved oxygen before and after incubation indicates the strength of the waste water. The BOD is normally performed at 20°C and usually for a period of 5 days (BOD_5). Some scientists have adopted a 20-day BOD test (BOD_{20}) for certain industrial effluents.

The BOD test oxidizes organic substances which are decomposable by bacterial action and it is therefore a very suitable test for domestic waste (sewage) to be treated in a conventional plant. As the oxidation of some degradable substances would not have gone to completion at the end of the 5 days' test; the use of numerical results of a test must be based on experience and knowledge of the details of the test. The test calls for the involvement of a skilled chemist and the specialized apparatus and consumes 5 (or 20) days to go to completion. It is suitable neither for snap checks, nor for use at small works where proper laboratory facilities are lacking. An indication of the oxygen demand can, however, be got from a four-hour test employing potassium permanganate.

3.2 OXYGEN ABSORBED FROM PERMANGANATE

Sometimes referred to as the OA test, this test is presently abbreviated to Permanganate Value (PV). It is perhaps the most widely employed test world over for evaluating the oxygen demand of waste water and is the oldest of the Chemical Oxygen Demand (COD) tests.

The standard PV test employs acid N/80 permanganate (potassium) and is a measure of the oxygen absorbed from the permanganate during a period of 4 hours. Oxidation during the 4-hour PV test is not complete but it is a quick and handy test to evaluate the quantity of organic carbon in the effluent stream. There is also a general relationship between the PV and BOD values for domestic sewage. An average sewage will usually have a PV value of 75 to 110 mg/l as crude sewage and 50 to 80 mg/l after settlement. The relationship of PV and BOD for both crude and settled domestic sewage is of the order of 1:4.

3.3 SUSPENDED SOLIDS

After the measurement of BOD, the suspended solid content is probably the next most important test for both crude sewage and treated effluents. The amount of suspended and soluble solids in a sewage will govern the design details of both the sedimentation tanks and the sludge drying process. Measurement of suspended solids is also employed to control the activated sludge treatment process.

The determination of the concentration of suspended solids is indicative of the contaminating strength of the sewage. It is normal to distinguish between organic (fixed) and organic (volatile) suspended solids as it is the latter that are putrescible. The quantity of volatile solids is measured by the loss on ignition at 600°C and it will normally be found that the mineral content of a domestic sewage is quite small. The tests call for the use of laboratory equipments and hence, are unsuitable for smaller works. The quantity of total suspended solids (fixed and volatile) will be indicative of the insoluble content of the sewage (both organic and mineral) while the quantity of settleable solids is an indicator of the quantity of sludge to be treated and the amount of solids which may be carried forward to the secondary (biological) process.

Average strength of domestic sewage will consist of 300–400 mg/l suspended solids before the settling process is effected. With efficient settlement this figure will normally be brought down to a maximum of about 120 to 150 mg/l.

3.4 NITROGEN

The nitrogen in wastes is evaluated as albuminoid nitrogen, as ammoniacal nitrogen or as organic nitrogen. The first is a measure of the strength and amount of nitrogenous organic matter; the reduction in this figure between that for the crude sewage and that of the final effluent will be indicative of the degree of purification obtained through a work. The measurement of ammoniacal nitrogen indicates the amount of nitrogenous organic matter which has been converted to ammonia and is therefore more relevant to the analysis of effluents after treatment.

As there is some degree of uncertainty as to how much of the albuminoid nitrogen is organic nitrogen, the analyst will normally determine the organic nitrogen figure separately. This gives him an idea of the impurity content of sewage and the extent of treatment of an effluent. An average domestic sewage will have a combined nitrogen content of 40 to 60 mg/l or higher. Of that figure about 10 to 15 mg/l or higher will be in the form of albuminoid nitrogen while the remainder will be ammoniacal nitrogen.

Nitrogen must be present in an effluent stream in adequate quantity if treatment is to be carried out by a biological process. The removal of BOD by biological treatment is accomplished partly by the breakdown of organic matter to carbon dioxide and water and partly by conversion to bacterial matter which is removed in the final settling tanks. This latter process is dependent on a supply of nutrients such as nitrogen and phosphorus. When these are falling short in an industrial waste, they must be added if the waste is to be treated without admixture with domestic sewage. It is normally considered that for effluent biological treatment, the waste should contain nitrogen equal to at least 3% of the BOD (6% or more preferable) and phosphorus about 1% of the BOD that is for a waste with a BOD of 400 mg/l, the total nitrogen content should be at least 12 mg/l while the phosphorus content should be about 4 mg/l.

3.5 pH VALUE

The pH value of a liquid is an indication of its degree of acidity or alkalinity (but not of the total quantity). Domestic sewage is normally slightly alkaline (about pH 7.2) as its basic constituent (tap water) is itself often slightly alkaline; in addition, soaps and other cleaning materials are also alkaline. Higher pH value may be caused by caustic discharges to the sewers. Acidity will generally be owing to the presence of mineral acids from industrial wastes or to organic acids from breweries, dairies and similar establishments, but acidity may also be due to septic conditions.

A simple field test uses a drop of a universal indicator in 0.003 litres of the liquid to be tested. The resultant colour compared with a standard table varies from deep red (acid) to violet alkaline with yellow-green at a natural pH of 7.0.

3.6 TEMPERATURE

The normal range of temperature of domestic sewage has no appreciable effect on the traditional treatment process. If the sewage contains high temperature industrial wastes (or cooling waters), the higher temperatures can accelerate corrosion of ferrous metals and concrete under certain conditions and they can also cause conditions in a sewer which are dangerous to workmen.

3.7 TOXIC SUBSTANCES

Toxic metals and chemicals may be present in many industrial wastes and in domestic sewage which contains those wastes. These substances will inhibit biological treatment and will normally affect aquatic life in a river. The effects of toxic metal contained in electroplating effluents are quite significant. The standards of industrial effluents to be discharged to sewers are frequently set so that the total concentration of toxic metals reaching a sewage treatment works will not be more than 1 mg/l. In some situations it may then be possible to allow an individual effluent to discharge to a sewer with a total toxic metal content of up to 40 or 50 mg/l, provided the proportion of effluent to domestic sewage is small.

3.8 CHLORIDES

Chlorides are always present in domestic sewage and their concentrations are not reduced by the normal treatment processes. They are therefore a useful indication of how much the sewage has been diluted by surface water or of the dilution afforded by a stream receiving an effluent. The concentration of chlorides varies throughout the 24 hours and this can be employed to check the extent to which samples correspond one with the other when they have been taken at various stages in the treatment process. Natural unpolluted water may contain upto 10 or 20 mg/l chlorides while domestic sewage and effluent will contain about 70 to 90 mg/l. These figures may be doubled when certain industrial wastes are present.

3.9 PHOSPHORUS

Phosphorus occurs in domestic sewage as organic phosphorus derived from food; as complex inorganic phosphates which are in turn derived from polyphosphates found in synthetic detergents and as soluble orthophosphate resulting from the biological breakdown of phosphorus. The usual concentration of phosphorus in crude sewage has been in the range of 3.0 to 22 mg/l. While sewage treatment works in a few countries are not usually designed to remove phosphorus, the processes normally employed will be, in general, bringing about some reduction. Some works in the USA and Europe have been designed specifically for the removal of phosphorus and other nutrients.

As with nitrogen, phosphorus is one of the nutrients which must be present in waste if it is to be treated successfully by a biological process. For efficient biological treatment the phosphorus content should be around 1 per cent of the BOD.

3.10 SAMPLING

If an analysis is to give a true picture of a sewage or effluent it must be made on a sample of typical dry weather flow, unless an analysis under any other special conditions is specifically needed. While spot samples may be taken to determine variations in constituents or the extreme values of pH, etc. It is common to take samples at regular time intervals throughout the day (normally, after every one hour), and to apportion these in proportion to the rate of flow to obtain a typical average sample. For general chemical examination, a sample of at least two litres is needed. It should be collected in a chemically clean bottle of either glass or polythene. The sample should be stored in a cool place and subjected to the test within 24 hours.

Automatic samples are available using either scoops or pumps. Samplers are available for coupling to flow recorders so that a given quantity of liquid is sampled each time. Successive predetermined volume increments are marked up on the integrator. The more usual types collect equal sub-samples at fixed time intervals. These must be subsequently mixed as per the recorded flows for the corresponding periods.

Whether sampling is by hand or by automatic sampler, necessary precautions must be taken to make sure that the sample is truly representative and that it does not contain more or less solids, etc. than the flow at the time of sampling. In case an automatic sampler is employed it must be robust and easily maintained; the first flush of each sub-sample should be discharged to waste, so that only fresh liquid is included and the containers must not be exposed to light or to the elements.

Except for samples taken for dissolved oxygen determination, the bottle containing the final sample should be filled so that a small air bubble is present after closure. This will prevent leakage in transit due to changes in temperature. The stopper should be firmly inserted and preferably tied down. Labels should be of the tie-on type. While prompt analysis is always desirable, this is not possible when samples are collected automatically over a period of time. Various possible methods of preserving the samples, to prevent bacterial action prior to the analysis include refrigeration and treatment with either acid or alkali or other chemical treatment.

3.11 INDUSTRIAL LIQUID EFFLUENTS—GENERAL CONSIDERATIONS

The quantity of water employed by industry for process purposes is continuously increasing and there are now relatively few industrial units which produce an effluent that does not need any form of treatment. The basic choice of disposal of a liquid industrial waste lies between discharge to a public sewer and discharge to a water course. Either method will usually call for some treatment before discharge. The treatment of industrial wastes along with domestic sewage ensures that the effluent will be subjected to treatment before being discharged to a river, but this may in some situations bring about a deterioration in the quality of the effluent from the sewage treatment works.

Before any decision can be made regarding the disposal of an industrial effluent, a complete survey of the effluent must be carried out to furnish information on the quantity and rates of discharge along with the composition of the effluent and any probable variations in quantity or quality which are likely to occur in the near future. Knowledge of the type of industry should be given, and indication of the types of wastes is to be anticipated. Snap samples of industrial units' effluents will rarely afford any dependable guide and the samples taken should be representative and should be collected where needed, over a 24-hour period relative to the rates of flow. A sample should be at least two litres in volume.

An automatic sampler may be the most convenient method of obtaining these samples. The laboratory analyst conducting the analysis of an effluent will prefer to have information on the original source of the waste water (that is public supply, private well, river water, etc.) as this helps the analyst to identify of suitable methods of analysis. He should also be given details regarding the type of sample (whether representative or snap), the place and time at which the sampling was carried out, and the temperature of the sample when it was taken.

The initial programme falls into three fairly distinct stages.

1. A survey of the drainage and waste discharges of the industrial unit (with analyses) to provide full information on the type of wastes, any variations in the flow throughout the 24 hours, any inclusion of cooling water or surface water, the source of the factory water supply and any proposals for new processes or extensions to the factory.
2. A decision as to whether the effluent should be treated before being discharged and whether discharge is to be to the local authority sewer or to a water course. This will be governed by the capacity of any available sewer and of the sewage treatment works and the relative economics of the alternative proposals.
3. Legal agreement on the basis of discharge to either sewer or water course.

3.11.1 The Layout of the Drainage System

When a new industrial unit is being constructed, suitable provision can be made for the disposal of waste waters during the design stage with the provision of drains for both present and probable future flow together with adequate space for any treatment plant. This is, however, rarely the case and more usually the problem is to provide drains and treatment plant for an existing establishment.

In these situations, a thorough survey of the existing drainage and factory layout is required to obtain details of quantities and qualities of waste waters at each part of the installation to establish whether water is now being wasted that is whether the flow could be brought down. In this respect, it is often preferable to isolate the industrial waste from any other wastes and particularly from rainfall run-off. This is, of course, very important if the industrial waste is going to be discharged to a public sewer and if the charge to be made by the local authority is based on the volume or the rate of discharge. It is often convenient to prepare a flow diagram to show these various aspects diagrammatically.

Where the industrial unit is new or where the existing drainage system is being re-laid, it will generally be advisable to segregate the various types of waste so that cooling water and storm water requiring minimum treatment are kept separate from polluted waters. If spent caustic soda solutions are collected separately from acid solutions, these can sometimes be mixed under controlled conditions to obtain neutral solution.

Some suggestions of points which the manufacturer is expected to answer include the following:

1. The source of the effluent
2. The nature of the effluent (e.g. acidic, alkaline, oily, sludgy)
3. Analysis of the effluent (maximum and minimum values of temperature, chemical composition, BOD, quantity and nature of the suspended solids' concentration and type of acid or alkali)
4. The quantity to be treated (maximum hourly discharge, intermittent or continuous quantity per day)
5. The point of discharge for treated effluent (river, local sewerage system, local main drainage)
6. Conditions required in the treated effluent, suspended solids, BOD, pH, others such as toxic metal limits)
7. Present method of treatment (if any)
8. Services available (clean water source and pressure electricity supply, compressed air)
9. Space available, location area, headroom
10. Other information which might be pertinent

3.12 CONTROL AND CHARGES

Before the design of any treatment plant can be commenced, the required standards of the final effluent must be known. This will necessitate discussions with the river authority and/or the local authority. Neither of these authorities can prepare general conditions. Any condition attached to any consent by an authority or any charges proposed by a local authority will have a bearing on the volume and rate of flow and the composition of the particular effluent. The obvious aim is to arrive at terms which are mutually acceptable to both the authority and the industrialist.

While an industrialist will no doubt wish to look at the matter from a cost/benefit point of view, this cannot be the approach of a public authority. A river authority is concerned with the maintenance of a certain standard in the water course and any consent must therefore have that in mind. A local authority must take into account the possible effects of the quality and the treatment works. The considerable expenses and difficulties of re-drainage in any industry are more logically met if the re-drainage can be phased into any factory redevelopment. To be equitable, this aspect should be taken into consideration by the relevant authority.

From any viewpoint it is essential that all wastes (domestic and industrial) are treated sufficiently to safeguard the nation's water resources and to maintain the water course in a satisfactory condition for its many uses—water supply, recreation, etc.

The quality of river water needs to be considered with the following seven points in mind:

1. A sufficient level of dissolved oxygen to avoid the development of anaerobic conditions.
2. Freedom from organic matter, which encourages the production of sewage fungus.
3. Lack of turbidity or liability to form substantial sludge deposits.
4. Freedom from oil, grease, etc.

5. Toxic substances not to be in sufficient quantity so as to affect fish life or the self-purification of the river.
6. Freedom from unnatural discolouration.
7. Temperature increase not to be excessive.

Discharge to a public sewer will normally be subject to less severe consent conditions than for discharge direct to a river and in accepting the effluent into the sewers, the local authority accepts responsibility for any poisonous, noxious or polluting matter which may ultimately be passed to the river. On the other hand, a charge will normally be levied by the local authority to cover the costs incurred and the industrialists must, therefore, select between paying this charge and being absolved from responsibility for the ultimate discharge or of accepting the cost of providing and maintaining a separate treatment plant at his own works.

Acceptance of industrial waste at the sewage treatment works will normally make it more difficult for that works to produce an effluent within the river authority's standard particularly if (as is usually the case) the majority of the industrial wastes are discharged to the sewers during working hours for five days each week.

3.13 THE PROBLEMS FACED BY INDUSTRIALISTS

The expense incurred in any treatment method of industrial waste is of prime consideration to a manufacturer. The capital costs of any re-drainage plus the capital and annual cost of treatment will almost certainly show no financial return and these costs must, therefore, be borne by the industrial unit as an addition to its normal cost. It may, however, be an economical proposition for the manufacturer to install specially designed separators for the recovery of certain solids; this form of pre-treatment of liquid wastes is particularly applicable to food processing and to the pulp and paper industry. The possibility of any charges being incurred for effluent treatment may in itself entail an improvement in the standards of housekeeping within an industrial unit with a consequent saving in effluent treatment costs and often a related saving in costs of water and other raw materials. The confederation of industries may stress the need for consideration to be given first to the modification of manufacturing process where applicable, and recommend that the processes should be examined to see if the total polluting matter discharged to the river can be reduced, and, in a few situations, eliminate the discharge of their effluent to rivers by treating and reusing them in a closed cycle system in the factory.

While pre-treatment of the effluent may result in the recovery of some materials for reuse or sale, the net result of installing an effluent treatment plant is normally an extra cost to the manufacturer. It is, therefore, essential from his standpoint of view that any treatment plant is adequately designed so that he avoids spending money on plant, which does not ultimately produce the desired result. It is also usually more economical for the industrialist to be able to make improvements to his effluent disposal arrangements over a period of time. Gradual and continuous programme of re-drainage will, in general, be more acceptable and will cause less interruption to the day-to-day running of the industrial unit concerned.

The impurities in industrial wastes may be organic or they may be toxic; with the exception of some toxic organic chemicals, the former are similar to the impurities in domestic sewage and are therefore more easily treated by accepted methods of sewage treatment; it may be more economical to arrange for these to be discharged to a public sewer where that is feasible. In situations where the effluent is from a food factory, it may be preferable for it to be discharged to the sewer without any treatment to avoid the possibility of causing any smells or of contamination of the product from the waste waters. In such a case, the manufacturer may be prepared to accept the local authority's charges in preference to providing treatment at the factory. Toxic substances such as chromium or cyanide are usually better dealt with the factory premises and a local authority will normally specify the limiting concentrations which can be discharged to any sewer.

Where a public sewer of sufficient capacity is available, the decision of the industrialist will be governed by the type of the effluent, where the effluent is of a type which can be satisfactorily and economically treated without admixture with domestic sewage. It will normally be in the industrialist's interests to negotiate with the river authority for the discharge of treated effluent to a water course if one is conveniently available. Where the source of the industrial water is a small stream, it may in fact be necessary to return the treated effluent to that stream. When the effluent is to be discharged to a river, the condition imposed by the river authority will take into consideration, the type and composition of the effluent and the size and standard of receiving water course.

The installation of any treatment plant for effluent before discharge to a river (or the installation of pre-treatment plant before discharge to a sewer) will include provisions for the settling of suspended solids and their removal in the form of sludge. The disposal of sludge at sewage treatment works is now a considerable problem to many local authorities; its production at an industrial site will usually cause equal if not more serious difficulties.

3.14 IMPURITIES IN EFFLUENTS AND UNDESIRABLE WASTE CHARACTERISTICS

Before considering the various forms of treatment available for industrial effluents it is convenient to consider first, the forms of impurities to be treated. Organic matter is measured in terms of oxygen demand or oxygen absorption. In addition, it is a common practice to measure the solids in suspension, the pH value, the temperature, and probably the colour. Depending on the type of effluent, separate assessments will also be made of toxic substances like metals, phenols, cyanides, etc. and, of oil and grease.

When discharged to a river, suspended solids can cause deposits of sludge behind weirs and in the more sluggish reaches of the river. If the solids are organic they can take up further oxygen from the river. Such sludge deposits will putrefy and then may rise to the surface as unsightly scum during the summer. The limiting figure of 15 mg/l for suspended solids in satisfactory river water has been suggested and a standard of 30 mg/l for the effluent from a sewage treatment works has been set. The limiting figures ranging from 40 to 60 mg/l for effluents from various industries have also been suggested. It is usually possible to treat an industrial effluent by sedimentation to bring it within these limits.

For the discharge of an effluent to sewers it is a common practice to set a limit at about 400 to 500 mg/l of suspended solids as this is about the same as that of average strength sewage. Limits of up to 1500 mg/l have been allowed for finely divided solids of vegetable origin without difficulty being experienced either in the sewers or at the sewage treatment works.

The pH value of a liquid is a measure of its acidity or alkalinity; the usual limiting values for effluent discharge either to sewers or rivers are in the range 5 to 9.

A river authority will normally set a maximum temperature for an effluent so that temperature of the receiving water course is maintained at a reasonable level to sustain fish life. It is normal to aim at a maximum river water temperature of 20 to 25°C and with this mind the authority might fix a maximum effluent temperature of 30 degrees Celsius where the boiler blow down will occur outside the normal hours of effluent discharge from an industrial unit, it may be necessary to provide some means for cooling this water before discharge to either a sewer or a water course.

When compounds of heavy metals and other toxic substances are present in an industrial effluent they may bring about suppression of the biochemical self-purification of a river. They may also have an adverse effect on both the biological treatment plant at a sewage treatment works and on the sludge if this is subsequently treated by digestion or used on the land. Some authorities require the concentrations to be limited to 5 mg/l of each metal, when a waste is discharged to a sewer. Many river authorities require treatment of an industrial waste if the toxic constituents such as ions of copper, chromium, nickel and zinc are likely to go beyond 1 mg/l. Concentrations in excess of 1 mg/l of metals in river water have been known to affect the fish life of a river; the effect of a discharge is dependent on other factors such as dissolved oxygen, pH, hardness and temperature.

The concentration of cyanide in an effluent to be discharged to a sewer is normally restricted to 1 mg/l because of possible toxic hazards from hydrogen cyanide gas in the sewers. Some river authorities have set a maximum limit for cyanide in an effluent at 0.1 mg/l. Phenols can be accepted for biological treatment at a sewage treatment works in reasonable concentrations. These will be governed by local circumstances and may vary from 500 mg/l to 2000 mg/l. Standards set by river authorities vary from about 0.5 to 1.0 mg/l.

Oil and grease may be present in an industrial effluent from engineering works or vehicle maintenance. For discharge to sewers of larger authorities, the amount of oil and grease together have sometimes been limited to 400 mg/l, but much lower figures would be used where the sewers discharge to small works. Petrol must be completely eliminated as far as possible in view of the danger to men working in the sewers. Oil used in an industrial unit may include those of mineral, animal or vegetable origin and also neat cutting oil and soluble oil. Any method of treatment will depend on the type of oil concerned. While soluble cutting oils are valueless when discarded, other waste oils provided they do not contain metal particles, etc. can often be disposed of at a profit to the industrialist.

When discharged to a water course, the presence of oil on the surface will be aesthetically objectionable and a surface film will inhibit the biological self-purification of the stream. Limits for oil, grease and scum in these situations may vary from no visible floating oil to 5 or 10 mg/l of free oil.

3.14.1 Treatment

Various proposals have been made from time to time for the assessment of the '*treatability*' of an industrial waste as compared with domestic sewage; but these wastes vary so widely from one industry to another and from time to time throughout the day so that no single formula or ratio would be suitable for all situations or cases. In any event, most are based on biological treatment and/or sludge treatment, as distinct from chemical treatment. The development of respirometer, technique seems to be a step in the right direction, particularly for evaluating new discharges, an automatic recording respirator can be a research tool for investigating biological oxidation and a means of evaluating the treatability of industrial effluents. While the treatment of most industrial effluents is feasible technically, any cost must be economical to the industrialist, whether this brings about full treatment to reach standards imposed by a river authority or any partial treatment before discharge to a public sewer. Estimates of costs must also include the cost of the ultimate disposal of any sludge from the plant.

In situations wherein a waste is acidic or alkaline, it may be necessary to correct this by the addition of suitable chemicals before any other treatment is given. Sulphuric acid is generally used for correction of alkaline effluents, lime, caustic soda or soda ash will be employed for acid wastes. Some metals (e.g. zinc and iron) can be precipitated as hydroxides by increasing the pH of a solution. By adjustments of the pH value, hexavalent chromium can be broken down and precipitated as a hydroxide. The amount of removal of the solid hydroxide will be dependent on the efficiency of the subsequent treatment processes.

Cyanide from plating shop wastes, etc. must be broken down before any conventional treatment is given. By maintaining the pH value at between 10.5 and 11.0 and by the addition of chlorine, cyanides can be oxidized to cyanates or they can be broken down to carbon dioxide and nitrogen. Provided the plant is suitably designed and the chemical reaction is complete, there are no basic difficulties (other than cost) in accomplishing a cyanide destruction down to 1.0 mg/l or less.

Often the only treatment or pre-treatment necessary for an industrial effluent may be the removal of suspended solids by settlement. The design of settling tanks to yield the requisite period of settlement and overflow rate will normally be based on a laboratory analysis of the effluent. To obtain efficient settlement of wastes containing emulsified oils or finely divided particles, chemical coagulation may be necessary; this must be followed by settlement for the removal of the precipitated solids.

A well-designed settlement tank will remove about 70 or 80 per cent of the suspended solids, but its efficiency will be governed to some extent by any variations in flow and also in solids content. The selection of type of settling tank between a radial-flow or an upward flow tank or in larger installations a rectangular tank, will depend to a larger extent on site conditions. The removal of solids from an effluent will give rise to subsequent problems of the disposal of the settled sludge and the possibility of offensive odours, these aspects must be taken into account when siting settling tanks and also when weighing up the relative merits of treatment against any local authority charges for discharge to the sewers without treatment.

Mechanical separators have been developed for use in many types of industrial effluents. While special filtering devices have been employed for the filtration of liquid wastes produced in the paper, food and chemical industries.

Oil and grease removal will have a bearing on the standards suggested for the effluent and also on the economics of recovery of these materials. Recovery by floatation is generally possible using a skimmer to remove the bulk of the oil; the residues can then be broken down with detergents. The rate of oil recovery will be dependent on the type, physical characteristics and the thickness of oil film.

If the waste is of the biodegradable type, it may be economical and more convenient to discharge this to a public sewer in preference to the construction of a separate biological treatment plant at the industrial unit. If treated separately, the methods employed will be identical to those for domestic sewage that is biological filters or the activated sludge process. The choice between these two processes will depend on various factors including both capital and annual costs and the availability of suitable maintenance staff. Generally, an activated sludge plant will give rise to higher running costs and a higher standard of maintenance, but on the other hand, it is more compact and requires less available head through the plant. Unless knowledge is available from prior experience, it is usual to consider each effluent individually, based on laboratory tests and possibly after the operation of pilot-scale plant.

Biological filtration may be either by *straight*, filtration or may include recirculation or alternating double filtration. For larger works one of the traditional methods of activated sludge treatment may be suitable, while for smaller discharges a more compact extended aeration plant will probably be more convenient. This latter term is used to include the many types of compact factory-assembled plants now in the market and also the oxidation ditch. These have been used successfully for a number of industrial effluents amenable to biological treatments.

3.14.2 Undesirable Waste Characteristics

Depending on the nature of the industry and the projected uses of the water of the receiving stream various constituents may have to be removed before the waste is discharged. These may be summarized as follows:

Soluble organic compounds causing depletion of dissolved oxygen

Since most receiving waters require maintenance of minimum dissolved oxygen, the quantity of soluble organic compounds is correspondingly restricted to the capacity of the receiving waters for assimilation or by specified effluent limitations.

Suspended solids

Deposition of solids in quiescent stretches of a stream will impair the normal aquatic life of the stream. Sludge blankets containing organic solids will undergo progressive decompositions resulting in oxygen depletion and the production of noxious gases.

Trace organic compounds

When a receiving water is to be used as a potable water supply, phenol and other organic compounds discharged in industrial wastes will cause taste and odour in the water. If these contaminants are not removed before discharge, additional water treatment will be needed.

Heavy metals

These are cyanides and toxic organic compounds. The Environmental Protection Agency of the USA has defined a list to toxic organic and inorganic chemicals that appear as specific limitations in most permits.

Colour and turbidity

These present aesthetic problems even though they may not be particularly deleterious for most water uses. In some industries, such as pulp and paper, economical methods are not presently available for colour removal.

Nitrogen and phosphorus

When effluents are discharged to lakes, ponds and other recreational areas the presence of nitrogen and phosphorus is particularly undesirable since it enhances eutrophication and stimulates undesirable algae growth.

Refractory substances resistant to biodegradation

These may be undesirable for certain water-quality requirements. ABS (Alkyl Benzene Sulphonate) from detergents is substantially non-degradable and frequently leads to a persistence of foam in a water course. Some refractory organics are toxic to aquatic life.

Oil and floating materials

These give rise to unsightly conditions and in most cases are restricted by regulations.

Volatile organic compounds

Hydrogen sulphide and other volatile organic compounds will create air pollution problems and are usually restricted by regulations.

3.15 AN OVERVIEW OF THE WASTE WATER TREATMENT SYSTEMS

Waste water also termed sewage originates from household wastes, human and animal wastes, industrial effluents, storm run-off and ground water infiltration. Waste water basically is the flow of used water from a community. It is 99.64 per cent water by weight. The balance quantity of 0.06 per cent is material dissolved or suspended in water.

An understanding of physical, chemical and biological characteristics of waste water is of paramount importance in design, operation and management of collection, treatment and disposal of waste water. The nature of waste water includes physical, chemical and biological characteristics which have a bearing on the water usage in the community, the industrial and commercial contributions, weather and infiltration/inflow.

When fresh, waste water is grey in colour and has a musty and not unpleasant odour. The colour gradually changes with passage of time from grey to black. Foul and unpleasant odours may then develop as a result of septic sewage. The most important physical characteristics of waste water are its temperature and its solid concentration.

Temperature and solid contents in waste water are very important factors for its treatment processes. Temperature affects chemical reaction and biological activities. Solids include Total Suspended Solids (TSS), Volatile Suspended Solids (VSS) and settleable solids after the operation and sizing of treatment units.

3.15.1 Chemical Constituents of Waste Water

The dissolved and suspended solids in waste water contain organic and inorganic material. Organic matter may include carbohydrates, fats, oils, grease, surfactants, proteins, pesticides and agricultural chemicals, volatile organic compounds and other toxic chemicals. Inorganic compounds may include heavy metals, nutrients (nitrogen and phosphorus) pH, alkalinity, chlorides, sulphur and a few other inorganic contaminants, gases such as carbon dioxide, nitrogen, oxygen, hydrogen sulphide and methane may be present in waste water.

3.15.2 Biological Characteristics of Waste Water

The principal groups of microorganisms found in waste water streams are bacteria, fungi, protozoa, microscopic plants and animals, and viruses. Most microorganisms (bacteria and protozoa) are responsible and are beneficial for biological treatment processes of waste water. However, some pathogenic bacteria and protozoa are responsible and are beneficial for biological treatment process of waste water. However, some pathogenic bacteria, fungi, protozoa and viruses found in waste water are of public concern.

3.15.3 Waste Water Treatment Systems

The natural waters in streams, rivers, lakes and reservoirs have a natural waste assimilative capacity to remove solids, organic matter, even toxic chemicals in the waste water. However, it is a long drawn process.

Waste water treatment facilities are designed to speed up the natural purification process that takes place in natural waters and to remove contaminants in waste water that might otherwise interfere with the natural process in the receiving waters.

Waste water contains varying quantities of suspended and floating solids, organic matter and fragments of debris. Conventional waste water treatment systems are combinations of physical and biological processes (sometimes with chemical processes) to remove the impurities.

The alternative methods for municipal waste water treatment are categorized into three major groups

1. Primary (physical process) treatment
2. Secondary (biological process) treatment and
3. Tertiary (combination of physical, chemical and biological processes) or advanced treatment

As can be seen in Figure 3.1, each group should include previous treatment devices (preliminary) disinfection and sludge management (treatment and disposal). The treatment devices shown in the preliminary treatment are not necessarily to be included, depending on the waste water characteristics and regulatory requirements.

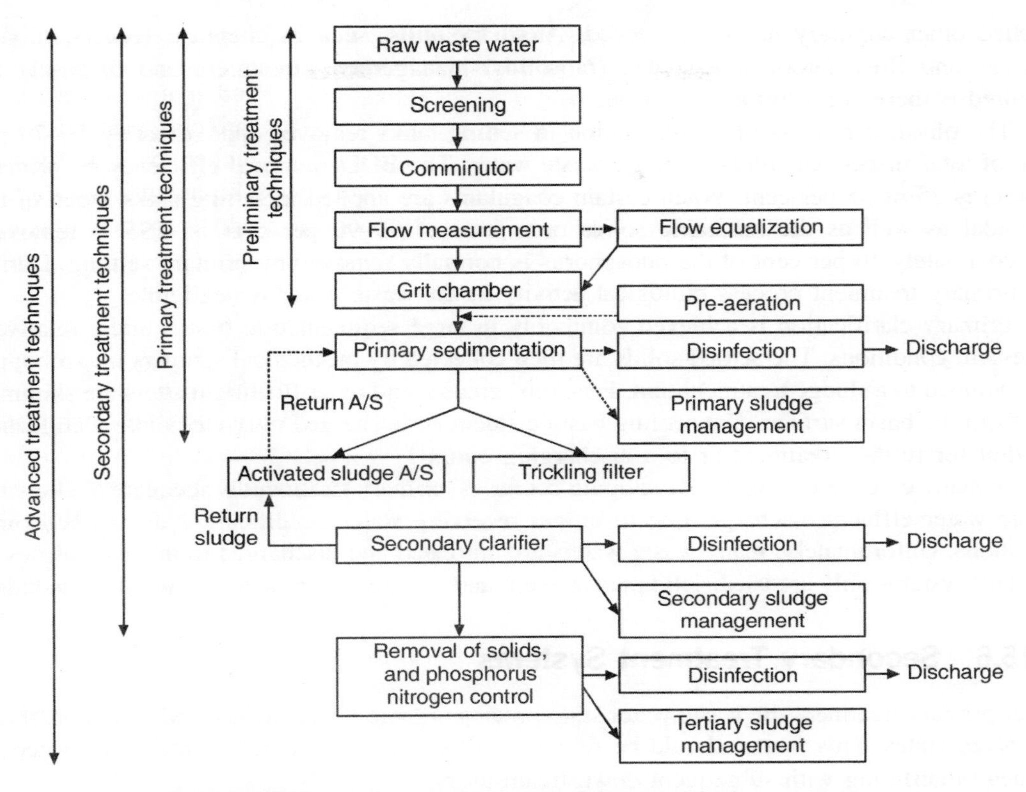

Figure 3.1 Flow diagram for waste water treatment processes.

Adapted from Loe, C.C. and Shun Darlin, *Handbook of Environmental Engineering Calculations,* McGraw Hill, New York, 2000.

3.15.4 Preliminary Treatment Systems

Preliminary systems are designed to remove or cut up the larger suspended and floating materials and to remove the heavy inorganic solids and excessive amounts of oil and grease. The purpose of preliminary treatment is to protect pumping equipment and the subsequent treatment units. Preliminary systems consist of flow measurement devices and regulators (flow equalization), racks and screens, comminuting devices (grinders, cutters and shredders), flow equalization devices, grit chambers pre-aeration tanks and possibly chlorination units. The quality of waste water is not substantially improved by preliminary treatment.

3.15.5 Primary Treatment Systems

The purpose of primary treatment is to bring down the flow velocity of the waste water sufficiently, to allow suspended solids to settle, that is, to remove settleable solids/materials. Floating materials are also removed by skimming. Thus, a primary treatment device may be called as a **settling tank** (or basin). Due to variations in design and operation, settling tanks can be categorized into four groups; plain sedimentation with mechanical sludge removal, two-storey tanks, up flow clarifiers with mechanical sludge removal and septic tanks. When chemicals are applied other auxiliary units are needed. Auxiliary units, such as chemical feeders, mixing devices and flocculators, and sludge (biosolids) management (treatment and disposal) are required if there is no further treatment.

The physical process of sedimentation in setting tanks removes approximately 50–70 per cent of total suspended solids from the waste water. The BOD_5 removal efficiency by primary system is 25 of 35 per cent. When certain coagulants are applied in setting tanks much of the colloidal as well as the settleable solids or a total of 80–90 per cent of TSS is removed. Approximately 10 per cent of the phosphorus is normally removed by primary settling. During the primary treatment process biological activity in the waste water is negligible.

Primary clarification is achieved commonly in large sedimentation basins under relatively quiescent conditions. The settled solids are then collected by mechanical scrapers into a hopper and pumped to a sludge treatment unit. Fats, oils, greases and other floating matters are skimmed off from the basin surface. The settling basin effluent is discharged over weirs into a collection conduit for further treatment or to a discharging outfall.

In many cases, especially in developing countries, primary treatment is adequate to allow the waste water effluent discharge, due to proper receiving water conditions or due to economic situations. Unfortunately, many waste waters are untreated and discharged in many countries. If primary systems only are used, solid management and disinfection processes should be included.

3.15.6 Secondary Treatment Systems

After primary treatment the waste water still contains organic matter in suspended, colloidal and dissolved states. This matter should be removed before it is discharged to receiving waters to eschew, interfering with subsequent downstream users.

Secondary treatment is used to the soluble and colloidal organic matter which remains after primary treatment. Although the removal of those materials can be effected by physico-chemical means, providing further removal of suspended solids, secondary treatment is commonly referred to as the biological process.

Biological treatment consists of application of a controlled natural process in which a very large number of microorganisms consume soluble and colloidal matter from the waste water in a relatively small container over a reasonable time. It is comparable to biological reactions that would occur in the zone of recovery during the self-purification of a stream.

Secondary treatment devices may be divided into two groups; attached and suspended growth processes. The attached (film) growth processes are trickling filters, rotating biological contents (RBC) and intermittent sand filters. The suspended growth processes include activated sludge and its modifications, such as contact stabilization (aeration tanks), sequencing batch reactors, aerobic and anaerobic digesters, anaerobic filters, stabilization ponds and aerated lagoons. Secondary treatment can also be achieved by physico-chemical or land application systems.

Secondary treatment process may remove more than 85 per cent of BOD_5 and TSS. However, they are not effective for the removal of nutrients (N and P), heavy metals, non-biodegradable organic matter, bacteria, viruses and other microorganisms. Disinfection is needed for this purpose. In addition, a secondary clarifier is needed to remove solids from the secondary processes. Sludges generated from the primary and secondary clarifiers need to undergo treatment and proper disposal.

3.15.7 Advanced Treatment Systems

Advanced waste water treatment is defined as the methods and processes that remove more contaminants from waste water than the conventional treatment. The term advanced treatment may be applied to any system that follows the secondary or that modifies or replaces a step in the conventional process. The term tertiary treatment is frequently used as a synonym; however, the two are not synonymous. A tertiary system is the third treatment step that is employed after primary and secondary treatment processes.

The goals of most of the advanced waste water treatment facilities are to remove nitrogen, phosphorus and suspended solids (including BOD_5) and to meet certain regulations for specific conditions. In some areas where water supply sources are limited, reuse of waste water is becoming more important. Also there are strict rules and regulations regarding the removal of suspended solids, organic matters, nutrients, specific toxic compounds and refractory organic compounds that cannot be achieved by conventional (traditional) secondary treatment systems; thus advanced waste water treatment processes are required.

In the US, federal standards for secondary effluents are BOD 30 mg/l and TSS 30 mg/l. In the European Community Commission for Environmental Protection has drafted the minimum effluent standards for large waste water treatment plants. The standards include BOD_5 <25 mg/l, COD <125 mg/l suspended solids <35 mg/l, total nitrogen <10 mg/l and phosphorus <1 mg/l. Stricter standards are presented in various countries.

TSS concentrations less than 20 mg/l are difficult to achieve by sedimentation through the primary and secondary systems. The objective of advanced waste water treatment techniques is specifically to bring down TSS, BOD, TDS, organic nitrogen, ammonia nitrogen, and total nitrogen or phosphorus. Biological nutrient removal processes can get rid of nitrogen or phosphorus and any combination.

Advanced process includes chemical coagulation of waste water, wedge-weir screens, granular media filters, diatomaceous earth filters, micro-screening and ultrafiltration and nanofiltration which are used to remove colloidal and fine-sized suspended solids.

For nitrogen, control techniques, such as biological assimilation, nitrification (conversion of ammonia to nitrogen and nitrate) and denitrification, ion exchange, break point chlorination, air stripping are employed. Soluble phosphorus may be removed from waste water by chemical precipitation and biological (bacteria and algae) uptake for normal cell growth in a control system. Filtration is required after chemical and biological processes. Physical processes, such as reverse osmosis and ultrafiltration also help to achieve phosphorus reduction, but these are primarily for overall dissolved inorganic solids reduction-oxidation ditch, anaerobic/oxidation processes.

The use of lagoons, aerated lagoons and natural and constructed wetlands is an effective method for nutrients (N and P) removal.

Removal of some species of groups of toxic compounds and refractory organics can be accomplished by activated carbon adsorption, air stripping, activated sludge, powder activated carbon processes and chemical oxidation. Conventional coagulation, sedimentation, filtration and biological treatment (trickling filter, RBC and activated sludge) processes are also used to remove the priority pollutants and some refractory organic compounds.

Sedimentation

Sedimentation is the process of removing solid particles heavier than water by gravity settling. It is the oldest and most widely employed unit operation in water and waste water treatment. The terms sedimentation, settling and clarification are used interchangeably. The unit sedimentation basin may also be referred to as a sedimentation tank, clarifier, settling basin or settling tank.

In waste water treatment, sedimentation is used to remove both inorganic and organic materials which are settleable in continuous flow conditions. It removes grit, particulate matter in the primary settling tank and chemical flocs from a chemical precipitation unit. Sedimentation is also used for solids' concentration in sludge thickeners.

Based on the solids' concentration and the tendency of particle interaction, there are four types of settling which may occur in waste water settling operations. The four categories are discrete, flocculant, hindered (also called **zone**) and compression settlings.

Biological (secondary) treatment systems

The purpose of primary treatment is to remove suspended solids and floating materials. In many situations in some countries, primary treatment with the resulting removal of approximately 40 to 60 per cent of the suspended solids and 25 to 35 per cent of BOD_5 together with removal of material from the waste water is adequate to meet the requirement of the receiving water body. If primary treatment is not sufficient to meet the regulating effluent standards, secondary treatment using a biological process is mostly used for further treatment due to its greater removal efficiency and less cost than chemical coagulation. Secondary treatment processes are meant to remove the soluble and colloidal organics (BOD) which remain after primary treatment, and to accomplish further removal of suspended solids and in some cases, also to remove nutrients such as phosphorus and nitrogen. Biological treatment processes provide similar biological activities to waste assimilation which would take place in the receiving waters, but in a reasonably shorter time. Secondary treatment which may remove more than 85 per cent of BOD_5 and suspended matter is not quite effective for removing non-biodegradable organic compounds, heavy metals and microorganisms.

Biological treatment systems are designed to maintain a large active mass and a variety of microorganisms, principally bacteria (and fungi, protozoa, rotifers, algae, etc.) within the confined system under favourable environmental conditions, such as dissolved oxygen, nutrient, etc. Biological treatment processes are generally classified mainly as suspended growth processes, attached (film) growth processes (trickling filter and rotating biological contactor RBC) and dual process systems (combined). Other biological waste water treatment processes include the stabilization pond, aerated lagoon, contaminant pond, oxidation ditch, activated sludge, biological nitrification, denitrification and phosphorus removal units.

In the suspended biological treatment process, under continuous supply of air or oxygen, living aerobic microorganisms are mixed thoroughly with the organic compounds in the waste water and use the organic compounds as food for their growth. As they grow, they clump or flocculate to form an active mass of microbes. This is the so called biological floc or activated sludge.

Advanced waste water treatment

Advanced Waste Water Treatment (AWT) refers to those additional treatment techniques needed to further reduce suspended and dissolved substances remaining after secondary treatment. It is also called **tertiary treatment**. Secondary treatment removes 85 to 95 per cent of BOD and TSS and minor portions of nitrogen, phosphorus and heavy metals. The purposes of AWT are to improve the effluent quality to meet stringent effluent standards and to reclaim waste water for reuse as a valuable water resource.

The targets for removal by the AWT process include suspended solids, organic matter, nutrients, dissolved solids, refractory organic compounds and specific toxic compounds. More specifically the common AWT processes are suspended solids' removal, nitrogen control and phosphorus removal. Suspended solids are removed by chemical coagulation and filtration. Phosphorus removal is carried out to reduce eutrophication of receiving waters from waste water discharge. Ammonia is oxidized to nitrate to reduce its toxicity to aquatic life and nitrogenous oxygen demand in the receiving water bodies.

3.16 WASTE WATER CHARACTERISTICS AND EFFLUENT STANDARDS

3.16.1 The Chemical Constituents of Waste Water

The dissolved and suspended solids in waste water contain organic and inorganic materials. Organic matter may include carbohydrates, fats, oils, grease, surfactants, proteins, pesticides and agricultural chemicals, volatile organic compounds and other toxic chemicals. Inorganic compounds may cover heavy metals (nutrients, nitrogen and phosphorus) pH, alkalinity, chlorides, sulphur and other inorganic pollutants. Gases, such as carbon dioxide, nitrogen, oxygen, hydrogen sulphide and methane may be present in waste water.

Normal ranges of nitrogen levels in domestic raw waste water are 25–85 mg/l for total nitrogen (the sum of ammonia, nitrate, nitrite and organic nitrogen); 12–50 mg/l ammonia, nitrogen and 8–35 mg/l organic nitrogen.

Typical total phosphorus concentrations of raw waste water range from 2 to 20 mg/l which includes 1–5 mg/l of organic phosphorus and 1–15 mg/l of inorganic phosphorus. Both nitrogen and phosphorus in waste water serve as essential elements for biological growth and reproduction during waste water treatment processes.

The strength (organic content) of waste water is usually measured as 5 days Biochemical Oxygen Demand (BOD_5), chemical oxygen demand and total organic carbon. The BOD_5 test estimates the amount of oxygen required to oxidize the organic matter in the sample during 5 days of biological stabilization at 20°C. This is commonly referred to as the first stage of carbonaceous BOD (not nitrification, which is second phase). Secondary waste water treatment plants are typically designed to remove carbonaceous BOD, not for nitrogenous BOD (except for advanced treatment). The BOD_5 of raw domestic waste water is around 250 mg/l or even more.

The ratio of carbon, nitrogen and phosphorus in waste water is very important for biological treatment process where there is normally a surplus of nutrients. The commonly accepted BOD/N/P weight ratio for biological treatment is 100/5/1, that is 100 mg/l BOD to 5 mg/l nitrogen to 1 mg/l phosphorus.

The main groups of microorganisms found in waste water are bacteria, fungi, protozoa, microscopic plants and animals, and viruses. Most microorganisms (bacteria, protozoa) are responsible and are beneficial for biological treatment processes of waste water. However, some pathogenic bacteria, fungi, protozoa and viruses found in waste water are of societal concern.

Pathogenic organisms are usually excreted by humans from gastrointestinal tract and discharged to waste water. Water-borne diseases include cholera, typhoid, paratyphoid fever, diarrhoea and dysentery. The number of pathogenic organisms in waste waters is generally low in density and they are difficult to isolate and identify. Therefore, indicator bacteria, such as total coliform, faecal coliform and faecal streptococcus are used as indicator organisms.

3.16.2 Effluent Standards

Effluent discharge standards are the most common and very useful form of direct regulation. The standards can be in the form of across the board standards which require that effluent of all the industrial units meet the same criteria or they may be individually developed by each industrial unit. The advantages of an across the board type of approach are that, it is easy to administer, it appears fair to all the industrial units and it provides the most rigid control over the quality of environment. The demerits or disadvantages are that, it may be uneconomical and therefore impractical, to insist that all the manufacturing units meet the same effluent standards. Some industrial units may easily meet standards that others will not be able to meet at all, or only at a prohibitively high cost. The different assimilative capacities of the environment in different locations can be taken into consideration only on a case-by-case basis. For instance, a large fast-moving river can accept a much larger amount of organic pollution than a small creek. Nevertheless, most jurisdictions prefer to set common effluent discharge guidelines, which must be fulfilled unless the contributor is specifically exempted.

Regulation through zoning and effluent standards has been the basis for controlling the pollution that is thrown into our environment. However, the legislation has not been adequate to bring an end to pollution. For the controls to be effective and successful, the polluter must be willing to abide by the law. If the industrial unit considers the law unjust, economic pressure in the form of punitive measures, such as levying a fine, may force compliance with effluent requirements. However, fines have been traditionally low providing no economic incentive to the industrial organizations to comply with the regulations. Other enforcement measures like the withholding of various permits or licences without which the organization concerned cannot lawfully operate, can also be adopted. Using a court injunction to force an industrial unit to cease operations is a drastic measure that is very rarely used.

Enforcement is the major problem with any form of direct regulation. Those who have been discharging wastes freely, naturally resent and protest any curtailing of their rights. Industries often threaten to close down if regulations are enforced. This will give rise to the local unemployment problem, and is politically unacceptable. It is likely that municipal corporations forced to upgrade their waste disposal systems often come out saying that they do not have money to do. It is difficult for any level of government to force polluters into actions that may have politically unacceptable side effects.

EXERCISES

3.1 What are the objectives in carrying out the analyses of waste waters?

3.2 Discuss about the commonly practiced analyses of waste water.

3.3 Write a short note on

(a) the layout of a drainage system
(b) impurities and undesirable characteristics of liquid effluents.
(c) effluent standards

3.4 What are chemical constituents and biological characteristics of waste waters?

4

Water Quality

The cleanest available source of groundwater and surface water should be protected and maintained for drinking water supply purpose. Several parameters are employed to ascertain and determine the suitability of water and the impact of the contaminants on human health, that may be found in untreated and treated water. It may be mentioned in this context that watershed and well head protection regulations should be a primary consideration.

Microbiological, physical, chemical and microscopic examinations are of paramount importance in assessing the quality of water. Water quality can be best assured by maintaining water clarity, chlorine residual in the distribution system and low bacterial population in the distributed water.

The physical water quality parameters which include turbidity, colour, suspended solids, taste, odour, and temperature are presented in the following section.

4.1 TURBIDITY

Turbidity is owing to suspended materials such as clay, silt or organic and inorganic materials. Stringent surface water regulations require that the maximum contaminant level for turbidity must not exceed 0.5 NTU (Nephelometric Turbidity Units) in 95 per cent of the samples taken every month and must never exceed 1 NTU. In addition to this, the utility must maintain a minimum of 0.2 mg/l free chlorine residual at representative points within the distribution system. Turbidity measurements are made in terms of Nephelometric Turbidity Units (NTU), Formazin Turbidity Units (FTU), and Jackson Turbidity Units (JTU). The lowest turbidity value that can be measured directly on the Jackson candle turbid meter is 25 units. There is no direct relationship between the aforementioned three readings. The NTU is the standard measure requiring the use of a nephelometer which measures the amount of light scattered, usually at 90° from the light direction by suspended solids/particles in the test (water) sample. It can measure turbidities of less than 1 unit and differences of 0.02 units.

The expectation of the public is sparkling clear water. This implies a turbidity of less than 1 unit; a level of 0.1 unit which is obtainable when water is coagulated, flocculated, settled and filtered, is practically feasible. Turbidity is a good measure of sedimentation, filtration and storage efficiency, particularly if supplemented by the total microscopic and particle count. Enhanced chlorine residual, bacteriological sampling and main flushing is indicated when the maximum contaminant level for turbidity is exceeded in the distribution system until the cause is ascertained and eliminated. Turbidity will interfere with proper disinfection of water, harbour microorganisms and give rise to taste and odour problem.

An increase in the turbidity of well water after heavy downpours may indicate the entrance of inadequately purified groundwater.

4.2 COLOUR

Colour should be less than 15 true colour units (cobalt platinum units). In the measurement of colour of water, the sample is first filtered. The persons accustomed to clear water may, however, notice a colour of only 5 units. The objective is less than 3 units. Waters for industrial applications should generally have a colour of 5 to 10 units or less. Colour caused by substances in solution is known as **true colour**. The substances in the suspended form (mostly organics), cause apparent or organic colour. Iron, copper, manganese and industrial wastes may also cause colour in water.

Water that has drained through peat bogs, swamps, forests or decomposing organic matter may contain a brownish or reddish stain owing to tannates and organic acids dissolved from leaves, barks and plants. Excessive growths of algae or microorganisms may also cause colour.

Colour resulting from the presence of organics in water may also cause taste, interfere with chlorination, induce bacterial growth, make water unusable by certain industries without further treatment. Foul anion exchange resins, interfere with colorimetric measurements, cause limitations on aquatic productivity by absorbing photosynthetic light, render lead in pipes soluble, hold iron and manganese in solution causing colour and staining of laundry and plumbing fixtures, and interfere with chemical coagulation. Chlorination of natural waters containing organic water colour (and humic acid) results in the formation of trihalomethanes including chloroform.

Colour can be controlled at the source itself, by watershed management. The major steps are, identifying water from sources contributing natural organic and inorganic colour and excluding them, controlling beaver populations, increasing water flow gradients using settling basins at inlets to reservoirs and blending water. Coagulation, flocculation, settling and rapid sand filtration should reduce colour-causing substances in solution to less than 5 units, with coagulation as major factor. Slow sand filters should get rid of about 40 per cent of the total colour. Removal of true colour is expensive. Oxidation (chlorine, ozone) or carbon adsorption also brings about a reduction in colour.

4.3 TEMPERATURE

The water temperature should preferably be less than 25°C. Groundwaters and surface waters from mountainous areas are generally in the temperature range of 12–18°C. Design and construction of water systems should provide for burying or covering of transmission mains to keep the drinking water cool and prevent freezing in cold climates or leaks due to vehicular traffic. High water temperatures accelerate the growth of nuisance organisms, and taste and odour problems are intensified. Low temperatures decrease the disinfections efficiency, to some extent.

4.4 TASTE AND ODOUR

4.4.1 Taste

The taste of water should not be objectionable. Otherwise, the consumer will resort to other sources of water that might not be of satisfactory sanitary quality. Algae, decomposing organic matter, dissolved gases, high concentrations of sulphates, chlorides and iron or industrial wastes may cause tastes and odours. Bone and fish oil and petroleum products, such as kerosene and gasoline are particularly objectionable. Phenols in concentrations of 0.2 ppb in combination with chlorine will impart a phenolic or medicinal taste to drinking water. The taste test like the odour test, is very subjective and may be dangerous to laboratory personnel. As in odour control, emphasis should be placed on the removal of potential causes of taste problems.

4.4.2 Odour

Odour should be absent or very faint for water to be acceptable, less than 3 Threshold Odour Number (TON). Water for food processing beverages and pharmaceutical manufacture should be essentially free of taste and odour. The test is very subjective, being dependent on the individual senses of smell and taste. The cause may be decaying organic matter, waste waters including industrial effluents, dissolved gases and chlorine in combination with certain organic compounds, such as phenols. Odours are sometimes confused with tastes. The sense of smell is more sensitive than taste. Activated carbon adsorption, aeration, chemical oxidation (chlorine, chlorine dioxide, ozone, potassium permanganate) and coagulation and filtration will usually remove odours and tastes. Priority should first be given to a sanitary survey of the watershed drainage area and the removal of potential sources or causes of odours and tastes.

A technique for determining the concentration of odour compounds from a water sample to anticipate consumer complaints involves the *stripping* of odour compounds from a water sample that is adsorbed on to a carbon filter. The compounds are extracted from the filter and injected into a gas chromatograph-mass spectrometer for identification and quantification.

4.5 SUSPENDED SOLIDS

Suspended solids in water may comprise inorganic or organic particles or immiscible liquids. Inorganic solids, such as clay, silt and other soil constituents are common in surface water. Organic materials, such as plant fibres and biological solids (algal cells, bacteria, etc.) are also common constituents of surface waters. These materials are often natural contaminants resulting from the erosive action of water flowing over surfaces. Owing to the filtering capacity of the soil, suspended material is seldom a constituent of groundwater.

Other suspended material may result from human use of the water. Domestic waste water usually contains large quantities of suspended solids that are mostly organic in nature. Industrial use of water may bring about a wide variety of suspended impurities of either organic or inorganic nature. Immiscible liquids, such as oils and grease are often constituents of waste water. Suspended material may be objectionable to water for many reasons. It is aesthetically displeasing and provides adsorption sites for chemical and biological agents. Suspended organic solids may be degraded biologically, resulting in objectionable by-products. Biologically active (live) suspended solids may include disease-causing organisms as well as other organisms, such as toxin-producing strains of algae.

Many tests for measuring solids are being used. Most are gravimetric tests involving the mass of residues. The total solids' test quantifies all the solids in the water, suspended and dissolved, organic and inorganic. This parameter is measured by evaporating a sample to dryness and weighing the residue. The total quantity of residue is expressed as milligrams/litre (mg/l) on a dry-mass-of solids, basis. A drying temperature slightly above boiling (105°C) is adequate to drive off the liquid and the water adsorbed to the surface of the particles while a temperature of about 180°C may be necessary to evaporate the occluded water.

Most suspended solids can be got rid of water by filtration. Thus, the suspended fraction of the solids in a water sample can be approximated by filtering the water, drying the residue and filtering it to a constant weight at 104°C or 105°C and determining the mass of the residue retained on the filter. The results of this suspended solid test are also expressed as **dry mass per volume** (milligrams per litre). The amount of dissolved solids passing through the filters, also expressed as **milligrams per litre**, is the difference between total solids' and suspended solids' content of a water sample.

It should be pointed out in this context that filtration of a water sample does not exactly divide the solids into suspended and dissolved fractions according to the definition given earlier. Some colloids may as well pass through the filter and be measured along with dissolved fraction while some of the dissolved solids adsorb to the filter material. The extent to which this takes place has a bearing on the size and nature of the solids and on the pore size and surface characteristics of the filter material. For this reason the terms filterable residues and non-filterable residues are frequently used. Filterable residues pass through the filter along with water and relate more closely to dissolved solids while non-filterable residues are retained on the filter and relate more closely to suspended solids. The distinction between filterable residues and non-filterable residues, and dissolved solids and suspended solids is not necessary.

Once the samples have been dried and measured, the organic content of both total and suspended solids can be determined by firing the residues at 600°C for 1 hour. The organic fraction of the residues will be converted to carbon dioxide, water vapour and other gases and will escape. The remaining material will represent the inorganic or fixed residue. When organic suspended solids are being estimated a filter made of glass fibre or some other materials that will not decompose at the elevated temperature needs to be used.

The chemical water quality parameters which include total dissolved solids, alkalinity, hardness, fluoride, metals (like lead, iron, mercury), organic compounds and nutrients. These are briefly presented hereinafter.

4.6 TOTAL DISSOLVED SOLIDS (TDS)

The total solid content in water should be less than 500 mg/l; however, this is based on the industrial applications of public water supplies and not on public health factors. Higher concentrations cause physiological effects and make drinking water less palatable. Dissolved solids, such as calcium, bicarbonates, magnesium, sodium sulphates and chlorides cause scaling in plumbing above 200 mg/l. The TDS can be reduced by distillation, reverse osmosis, electrodialysis, evaporation, ion-exchange and chemical precipitation (in some cases). Water, with more than 1000 mg/l of dissolved solids is classified as saline, irrespective of the nature of the minerals present. The United States Geological Survey, classifies water with less than 1000 mg/l dissolved solids as fresh, 1000 to 3000 mg/l dissolved solids as slightly saline, 3000 to 10000 mg/l dissolved solids as moderately saline, 10,000 to 35,000 mg/l dissolved solids as very saline and more than 35,000 mg/l dissolved solids as briny.

4.7 ALKALINITY

The alkalinity of water passing through distribution systems with iron pipe should lie in the range of 30 to 100 mg/l as $CaCO_3$ to prevent the occurrence of serious corrosions; up to 500 mg/l is acceptable although this factor must be apprised from the standpoint of pH, hardness, carbon dioxide and dissolved oxygen content. Corrosion of iron pipe is prevented by the maintenance of calcium carbonate stability. Under saturation will result in corrosive action in iron water mains and bring about red colour in water. Over saturation will result in carbonate deposition in piping, water heater and on utensils. Potassium carbonate, potassium bicarbonate, sodium carbonate, sodium bicarbonates, phosphates and hydroxides cause alkalinity in natural water. Calcium carbonate, calcium bicarbonate, magnesium carbonate and magnesium bicarbonate cause hardness as well as alkalinity. Sufficient alkalinity is needed in water to react with added alum to form a floc in water coagulation. Insufficient alkalinity will cause alum to remain in solution. Bathing or washing in water of excessive alkalinity can cause change in the value of pH of the lacrimal fluid around the eye, causing eye irritation.

4.8 HARDNESS

Hardness is primarily due to calcium, and secondarily due to magnesium carbonate and bicarbonates (carbonate or temporary hardness that can be removed by heating), and calcium sulphate, calcium chloride, magnesium sulphate and magnesium chloride (non-carbonate or permanent hardness which cannot be removed by heating), the sum is the total hardness expressed as calcium carbonate. In general, water softer than 50 mg/l as $CaCO_3$ is corrosive, whereas waters harder than about 80 mg/l lead to the use of more soap and above 200 mg/l may bring about incrustation in pipes. Lead, cadmium, zinc and copper in solution are usually caused by pipe corrosion associated with soft water. Desirable hardness values, therefore, should be 50 to 80 mg/l, with 80 to 150 mg/l as passable, over 150 mg/l as undesirable and greater than 500 as unacceptable. The U.S Geological Survey (USGS) and WHO classify hardness, in milligram per litre as $CaCO_3$, as 0 to 60 soft, 61 to 120 moderately hard, 121 to 180 hard and more than 180 very hard. Water high in sulphate (above 600 to 800 mg/l calcium sulphate, 300 mg/l sodium sulphate or 390 mg/l magnesium sulphate) are laxative to those not accustomed to the water. Depending on alkalinity, pH and other factors, hardness above 200 mg/l may cause the build-up of scale and flow reduction in pipes. Besides its being objectionable for laundry and other washing purposes due to soap curdling, excessive hardness contributes to the deterioration of fabrics. Hard water is not suitable for the production of ice, soft drinks, felts, or textiles. Satisfactory cleansing of laundry, dishes and utensils turns out to be difficult and impractical. When heated, bicarbonates precipitate as carbonates and adhere to the pipes or vessels. In boilers and hot water tanks, the scale resulting from hardness reduces the thermal efficiency and eventually causes restriction of the flow or plugging in the pipes. Calcium chloride when heated becomes acidic and pits boiler tubes. Hardness can be reduced by lime soda ash chemical treatment or the ion-exchange process, but the sodium concentration will be increased. Desalination will also remove water hardness.

There seems to be higher mortality rates from cardiovascular diseases in people provided with soft water than those provided with hard water. Water softened by the ion-exchange process increases the sodium content of the finished water. The high concentration of sodium and the low concentration of magnesium have been implicated, but low concentrations of chromium and high concentrations of copper have also been suggested as being responsible. High concentration of cadmium is believed to be associated with hypertension and the cause and effect for any of these is not established.

4.9 FLUORIDES

Fluorides are found in many groundwaters as a natural constituent, ranging from a trace to 5 mg/l or even more. And in some foods, fluorides, in concentration greater than 4 mg/l can cause the teeth of children to become mottled and discoloured, depending on the concentration and amount of water consumed. Mottling of teeth has been reported very occasionally above 1.5 mg/l according to WHO guidelines. Drinking water containing 0.7 to 1.2 mg/l natural or added fluoride is beneficial to children during the time they are developing permanent teeth. An optimum level is 1.0 mg/l in temperate climate. The maximum contaminant level in drinking water has been established at 4 mg/l as per drinking water regulations. The probable oral lethal dose for sodium fluoride is 70 to 140 mg/kg. Fluoride removal methods include reverse osmosis, lime softening, ion-exchange using bone char or activated alumina and tri-calcium phosphate adsorption. It has been found that it is not possible to reduce the fluoride level to 1 mg/l using only lime. The WHO report shows no evidence to support any association between fluoride addition (fluoridation) to drinking water and the occurrence of cancer.

4.10 METALS

4.10.1 Aluminium

Aluminium is not found naturally in the elemental form, although it is one of the most abundant metals on the earth's surface. It is found in all soils, plants and animal tissues. The Environmental Protection Agency (EPA) recommended goal is less than 0.05 mg/l, the WHO guideline is 0.2 mg/l. Aluminium-containing wastes concentrate can harm shellfish and bottom life. Alum as aluminium sulphate is commonly employed as a coagulant in water treatment, excessive aluminium may pass through the filter with improper pH control. Precipitation may occur in the distribution system or on standing when the water contains more than 0.5 mg/l. Its presence in filter plant effluent is used as a measure of filtration efficiency. Although, ingested aluminium does not appear to be harmful, aluminium compounds have been associated with neurological disorders in persons on kidney dialysis machines. Aluminium in the presence of the iron may bring about discolouration in water.

4.10.2 Arsenic

The maximum contaminant level for arsenic in drinking water was lowered from 0.05 mg/l to 0.01 mg/l by the EPA in January 2001. The WHO guideline is also 0.01 mg/l. The Occupational Safety and Health Administration (OSHA) standard is 10 $\mu g/m^3$ for occupational exposure to inorganic arsenic in air over 8-hr day; 2 $\mu g/m^3$ for 24-hr exposure to ambient air. Sources of arsenic are natural rock formations (phosphate rock), industrial wastes, arsenic pesticides, fertilizers and a few detergents. It is also found in foods including shellfish and tobacco, and in air of some locations.

4.10.3 Barium

Barium may be found naturally in groundwater in concentrations less than 0.1 mg/l and in surface water receiving industrial wastes; it is also found in air. It is a muscle stimulant and in large quantities may be harmful to the nervous system and heart. The Maximum Contaminant Level (MCL) is 2 mg/l in drinking water. Barium can be removed by weak acid ion-exchange.

4.10.4 Cadmium

The WHO guideline for MCL of cadmium in water is 0.005 mg/l. Common sources of cadmium are water mains and galvanized iron pipe, tanks, metal roof where cistern water is collected, industrial wastes (electroplating), tailings, pesticides, nickel plating solder, incandescent light filaments, photography wastes, paints, plastics, inks, nickel-cadmium batteries and cadmium plated utensils; salts of cadmium readily dissolve in water and can therefore be found in air pollutants, waste water, waste water sludge, fertilizers, land run-off and drinking water. They are also found in some food crops and tobacco.

4.10.5 Chromium

The MCL and WHO guideline for total chromium is 0.1 mg/l in drinking water. Chromium is found in cigarettes, some foods, industrial platings, paints and leather tanning wastes. Chromium deficiency is associated with altherosclerosis. Hexavalent chromium dust can cause cancer of the lungs and kidney damage.

4.10.6 Copper

The EPA action level for copper is 1.3 mg/l, the WHO guideline is 1.0 mg/l. The goal is less than 0.2 mg/l. Concentrations of this magnitude are not present in natural waters, but may be due to the corrosion of copper or brass piping; 0.5 to 1.0 mg/l in soft water stains laundry and plumbing fixtures. 1 mg/l reacts with soap to produce a green colour in water. Corrosion of galvanized iron and steel fittings is reported to be enhanced by copper in public water supplies. Copper salts are commonly used to control algal growth in reservoirs and slime growth in water systems. Copper can be removed by ion-exchange, conventional coagulation, sedimentation, filtration, softening or reverse osmosis; when caused by corrosion of copper pipes, it can be controlled by proper water treatment and pH control. Copper sulphate treatment of water source for algal control may contribute copper to the finished water. Electrical grounding to copper water pipe can add to the copper dissolution.

4.10.7 Iron

Iron is found naturally in groundwaters and in some surface waters and as the result of corrosion of iron pipes. Iron deposits and mining operations and distribution systems may be a source of iron and manganese. Water should have a soluble iron content of less than 0.1 mg/l to prevent reddish brown staining of laundry, fountains and plumbing fixtures and to prevent pipe deposits. The secondary MCL and WHO guideline value (level) is 0.3 mg/l; the goal should be less than 0.05 mg/l. Precipitated ferric hydroxide may cause a slight turbidity in water that is likely to be objectionable and may entail clogging of filters and softener resin beds. In combination with manganese, concentrations in excess of 0.3 mg/l cause problems. Precipitated iron may cause some turbidity. Iron in excess of 1.0 mg/l will cause an unpleasant taste. Conventional water treatment or ion exchange will remove iron.

Chloride or oxygen, will precipitate soluble iron. Iron is an essential element for human health.

4.10.8 Lead

The EPA of USA requires that when more than 10 per cent of tap water samples exceed 15 mg/l, the utility must institute corrosion control treatment. Concentrations exceeding this value occur when corrosive waters of low mineral content and softened waters are piped through lead pipes, zinc-galvanized iron pipes, copper pipes with lead-based solder joints and brass pipes. Faucets and fittings may also contribute lead. Lead should not exceed 5 mg/l in the distribution system.

Lead, cadmium, zinc and copper are dissolved by carbonate beverages which are also charged with carbon dioxide. Limestone, galena, water and food are natural sources of lead. Other sources are motor vehicle exhaust, certain industrial wastes, mines and smelters, lead paints, glazes, car battery salvage operations, soil, dust, tobacco, cosmetics and agricultural sprays. Fallout from airborne pollutants also contribute significant concentrations of lead to water supply, reservoirs and drainage basins. About one-fifth of the lead ingested in water is absorbed. The EPA estimates that in young children about 20 per cent of lead exposure comes from drinking water; dust contributes at least 30 per cent, air 5 to 20 per cent and food 30 to 45 per cent.

Water containing lead in excess of the standard should not be used for baby formula or for cooking or drinking. Water treatment or use of a corrosion inhibitor is advised. Conventional water treatment including coagulation will partially remove natural or man-made lead in raw water. Measures to prevent or minimize lead dissolution include maintenance of pH greater or equal to 8.0 and use of zinc orthophosphate or polyphosphates. Silicates may have a long beneficial effect.

Removal of lead service line is required if treatment is not adequate to reduce lead level.

4.10.9 Manganese

Manganese is found in gneisses, quartzites, marbles and metamorphic rocks and hence, in well waters from these formations. It is also found in many soils and sediments, such as deep lakes, reservoirs and surface water. Manganese concentrations should not be greater than 0.05 mg/l and preferably less than 0.01 mg/l (MCL), to avoid the black-brown staining of plumbing fixtures, and laundry when chlorine bleach is added. The WHO guideline value for manganese is 0.1 mg/l.

Concentrations greater than 0.5 to 1.0 mg/l may give a metallic taste to water. Concentrations above 0.05 mg/l or less than this can cause or build up coatings on sand filter media, glass parts of chlorinators and concrete structures, and in piping which may cause reduction in pipe capacity. When manganese in solution comes in contact with air or chlorine, it is converted to the insoluble manganic state which is very difficult to remove from materials on which it precipitates. Excess polyphosphate for sequestering manganese may prevent absorption of essential trace elements from the diet; it is also a source of sodium.

4.10.10 Mercury

Mercury is found in nature in the elemental and organic forms. Concentrations in unpolluted waters are normally less than 1.0 mg/l. The organic methyl mercury and other alkyl mercury compounds are highly toxic, affecting the central nervous system and kidneys. It is taken up by the aquatic food chain. The maximum permissible contaminant level in drinking water is 0.002 mg/l as total mercury. The WHO guideline is 0.001 mg/l.

4.10.11 Silver

The secondary MCL for silver in drinking water is 0.10 mg/l. Silver is sometimes used to disinfect small quantities of water and home faucet *purifiers*. Colloidal silver may cause permanent discolourations in the eyes and mucous membranes. A continuous dose of 400 mg of silver may produce the discolouration. Only about 10 per cent of the ingested silver is absorbed.

4.10.12 Sodium

Persons on a low sodium diet, because of heart, kidney or circulatory (hypertension) disease or pregnancy should use distilled water if the water supply contains more than 20 mg/l of sodium and be guided by a physician's advice. The consumption of 2.0 litres of water per day is assumed. Water containing more than 200 mg/l sodium should not be used for drinking by those on a moderately restricted sodium diet. It can be tasted at this concentration when combined with other anions. Many groundwater supplies and most home softened (using ion-exchange) well waters contain too much sodium for persons on sodium-restricted diets. If the well water is low in sodium (less than 20 mg/l), but the water is softened by the ion-exchange process because of excessive hardness the cold water system can be supplied by a line from the well that bypasses the softener and low sodium water can be made available at cold water taps. Sodium can be removed by reverse osmosis, distillation and cation exchange, but it is quite expensive. A laboratory analysis is necessary to determine the exact amount of sodium in water. The WHO guideline for sodium in drinking water is 200 mg/l. Common sources of sodium in addition to food are certain well waters, ion-exchange water-softening units, water treatment chemicals (sodium aluminates, lime-soda ash in softening sodium hydroxide, sodium bisulphite and sodium hypochlorites) and possibly industrial effluents.

4.10.13 Zinc

The concentration of zinc in drinking water should be less than 1.0 mg/l. The MCL and WHO guideline is 5.0 mg/l. Zinc is dissolved by surface water. A greasy film forms in surface water containing 5 mg/l or more of zinc upon boiling. More than 5.0 mg/l brings about a bitter metallic taste and 25 to 40 mg/l may cause nausea and vomiting. At high concentrations zinc may contribute to the corrosiveness of water. Common sources of zinc in drinking water are brass and galvanized pipe and natural waters where zinc has been mined. Zinc from zinc oxide in automobiles is a significant pollutant in urban run-off. The ratio of zinc to cadmium is also of public health importance. Zinc deficiency is associated with dwarfism and hypogonadism. Zinc is an essential nutrient. Zinc can be reduced by ion-exchange, softening, reverse osmosis and electrodialysis.

4.11 CHEMICALS

4.11.1 Benzene

This chemical is employed as a solvent and degreaser of metals. It is also a major component of gasoline. Drinking water contamination generally results from leaking under ground gasoline and petroleum tanks or improper waste disposal. Benzene has been associated with significantly increased risks of leukaemia among certain industrial workers exposed to relatively large amounts of this chemical during their working careers. This chemical has also been shown to cause cancer in laboratory animals when the animals are exposed to high levels over their lifetimes. Chemicals that cause increased risk of cancer among exposed industrial workers also may increase the risk of cancer in humans who are exposed at lower levels over long period of time. The EPA has set the enforceable drinking water standard for benzene at 0.005 mg/l to reduce the risk of cancer or other adverse health effects observed in humans. The Occupational Health and Safety Administration (OHSA) standard is 1 mg/l with 5 mg/l for short term (15 min exposure).

4.11.2 Carbon Tetrachloride

This chemical was once a popular household cleaning fluid. It generally gets into drinking water by improper disposal. This chemical has been shown to cause cancer in laboratory animals such as rats and mice when exposed to high levels over their lifetimes. Chemicals that cause cancer in laboratory animals may also increase the risk of cancer in humans exposed at lower levels over long period of time. The EPA has set the enforceable drinking water standard for carbon tetrachloride at 0.005 mg/l to reduce the risk of cancer or other adverse health effects observed in laboratory animals. The WHO tentative guideline value is 3 mg/l.

4.11.3 Cyanide

Cyanide is found naturally and in industrial wastes. Cyanide concentrations as low as 10 mg/l have been reported to cause adverse effects in fish. Long-term consumption of upto 4.7 mg/day has shown no injurious effect. The cyanide concentration in drinking water should not exceed 0.2 mg/l. The probable oral lethal dose is 1 mg/l. The WHO guideline is 0.1 mg/l. The Maximum Contaminant Level (MCL) and Maximum Contaminant Level Goal (MCLG) of 0.2 mg/l has been set by the EPA of USA. Cyanides can ultimately decompose to carbon dioxide and nitrogen gas. Cyanide is readily destroyed by conventional treatment process.

4.11.4 1, 1 Dichloroethylene

This chemical is used in industry and is found in drinking water as a result of the breakdown of related solvents. The solvents are used as cleaners and degreasers of metals and generally found their way into drinking water by improper waste disposal. This chemical has been found to cause liver and kidney damage in laboratory animals, such as rats and mice when exposed to high levels over their lifetimes. Chemicals that cause adverse effects in laboratory animals may also cause adverse effects in humans exposed at lower levels over a long duration of time. The EPA has set the enforceable drinking water standard for 1, 1 dichloroethylene at 0.007 mg/l to reduce the risk of the adverse health effects observed in the case of laboratory animals.

4.11.5 1, 2 Dichloroethane

This chemical is used as a cleaning fluid for fats, oils, waxes and resins. It generally gets into drinking water from improper waste disposal. This chemical has been shown to cause cancer in laboratory animals, such as rats and mice when exposed to high levels over their lifetimes. Chemicals that cause cancer in laboratory animals may also increase the risk of cancer in humans exposed at lower levels over long period of time. The EPA has set the enforceable drinking water standard for 1, 2 dichloroethane at 0.005 mg/l to reduce the risk of cancer or other adverse health effects observed in the case of laboratory animals. The WHO guideline is 10 mg/l.

4.11.6 Hydrogen Sulphide

Hydrogen sulphide is most frequently found in groundwaters as a natural constituent and is easily identified by the rotten egg odour. It is caused by microbial action or organic matter or the reduction of sulphate ions to sulphide. A concentration of 70 mg/l is an irritant, but 700 mg/l is highly poisonous. In high concentration it paralyzes the sense of smell, thereby making it more dangerous. Black stains on laundered clothes, and black deposits in piping and plumbing fixtures are caused by hydrogen sulphide in the presence of soluble iron. Hydrogen sulphide in drinking water should not be detectable by smell and its concentration should not exceed 0.05 mg/l. Hydrogen sulphide predominates at pH of 7.0 or less. It can be removed by aeration or chemical oxidation followed by filtration.

4.11.7 Para-dichlorobenzene

This chemical is a component of deodorizers, moth balls and pesticides. It generally gets into drinking water by improper waste disposal. This chemical has been shown to cause liver and kidney damage in laboratory animals, such as rats and mice exposed to high levels over their lifetimes. Chemicals that cause adverse effects in laboratory animals may also cause adverse health effects in humans exposed at lower levels over long period of time. The EPA has set the enforceable drinking water standard for para-dichlorobenzene at 0.075 mg/l to reduce the risk of the adverse effect on health, observed in the case of laboratory animals.

4.11.8 Pesticides

Pesticides include insecticides, herbicides, fungicides, rodenticides, regulators of plant growth, defoliants or desiccants. Sources of pesticides in drinking water are industrial wastes, spills and dumping of pesticides, run-off from fields, inhabited areas, and farms or orchards treated with pesticides. Surface and groundwater may be contaminated. Pesticides cannot be adequately removed by conventional water treatment. Powered or granular activated carbon treatment may also be needed. Maximum permissible contaminant levels of certain pesticides in drinking water and their uses and health effects have been given by Drinking Water Regulations Authority.

4.11.9 Phenols

The WHO guideline for individual phenols, chlorophenols and 2, 4, 6 trichlorophenols is not greater than 0.1 mg/l (0.1 ppb), as the taste and odour can be detected at or above that level after chlorination. The odour of some chlorophenols is detected at 1 mg/l. In addition, 2, 4, 6 trichlorophenol, found in biocides and chlorinated water containing phenol is considered as a chemical carcinogen based on animal studies. The guideline for pentachlorophenol in drinking water, a wood preservative, is 0.001 mg/l based on its toxicity. It also causes objectionable taste and odour. If the water is not chlorinated, phenols upto 100 mg/l are acceptable. Phenols are a group of organic compounds that are by-products of steel, coke distillation, petroleum refining and chemical operations. They should be removed prior to discharge to drinking water sources. Phenols are also associated with the natural decay of wood products, biocides and municipal waste water discharges. The presence of phenols in water can cause serious problems in food and beverage industries, and can taint fish. Chlorophenols can be removed by chlorine dioxide and ozone treatment and by activated carbon. It is desirable to have the phenol concentrations to be less than 2.0 mg/l at the point of chlorination. Chlorine dioxide, ozone or potassium permanganate pretreatment is preferred, where possible, to remove phenolic compounds.

4.11.10 Polychlorinated Biphenyls (PCBs)

Polychlorinated biphenyls give an indication of the presence of industrial effluents containing mixtures of chlorinated biphenyl compounds having various percentages of chlorine. Organochlorine pesticides have a similar chemical structure. The PCBs cause skin disorders in humans. They are stable and fire resistant and have good electrical insulation capabilities. The

Food and Drug Administration (FDA) action levels are 1.5 mg/l in fat of milk and dairy products. The MCL for drinking water is 0.0005 mg/l with zero as EPA Maximum Contaminant Level Goal. The PCB contamination of well water has been associated with leakage from old submersible well pumps containing PCB in the capacitors. Activated carbon adsorption and ozonation plus UV treatment are possible water treatments to remove PCBs. Polybrominated biphenyl, a derivative of PCB, is more toxic than PCB. Aroclor is the trade name for a PCB mixture used in a pesticide.

4.11.11 Polynuclear Aromatic Hydrocarbons

Polynuclear aromatic hydrocarbons, such as fluoranthene, 3.4 benz fluoranthene, etc. are known carcinogens, and are potentially hazardous to humans. The WHO set a limit of 0.2 μg/l for the sum of these chemicals in drinking water, comparable in quality with unpolluted groundwater. It was also recommended that the use of coal tar-based pipe linings be discontinued.

4.11.12 1, 1, 1 Trichloroethane

This chemical is used as a cleaner and degreaser of metals. It generally gets into drinking water by improper waste disposal. Some industrial workers who were exposed to relatively large amount of this chemical during their working careers suffered damage to liver, nervous system and circulatory system. Chemicals that cause adverse effects among exposed industrial workers and laboratory animals may also cause adverse health effects observed in humans exposed at lower levels over longer period of time. The EPA has set an enforceable drinking water standard for 1, 1, 1 trichloroethane at 0.2 mg/l to protect against the risk of adverse health effects in humans and laboratory animals.

4.11.13 Trichloroethylene

This chemical is a common metal cleaning and dry cleaning fluid. It generally gets into drinking water by improper waste disposal. The EPA, has set forth the enforceable drinking water standard for trichloroethylene at 0.005 mg/l to reduce the risk of cancer or other adverse health effects observed in laboratory animals.

4.11.14 Vinyl Chloride

This chemical is used in industry and is found in drinking water as a result of the breakdown of related solvents. The solvents are used as cleaners and degreasers of metals and generally get into drinking water by improper waste disposal. This chemical has been associated with significantly increased risks of cancer among certain industrial workers who were exposed to relatively large amount of this chemical during their working careers. The EPA has set the enforceable drinking water standard for vinyl chloride at 0.002 mg/l to reduce the risks of cancer or other adverse effects on health observed in humans and laboratory animals. Packed tower aeration removes vinyl chloride.

4.11.15 Nutrients

Nutrients are chemicals, such as nitrogen, phosphorus, carbon, sulphur, calcium, potassium, iron, manganese, boron and cobalt which are essential for the growth of living organisms. From the perspective of water quality, nutrients can be considered as pollutants when their concentrations are adequate to allow excessive growth of aquatic plants particularly algae. When nutrients stimulate the growth of algae, it is likely that the attractiveness of the body of water for recreational uses, as a drinking water supply and as a viable habitat for other living organisms gets adversely affected.

The enrichment of nutrients can lead to blooms of algae which eventually die and decompose. Their decomposition removes oxygen from the water, potentially leading to levels of dissolved oxygen that are inadequate to sustain normal life forms. Algae and decaying organic matter add colour, turbidity, odours and objectionable tastes to water that are hard to remove and that may greatly reduce its acceptability as a domestic water source. The process of nutrient enrichment is termed as **eutrophication**. It is of significance in lakes.

From a water quality perspective, the three most important nutrients are carbon, nitrogen and phosphorus. Plants need relatively large amounts of each of the above three nutrients and unless all the three are available, growth will be limited. The nutrient that is least available relative to the plant's needs is called the **limiting nutrient**. This suggests that algal growth can be controlled by identifying and reducing the supply of that particular nutrient. Carbon is usually available from a number of natural sources including alkalinity, dissolved carbon dioxide from the atmosphere and decaying organic matter; so it is not often the limiting nutrient. As a matter of fact, it is usually either nitrogen or phosphorus that controls algal growth rates. In general, sea water is most often limited by nitrogen, while freshwater lakes are most often limited by phosphorus.

Major sources of both nitrogen and phosphorus include chemical fertilizers run-off from animal feedlots and municipal waste water discharges. Besides this, certain bacteria and blue-green algae can obtain nitrogen directly from the atmosphere. These life forms are likely to be abundant in lakes that have high rates of biological productivity, making the control of nitrogen in such lakes is extremely difficult. Certain forms of acid rain can also contribute nitrogen to lakes. While nitrogen has several special sources, the only unusual source of phosphorus is from detergents. When phosphorus is the limiting nutrient in a lake that is experiencing an algal problem, it is particularly important to limit the nearby use of phosphate in detergents.

Nitrogen is not only capable of contributing to eutrophication problems, but when found in drinking water, a particular form of it can also pose a serious public health threat. Nitrogen in water is commonly found in the form of nitrate (NO_3) which itself is not particularly dangerous. However, certain bacteria commonly found in the intestinal tract of infants can convert nitrates (NO_3) to highly toxic nitrites (NO_2). Nitrites have a greater affinity for haemoglobin in the bloodstream than oxygen and when they replace that needed oxygen, a condition called **methemoglobinemia** results. The resulting oxygen starvation causes a bluish discolouration of the infant; hence it is commonly referred to as the 'blue baby' syndrome. In extreme cases, the victim may expire from suffocation. Usually, after the age of about 6 months, the digestive system of a child is sufficiently developed so that this syndrome does not take place.

4.12 BIOLOGICAL WATER QUALITY PARAMETERS (PATHOGENS)

It is a well-known fact that contaminated water is responsible for the spread of many contagious diseases.

Pathogens are disease-producing organisms that grow and multiply within the host. Examples of pathogens associated with water include bacteria—responsible for cholera, bacillary dysentery, typhoid and paratyphoid fever; viruses—responsible for infectious hepatitis and poliomyelitis; protozoa—which cause amoebic dysentery and giardiasis, and helminths or parasitic worms which cause disease known as **schistosomiasis**. The intestinal discharge of an infected person, a carrier, may contain billions of these pathogens which if allowed to enter the water supply, can cause epidemics of immense proportions. Carriers may not even necessarily exhibit symptoms of their disease, which makes it all more important to carefully protect all water supplies from any human waste contamination. Contaminated water caused by poor sanitation can lead to both water-borne and water-contact diseases. Water-borne diseases are those acquired by ingestion of pathogens not only in drinking water, but also from the water that makes it into a person's mouth from washing food, utensils and hands. Giardic cysts passed through the faeces of carriers, pose an unusual threat to surface water and even to municipal supply systems. They can be carried by wild animals as well as humans, may survive for months in the environment and are not easily destroyed by chlorination.

Water-contact diseases do not even require ingestion of the water. Schistosomiasis (bilharzia) is the most common water contact disease in the world. It is spread by free swimming larva in the water called **cercaria**, that attach themselves to human skin, penetrate it and enter the bloodstream. Water also plays an indirect role in other diseases which are common in developing countries. Insects that breed in water are responsible for the spread of malaria. Yellow fever, sleeping sickness and river blindness are spread in the same way. Inadequate supplies of water for personal hygiene results in skin diseases, such as scabies, leprosy as well as eye diseases, such as trachoma and conjunctivitis. Table 4.1 presents some of these water related problems.

Table 4.1 Water Related Health Problems

Category	*Spread by*	*Examples*
Water-borne	Drinking water contaminated by pathogens or washing hands, food or utensils in contaminated water	Typhoid, cholera, dysentery, diarrhoea, hepatitis, guinea, worm disease
Water-contact	Invertebrates living in water which act as carriers (vectors)	Schistomiasis (bilharzia), leptospirosis, tularemia
Water-hygiene	Inadequate supplies of water for personal hygiene	Skin diseases, scabies, leprosy, eye diseases, trachoma, conjunctivitis

Table 4.2 lists industrial waste water characteristics from a few industries.

Table 4.2 Industrial Waste Water Characteristics from a Few Selected Industries

Industry	*Nature of flow of waste water*	*BOD (Biological oxygen demand)*	*TSS (Total suspended solids)*	*COD (Chemical oxygen demand)*	*pH*	*N*	*P*
Meat products	Intermittent	High, extremely high	High	High, extremely high	Neutral	Present	Present
Milk handling	Intermittent	Average to high	Between low and average	Average to high	Acidic, alkaline	Adequate	Present
Cheese products	Intermittent	Very high	Between average and very high	Very high	Acidic, alkaline	Deficient	Present
Alcoholic beverages	Intermittent	Between high and very high	Between low and high	High to very high	Alkaline	Deficient	Deficient
Soft drinks	Intermittent	Between average and high	Low to high	Average to high	–	Deficient	Present
Textiles	Intermittent, continuous	High	High	High	Alkaline	Deficient	Present
Tanning and finishing	Intermittent	Very high	Very high	Very high	Acidic, alkaline	Adequate	Deficient
Metal finishing	Continuous, variable	Low	Between average and high	Low	Acidic	Present	Present
Paper and allied products	Continuous	Between average and very high	Between low and high	Between low and high	Neutral (mechanical pulping)	Deficient	Deficient
Pharmaceutical	Continuous, intermittent	High	Between low and high	High	Acidic, alkaline	Deficient	Deficient
Plastics and resins	Continuous, variable	Between low and high	Between low and high	Between average and high	Acidic, alkaline	–	–

The physical, chemical and biological characteristics of waste water and their sources are given in Table 4.3.

Table 4.3 Physical, Chemical and Biological Characteristics of Waste Waters and their Sources

Characteristics	*Sources*
Physical properties	
Colour	Domestic and industrial wastes, natural decay of organic materials
Odour	Decomposing waste water, industrial effluents
Solids	Domestic water supply, domestic and industrial wastes, soil erosion, inflow, infiltration
Temperature	Domestic and industrial wastes
Chemical constituents	
Organic Carbohydrates	Domestic, commercial and industrial wastes
Fats, oils and grease	Domestic, commercial and industrial wastes
Pesticides	Agricultural wastes
Phenols	Industrial effluents
Proteins	Domestic and commercial wastes
Surfactants	Domestic and industrial wastes
Others	Natural decay of organic materials
Inorganic	
Alkalinity	Domestic wastes, domestic water supply, groundwater infiltration
Chlorides	Domestic water supply, domestic wastes, groundwater infiltration, water softeners
Heavy metals	Industrial effluents
Nitrogen	Domestic and agricultural wastes
pH	Industrial effluents
Phosphorus	Domestic and industrial wastes, natural run-off
Sulphur	Domestic water supply, domestic and industrial wastes
Toxic compounds	Industrial effluents
Gases	
Hydrogen sulphide	Decomposition of domestic wastes
Methane	Decomposition of domestic wastes
Oxygen	Domestic water supply, surface water infiltration
Biological constituents	
Animals	Open water course and effluent treatment plants
Plants	Open water course and effluent treatment plants
Protista	Domestic wastes, effluent treatment plants
Viruses	Domestic wastes

EXERCISE

4.1 Discuss the physical and chemical water quality parameters of interest to environmental engineers.

5

Waste Water Treatment and Disposal

The objective of waste water treatment is to remove the contaminants from the waste water so that the treated water can meet the recommended quality standards. These quality standards will have bearing on whether the treated water will be reused or discharged into a receiving stream. The waste water (industrial and other effluents) treatment can be broadly categorized as physical, chemical or biological depending on the techniques used. These processes, comprising a series of unit operations and sequences depending upon the existing situations of influent concentrations and composition. The condition and specifications of the effluent have relevance on the choice of the sequence.

Physical processes are based on the exploitation of the physical properties of the pollutants and it is easier to carry out these treatment methods. The physical processes principally consist of techniques like screening, floatation, sedimentation and filtration. Chemical processes, on the other hand, make use of the chemical properties of the contaminants or of the reagents that are added. These include commonly employed processes such as precipitation, coagulation and disinfection.

Air stripping, adsorption using activated carbon, oxidation and reduction, ion-exchange also come under physical and chemical processes. In certain situtations membrane processes like reverse osmosis and electrodialysis assume importance. Biological processes which use biochemical reactions include biological filtration and activated sludge process.

The waste water processes are categorized in accordance with the water quality they are expected to give rise to. These processes are usually classified as the primary treatment, the secondary treatment and the tertiary or the advanced waste water treatment. In the primary treatment, identifiable suspended solids and floating matter are removed. In the secondary treatment, organic matter that is soluble or in the colloidal form, is removed. Physical, chemical or biological processes or their various combinations depending on the impurities to be removed, constitute advanced waste treatment methods. Residual soluble non-biodegradable organic compounds including surfactants, inorganic nutrients and salts, trace contaminants of various types and dissolved inorganic salts are removed by these processes.

5.1 WASTE WATER PRE-TREATMENT

By definition, pre-treatment is the process (in certain situations more than one process may be needed) that prepares the waste water in a condition that it can be subjected to further treatment by conventional secondary treatment involving biological processes. In municipal waste water, it means the removal of floating debris and grit and the removal of oily scums. These pollutants are likely to inhibit the biological process and there is a possibility of the pollutants damaging mechanical equipments. Ideal influent parameters for municipal activated sludge, the principal biological treatment processes, are in the range of 100 to 400 mg/l for BOD_5 and suspended solids. There may be situations when municipal waste water and industrial effluents may have a pH either too acidic or too alkaline for optimum biological degradation and may thus require pH correction or adjustment. This may be accomplished by the addition of sulphuric acid or lime. There may also be requirements when the flow rate is so inconsistent (e.g. five-day week industrial effluents) that flow balancing in a storage tank is to be provided. This balancing or equalization tank may also be employed to balance the organic loading if that varies substantially. If a waste water is deficient in nutrients, essential for biological treatment, then nutrients may be added in the pre-treatment stage. Pre-treatment for municipal waste waters is normally only physical that is flow balancing, screenings, removal of grit or oily scum removal. Industrial influents may additionally need chemical pre-treatment in the form of air stripping (ammonia removal), oxidation, reduction (heavy metal precipitation) and air floatation (oil removal). Figure 5.1 shows some of the pre-treatment processes required in the case of municipal effluents. If industrial effluents are further treated in a municipal plant they would usually first undergo the processes shown in part (a) of Figure 5.1.

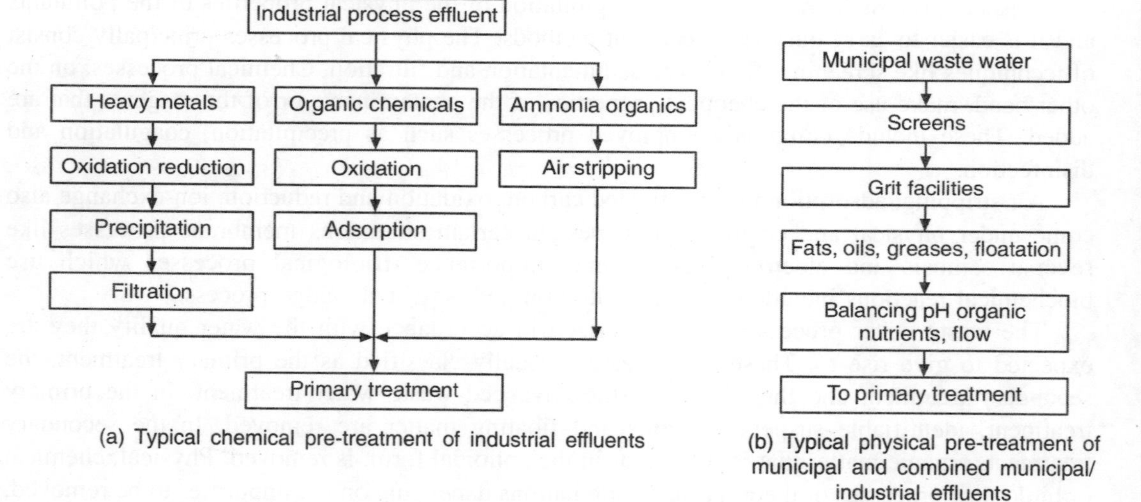

(a) Typical chemical pre-treatment of industrial effluents

(b) Typical physical pre-treatment of municipal and combined municipal/industrial effluents

Figure 5.1 Pre-treatment processes required in the municipal effluents.

5.2 PRIMARY TREATMENT

Primary treatment is often termed clarification, sedimentation and settling. This is the unit process where the waste water is allowed to settle for period not exceeding two hours in a settling tank and in that process produce a somewhat clarified liquid effluent in one stream and a liquid solid sludge (termed primary sludge) in a second stream. The objective is to produce a liquid effluent of fairly improved quality for the next treatment stage that is secondary biological treatment and to accomplish a solid separation resulting in a primary sludge that can be conveniently treated and disposed of. The benefits of primary treatment include

1. Reduction in suspended solids
2. Reduction in BOD_5
3. Reduction in the quantity of waste activated sludge in the activated sludge plant
4. Removal of floating material
5. Partial equilization of flow rates and organic load

Primary treatment is quiescent sedimentation with surface skimming of floating matter and grease, and bed level collection and removal of settled sludge. Sedimentation is carried out in different types of configurations including circular (which is most common), rectangular and square.

The sedimentation tanks may be flat bottomed or hopper bottomed. The waste water to be treated enters the tank, usually at the centre, through a well or diffusion box. The tank is so sized that the retention time is about 2 hours (the range can be 20 minutes to 3 hours). In this quiescent period, the suspended particles settle to the bottom as sludge and are raked towards a centre hopper from where the sludge is withdrawn. The clarified water is discharged over a perimeter weir at the surface of the tank at a rate known as the **basin overflow rate** or **surface overflow rate**. The units of the Surface Overflow Rate (SOR) are $m^3/day/m^2$. The last square metre is the plan area of the tank.

Primary sedimentation is among the oldest of waste water treatment processes. The amount of money spent on primary treatment invariably provides the greatest return on the investment in terms of Rupees per kg of pollutant removed.

Besides the usual design criteria, additional parameters termed as performance criteria have been established to monitor and improve the day-to-day performance. These criteria include

1. Influent flow rates and their variation (daily variation)
2. Influent waste strength rates and their variation
3. Recycle influent streams

From activated sludge or septage

- Supernatants from sludge dewatering
- Washings from tertiary filter processes

The aforementioned parameters may vary from hour to hour or from day to day. The flow rates may develop peaks several times; the daily average and waste strengths may vary accordingly. Recycle streams can come from several sources and in widely varying waste strengths. Septage, for instance, may have a BOD_5 value 30 times greater than municipal raw waste water. Supernatants from anaerobic digestion processes or filtrate back washings may also be very high in waste strength. As such, the performance of a primary clarification is not solely dependent on influent flow variations. For instance, plants that may have been over designed for flow may find that the retention time in the tank is not two hours, which is in accordance with the right design, but many times that quantity. Excessive retention time leads to septicity as there is no mixing in primary sedimentation. It is also likely that in situations wherein the operation and maintenance of primary tanks is poor (that is long retention times and infrequent sludge withdrawals), the quality of clarified water is no improvement on the influent waste water. However, with good performance management, removal rates in the range of 50 to 70 per cent for suspended solids and 25 to 40 per cent for BOD_5 can be accomplished. The depth of the sedimentation tanks varies from 2.5 to 5 metres.

The addition of coagulant chemicals (iron salts, lime, alum) before sedimentation, promotes flocculation of fine suspended matter into more readily settleable flocs. This enhances the efficiency very significantly of suspended solids and BOD_5 removal rates.

Chemical enhancement sustains the high removal efficiency over a wide range of removal rates. In conventional primary sedimentation tanks, as the surface overflow rate increases, the removal efficiency comes down. With chemical coagulants, the removal efficiencies are almost constant over an surface overflow rate range of 20 to 80 m^3/m^2/day. A demerit of the addition of chemical coagulants is an increase in primary sludge which is a chemical-type sludge, quite different from the biological sludge, from primary sedimentation. The mechanism of chemically enhanced primary sedimentation is to employ an aeration tank prior to the settling tank. The chemicals are added to the aeration tank.

In many plants, the total treatment process is pre-aeration with coagulant addition, followed by sedimentation. No secondary biological treatment follows. It is possible to meet most water quality standards using this process. This process merits consideration particularly for upgrading of existing plants, where space and cost may be the limiting conditions.

5.2.1 Screening and Communiting

The objective of screens is to remove large floating material (e.g., rags, plastic bottles, etc.) and thereby protect downstream mechanical equipments (pumps). There are four types of screens in normal use.

1. Coarse screens with opening greater than 6 mm that can remove large material.
2. Fine screens with openings in the range of 1.5 mm to 6 mm, which are sometimes used as a substitute for primary clarification (e.g. when activated sludge is used).
3. Very fine screens with openings in the range of 0.2 to 1.5 mm which reduce the suspended solids to primary clarification levels.
4. Micro-screens with opening in the range of 0.001 to 0.3 mm, which are used for effluent polishing as a final treatment step. These are not used in pre-treatment except as a single one step treatment process for predominately inorganic waste water, e.g. quarry washings.

Communiting is a traditional method of screening and shredding the retained material and then allowing it back into the flow. They are no longer recommended in view of the fact that items like plastic pieces can find their way to the biological plant creating inhibition conditions for the microbial population. Screens are designed to accommodate through flow velocities greater than 0.5 m/sec and less than 1.2 m/sec with maximum head losses of about 0.7 m.

Communities can be employed (as an alternative to racks or course screens) to grind up the coarse solids without removing them from flow. Communitors function to cut up (comminute) coarse solids to improve the downsteam operations and processes and to get rid of problems caused by the varied sizes of solids present in effluents. The solids are cut up into a smaller, more uniform size for return to the flow stream for removal in the subsequent downstream treatment operations and processes. Communitors can theoretically get rid of the messy and offensive task of screenings, handling and disposal. Their use is very advantageous in a pumping station to protect the pumps against clogging by rags and large objects, and to eliminate the need to handle and dispose of screenings.

There are conflicting views on the suitability of using communition devices at effluent treatment plants. Certain section of people maintain that once material has been removed from the effluent, it should not be returned, irrespective of the form. Certain other people are of the view that once cut up, the solids are more easily handled in the downstream processes.

A demerit of using communitors is that the communited solids often present downstream problems. The problems are particularly bad with rags which show the tendency to recombine after communition into ropelike stands if agitated. At treatment plants, this agitation is provided in grit chambers and aerated channels. These recombined rags can have a number of negative impacts like clogging pump impellers, sludge pipelines and heat exchangers, and accumulating on air diffusers.

5.2.2 Grit Removal

Grit removal may be accomplished in grit chambers or by the centrifugal separation of sludge. Grit chambers are designed to remove grit comprising sand, gravel cinders or other heavy solid materials that have subsiding velocities or specific gravities substantially greater than those of the organic putrescible solids in effluent.

Grit comprises sand, gravel, cinders or other heavy materials that possess specific gravities or settling velocities, considerably greater than those of orgánic putrescible solids. Besides these materials, grit includes eggshells, bone chips, seeds and large organic particles, such as food wastes. Generally, what is removed as grit is predominately inert and relatively dry. However, grit composition can be highly variable with moisture content ranging from 15 to 70 per cent and the volatile content from 1 to 60 per cent. The specific gravity of clean grit particles, reaches 2.8 for inerts, but can be as low as 1.5 when substantial organic material is agglomerated with inerts. A bulk density of 1600 kg/m^3 is commonly used for grit. Often, enough organics are present in the grit so that it quickly putrefies if not properly handled after removal from the waste water. Grit particles larger than 65 mesh (0.2 mm) have been cited as the cause of many of the downstream problems.

The actual size distribution of retained grit exhibits variation due to differences on collection system characteristics as well as variations in grit removal efficiency. Generally, most grit particles are retained on a No. 100 mesh (0.15 mm) sieve, reaching nearly 100 per cent retention in some instances. However, grit can be much finer.

Grit chambers are provided to serve the following purposes:

1. To protect moving mechanical equipment from abrasion and accompanying abnormal wear
2. To reduce formation of heavy deposits in pipelines channels and conduits, and

3. To bring about or effect a reduction in the frequency of cleaning necessitated by excessive accumulations of grit. The removal of grit is essential ahead of centrifuges—heat exchangers and high-pressure diaphragm pumps.

It is a common practice to locate the grit chambers after the bar racks and before the primary sedimentation tanks. In some installations, grit chambers precede the screening facilities. Generally, the installations of screening facilities ahead of grit chambers make the operation and maintenance of the grit removal facilities easier.

Locating grit chambers ahead of waste water pumps, when it is desirable to do so, would normally involve placing them at considerable depth at an additional expense. It is, therefore, usually considered more economical to pump the effluent, including the grit, to grit chambers located at a convenient position ahead of the treatment plant units, taking cognizance of the fact that pumps may require greater maintenance.

There are three general types of grit chambers. They are horizontal-flow: either of a rectangular or square configuration, aerated, or vortex type.

5.2.3 Primary Sedimentation

When a liquid, containing solids in the suspended state, is placed in a relatively quiescent state, those solids possessing a higher specific gravity than the liquid will show the tendency to settle and those with a lower specific gravity will tend to rise. These principles are employed in the design of sedimentation tanks for treatment of industrial effluents. The purpose of treatment by sedimentation is to remove readily settleable solids and floating materials and thus effect a reduction in the suspended solid contents.

Primary sedimentation tanks may provide the principal degree of waste water treatment or they may be employed as a preliminary step in the further processing of the waste water. When these tanks are used as the only means of treatment, they provide for the removal of

1. Settleable solids capable of forming sludge deposits in the receiving waters
2. Free oil and grease and other floating materials, and
3. A portion of the organic load discharged to the receiving waters

When primary sedimentation tanks are used ahead of biological treatment, their role or function is to effect a reduction on the load on the biological treatment units. A well-designed and operated primary sedimentation tank must be in a position to remove from 50 to 70 per cent of the suspended solids and from 25 to 40 per cent of the BOD_5.

Primary sedimentation tanks that precede biological processes may be designed to provide shorter detention time and a higher rate of surface loading than tanks serving as the only method of treatment except when waste activated sludge is returned to the primary sedimentation tanks for co-settling with primary sludge.

Sedimentation tanks have also been used as storm water retention tanks, which are designed to provide a moderate detention period (10 to 30 minutes) for overflows from either combined sewers or storm sewers. The objective is to get rid of a major portion of the organic solids that otherwise would be discharged directly to the receiving water and that could form offensive sludge deposits. Sedimentation tanks have also been used to provide sufficient detention periods for effective chlorination of such overflows.

5.3 SECONDARY TREATMENT

Under primary treatment, only the materials that could be removed by some type of physical or mechanical action, have been considered. The primary treatment processes will turn out to be largely ineffective in removing organic material in waste water which may be present in the colloidal or dissolved form. This organic matter (material) represents a high demand for oxygen which should be brought down further in order to make the effluent suitable for discharge into the water bodies.

Biological or secondary treatment, as it is commonly referred to is very similar in concept to the natural biodegradation of organic material by aerobic bacteria. The secondary treatment techniques are presented below.

5.3.1 Activated Sludge

The activated sludge process is based on the principle or has a bearing on a dense microbial population being in mixed suspension with the liquid effluent under aerobic conditions. With unlimited food and oxygen, extremely high rates of microbial growth and respiration can be accomplished, resulting in the utilization of the organic matter present to either oxidized end products, such as CO_2, NO_3, SO_4 and PO_4 or the biosynthesis of new microorganisms. The activated sludge process relies on the following five interrelated components presented in Table 5.1.

Table 5.1 Main Components of all Activated Sludge Systems

S. No.	*Components*	*Brief description*
1.	The reactor	This unit can be a tank, lagoon or ditch. The main criteria of a reactor are that the contents can be adequately mixed and aerated. The reactor is also termed the aeration tank or basin.
2.	Activated sludge	This is the microbial biomass within the reactor, which is constituted mainly of bacteria and other micro-fauna and flora. The sludge is a flocculant suspension of these organisms and is frequently referred to as the mixed liquor. The normal range concentration of mixed liquid, expressed as suspended solids is between 2000 and 5000 mg/l.
3.	Aeration/mixing system	Aeration and mixing of the activated sludge and incoming effluent are essential. While these tasks can be performed independently, they are usually carried out using a single system. Either surface aeration or diffused air is used.
4.	Sedimentation tank	Final settlement (or clarification) of the activated sludge displaced from the aeration tank by the incoming waste water is required. This effects the separation of the microbial biomass from the treated effluent.
5.	Returned sludge	The settled activated sludge in the sedimentation tank is recycled back to the reactor to maintain the microbial population at a required concentration so as to ensure continuation of treatment.

Removal of organic matter, the substrate, in the activated sludge process consists of three mechanisms.

1. Adsorption and also agglomeration onto microbial flocs.
2. Assimilation which is the conversion to new microbial cell material and eventually.
3. Mineralization which is complete oxidation.

The predominant removal mechanism can be selected by specific operating conditions. For instance, conditions of favouring assimilation removes substrates by precipitating it in the form of biomass which results in a higher proportion of the cost required for sludge separation and disposal (high rate activated sludge). Under circumstances favouring mineralization, the volume of biomass is brought down under endogenous respiratory conditions. This results in lower sludge handling costs, but higher aeration costs. Currently the higher cost of sludge treatment and disposal favours plants operating with low sludge production. The relationship between substrate (food) concentration and sludge biomass (microorganisms) concentration is a fundamental one in activated sludge operation. In the activated sludge plant, the mass of micro-organisms multiply rapidly in the presence of oxygen, food and nutrients. After maturation the microorganisms are developed to assimilate specific waste (log growth phase), which is the period of maximum removal. Then under substrate-limiting conditions, the microorganisms enter a declining growth phase leading eventually to auto-oxidation. In practice, activated sludge processes operate towards the end of the log phase and in the declining stationary growth phases.

5.3.2 Disinfection of Effluents

The term disinfection pertains to the selective destruction of disease-causing organisms. All the organisms are not normally destroyed during the process. This differentiates disinfection from sterilization, in which, all the organisms are destroyed. In the field of effluent treatment the three categories of human enteric organisms of the greatest consequence in producing disease are bacteria, viruses and amoebic cysts. Diseases caused by water-borne bacteria include typhoid, cholera, paratyphoid and bacillary dysentery. Diseases caused by water-borne viruses include poliomyelitis and infectious hepatitis.

The requirements for an ideal chemical disinfectant include toxicity to microorganisms, solubility, stability, non-toxicity to higher forms of life, toxicity at ambient temperatures, non-corrosiveness and non-staining characteristic, penetration, deodorizing ability, etc. It is also important that the disinfectant be safe to handle and apply, and that its strength or concentration in treated waters be measurable. Disinfection is most commonly accomplished by the use of techniques presented briefly as follows:

Chemical agents

Chemical agents that have been employed as disinfectants include

- Chlorine and its compounds
- Bromine
- Iodine
- Ozone
- Phenol and phenolic compounds
- Alcohols

- Heavy metals and related compounds
- Soaps and detergents
- Dyes
- Quaternary ammonium compounds
- Hydrogen peroxide, and
- A variety of alkalis and acids.

Of the aforementioned chemical agents, the most common disinfectants are the oxidizing chemicals and chlorine is the one which is most widely and universally used. Bromine and iodine have also been used for effluent disinfections. Ozone is a highly effective disinfectant and its application is increasing even though it leaves no residual. Highly acidic or alkaline water can also be used to destroy pathogenic bacteria because water with pH greater than 11 or less than 3 is relatively toxic to most bacteria.

Physical agents

Physical disinfectants, that can be used are heat and light. Heating water to the boiling point, for instance, will destroy the major disease producing non-spore forming bacteria. Heat is commonly used in the beverage and dairy industries, but it is not a feasible means of disinfecting large quantities of waste water in view of the high cost involved in this process. However, pasteurization of sludge is used extensively in some developed countries.

Sunlight is also a good disinfectant. In particular, ultraviolet radiation can be used. Special lamps that emit ultraviolet rays have been used successfully to sterilize small quantities of water. The efficacy of the process depends on the penetration of the rays in to water. The contact geometry between the ultraviolet source and the water is of paramount importance owing to the fact that suspended matter, dissolved organic molecules and water itself as well as the microorganisms will absorb the radiation. It is, hence, difficult to use ultraviolet radiation in aqueous systems especially when large amounts of particulate matter are present in the effluent.

Mechanical means

Bacteria and other organisms are also removed by mechanical means during the effluent treatment. Typical removal efficiencies for a variety of treatment operations and processes are presented in Table 5.2.

Table 5.2 Removal or Destruction of Bacteria by Different Treatment Processes

Process	*Per cent removal*
Coarse screen	0–5
Fine screen	10–20
Grit chambers	10–25
Plain sedimentation	25–75
Chemical sedimentation	40–80
Trickling filters	90–95
Activated sludges	90–98
Chlorination of treated waste water	98–99

The first four operations listed in Table 5.2 may be considered to be physical. The removals accomplished are a by-product of the primary function of the process.

Radiation

The major types of radiation are electromagnetic, acoustic and particle. Gamma rays are emitted from radioisotopes, such as cobalt 60. In view of their penetration power, gamma rays have been used to disinfect (sterilize) both water and waste water.

Four mechanisms that have been proposed to explain the action of disinfectants are:

- Damage to the cell wall
- Alteration of cell permeability
- Alteration of the colloidal nature of protoplasm, and
- Inhibition of enzyme activity

Damage or destruction of the cell wall will result in cell lysis and death.

Some agents, such as penicillin inhibit the synthesis of the bacterial cell wall.

Chemical agents, such as phenolic compounds and detergents alter the permeability of the cytoplasmic membranes. These substances destroy the selective permeability of the membrane and allow vital nutrients, such as nitrogen and phosphorus to escape.

Heat, radiation and highly acidic or alkaline agents alter the colloidal nature of the protoplasm. Heat will coagulate the cell protein, and acids or bases will denature proteins producing a lethal effect.

Another mode of disinfection is the inhibition of enzyme activity. Oxidizing agents such as chlorine, can bring about an alteration in the chemical arrangements of enzymes and deactivate the enzymes.

Ponds and lagoons

Besides the activated sludge processes, other suspended culture biological treatment systems are available for handling waste water with ponds and lagoons being the common ones. A waste water pond, also called **stabilization pond**, **oxidation pond** and **sewage lagoon** comprises a large shallow earthen basin in which waste water is retained long enough for natural purification processes to provide the necessary degree of treatment. At least a part of the system must be aerobic to produce an acceptable effluent. Even though some oxygen is provided by the diffusion from the air, the bulk of the oxygen in ponds is provided by photosynthesis. Lagoons are distinguished from ponds, as oxygen for lagoons is provided by artificial aeration. There is a wide variety of ponds and lagoons, each uniquely suited to specific applications.

Shallow ponds in which dissolved oxygen is present at all depths are termed aerobic ponds. Very often employed as additional treatment processes, aerobic ponds are referred to as polishing or tertiary ponds. Deep ponds in which oxygen is absent except for a relatively thin surface layer, are termed anaerobic ponds. Anaerobic ponds can be employed for partial treatment of a strong organic waste water, but must be followed by some form of aerobic treatment to produce acceptable end products. Under favourable conditions, facultative ponds in which both aerobic and anaerobic zones exist may be employed as the total treatment system for municipal waste water.

Lagoons are categorized on the basis of mechanical mixing product. When sufficient energy is supplied to keep the entire contents including the sewage solids mixed and aerated, the reactor is termed an aerobic lagoon. The effluent from an aerobic lagoon requires solids' removal in order to meet the suspended solids' effluent standards. When only adequate energy is supplied to mix the liquid portion of the lagoon, solids settle to the bottom in areas of low velocity gradients and proceed to degrade anaerobically. This facility is termed facultative lagoon and the process differs from the facultative pond only in the method by which oxygen is supplied.

The majority of ponds and lagoons serving municipal corporations are of the facultative type. Facultative ponds and lagoons are assumed to be completely mixed reactors without biomass recycle. Raw waste water is transported into the reactor and is released near the bottom. Waste water solids settle near the influent while biological solids and flocculated colloids form a thin sludge blanket over the rest of the bottom. Outlets are located so as to minimize short-circuiting.

5.3.3 Sludge Treatment, Utilization and Disposal

Some job is still remaining even after treating the waste water and discharging into a water course. The sludge is nothing, but the left behind solids which are suspended in water. At present, sludge treatment and disposal accounts for more than 50 per cent of the treatment costs in a typical secondary plant, making this non-glamorous operation, an essential aspect of waste water treatment.

This section is devoted to the problem of sludge treatment and disposal. The sources and qualities of sludge from various types of waste water treatment systems are presented first followed by a definition of sludge characteristics. Solid concentration techniques, such as thickening and dewatering are dealt with next. At the conclusion, the considerations for the ultimate disposal of the treated sludge are presented.

Sources of sludge

The first source of sludge is the suspended solids that enter the treatment plant and are partially removed in the primary settling tank or clarifier. Normally around 60 per cent of the suspended solids become raw primary sludge which is highly putrescent, contains pathogenic organisms and is very wet with around 96 per cent water content.

The removal of BOD is basically a method of wasting energy, and secondary waste water treatment plants are designed to bring about a reduction in this high energy material to low energy chemicals, typically accomplished by biological methods, using microorganisms (these are called **decomposers** in ecological parlance) that use the energy for their own life and procreation. Secondary treatment processes, such as the popular activated sludge system, are almost perfect systems except that the microorganisms convert too little of the high energy organics to CO_2 and H_2O and too much of it to new organisms. Thus, the system operates with an excess of these microorganisms as waste activated sludge. Normally, the mass of waste activated sludge per mass of BOD removed in secondary treatment is known as the yield, expressed as mass of Suspended Solids (SS), produced per mass of BOD removed. Typically, the yield of waste activated sludge is 0.5 kilogram of dry solids per kilogram of BOD reduced.

Phosphorus removal processes also invariably end up with excess solids. If lime is used, the calcium carbonates and calcium hydroxyapatites are formed and must be disposed of. Aluminium sulphate likewise produces solids, in the form of aluminium hydroxides and aluminium phosphates. Even the biological processes for phosphorus removal endup with solids. The use of an oxidation pond or marsh for phosphorus removal is possible only if some organisms (algae, water hyacinths, fish, etc.) are periodically harvested.

Sludge treatment

There is a potential for a lot of savings and troubles can be averted if sludge could be disposed of as it is drawn off the main process train. Unfortunately the sludges have three characteristics that make such a solution unlikely. They are aesthetically displeasing, they are potentially harmful and they have too much water.

The first two problems are frequently solved by stabilization, such as anaerobic or aerobic digestion. The third problem needs the removal of water by either thickening or dewatering. The following three sections deal with the topics of stabilization, thickening and dewatering and then ultimate disposal of the sludge.

Sludge stabilization

The objective of sludge stabilization is to entail reduction in the problems associated with two detrimental characteristics—sludge odour and putrescence, and the presence of pathogenic organisms. Sludge may be stabilized by using lime, by aerobic digestion or by anaerobic digestion.

Lime stabilization is accomplished by adding lime in the form of hydrated lime (calcium hydroxide) or as quicklime (CaO) to the sludge and thus increasing the pH to 11 or even above. This appreciably reduces the odour and helps in the destruction of pathogens. The major disadvantage of lime stabilization is that it is temporary with efflux of time and in terms of days the pH drops and the sludge once again becomes putrescent.

Aerobic digestion is a logical extension of the activated sludge system. Waste activated sludge is placed in dedicated tanks and the concentrated solids are allowed to continue their decomposition. The food for microorganisms is available only by the destruction of other viable organisms and both total and volatile solids are thereby reduced. However, aerobically digested sludges are more difficult to dewater than are anaerobic sludges and are not as effective in the reduction of pathogens as anaerobic digestion, a process shown in the Figure 5.2.

The biochemistry of anaerobic digestion is a step wise or staged process. Solution of organic compounds by extracellular enzymes is followed by the production of organic acids by a large and hearty group of anaerobic microorganisms known as the **acid formers**. The organic acids are in turn degraded further by a group of strict anaerobes termed **methane formers**. These microorganisms are the prima donnas of waste water treatment, becoming upset at the least change in their environment, and the success of anaerobic treatment depends on maintenance of suitable conditions for the methane formers. In view of the fact that they are strict anaerobes, they are unable to function in the presence of oxygen and are very sensitive to environmental conditions, such as pH, temperatures and the presence of toxins. A digester goes *sour* when the

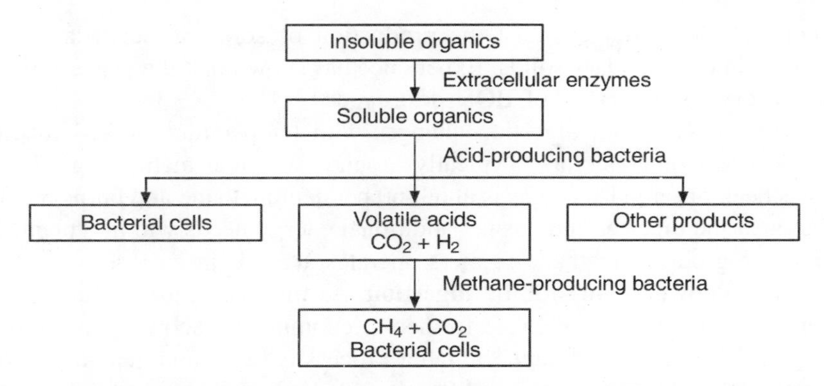

Figure 5.2 Generalized biochemical reactions in an aerobic sludge digestion.

methane formers have been inhibited in some way and the acid formers keep chugging away, making more organic acids, further lowering the pH and making conditions even worse for the methane formers. Curing a sick digester requires suspension of feeding and, often, massive doses of lime or other antacids.

Many of the treatment plants have both a primary and a secondary anaerobic digester like the one shown in Figure 5.3.

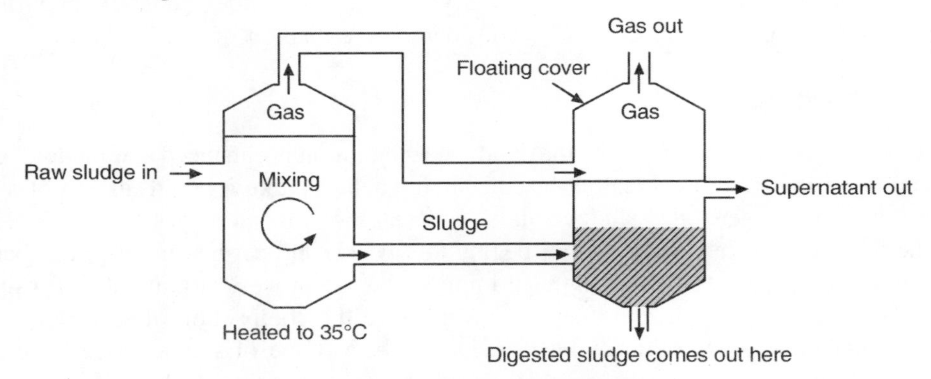

Figure 5.3 Anaerobic sludge digesters.

The primary digester is covered, heated and mixed to enhance the reaction rate. The temperature of the sludge is usually around 35°C. Secondary digesters are not mixed or heated and are used for storage of gas and for concentrating the sludge by settling. As the solids settle down, the liquid supernatant is pumped back to the main plant for further treatment. The cover of the secondary digester often floats up and down depending on the amount of gas stored. The gas is high enough in methane to be used as a fuel and in fact is usually employed to heat the primary digester.

Anaerobic digesters are commonly designed on the basis of solids holding. Experience has indicated that domestic waste waters contain about 120 grams of suspended solids per day per capita. This may be translated, knowing the population served into total suspended solids to be handled. Of course, added to this, must the production of solids in secondary treatment be expressed as the yield of secondary solids. Experience has shown that the waste activated sludge yield is 0.2 kg of dry solids per kg of BOD destroyed.

The production of gas from digestion changes with temperature, solids' loading, solids' volatility besides a few other factors. Typically around 0.6 cubic metre of gas/kg of volatile solids added, has been observed. The gas is about 60 per cent methane and burns readily, usually is being used to heat the digester and answer additional energy needs within a plant. It has been found that an active group of methane formers operates at 35°C in common practice, and this process has become known as **mesophilic digestion**. As the temperature is increased to about 45°C, however, another group of methane formers predominantly show up and this process is termed thermophilic digestion. Although the latter process is faster and generates more gas, the necessary elevated temperatures are more difficult and expensive to maintain.

All the three stabilization processes entail a reduction in the concentration of pathogenic organisms, but to varying extents. Lime stabilization accomplishes a high degree of sterilization owing to the high pH value. Further if quicklime (CaO) is employed the reaction turns out to be exothermic and the elevated temperatures assist in the destruction of pathogens. Although aerobic digestion at ambient temperatures is not very effective in the destruction of pathogens, anaerobic digesters have been well studied from the standpoint of pathogen viability, since the elevated temperatures should result in substantial sterilization. Unfortunately, many pathogens can survive digestion and polio viruses similarly survive with little reduction in virulence. Therefore, an anaerobic digester cannot be considered as a method of sterilization.

Sludge thickening

Sludge thickening is a process in which the solids' concentration is enhanced and the total sludge volumes are correspondingly decreased, but the sludge behaves like a liquid instead of a solid. Thickening commonly generates sludge solids' concentrations in the 3 per cent to 5 per cent range, whereas the point at which sludge begins to have the properties of a solid is between 15 per cent and 20 per cent. Thickening also implies that the process is gravitational using the difference between particle and fluid densities to achieve the compaction of solids.

The advantages of sludge thickening in reducing the volume of sludge to be handled are plenty. With reference to Figure 5.4, a sludge with 1 per cent solids thickened to 5 per cent, results in an 80 per cent volume reduction.

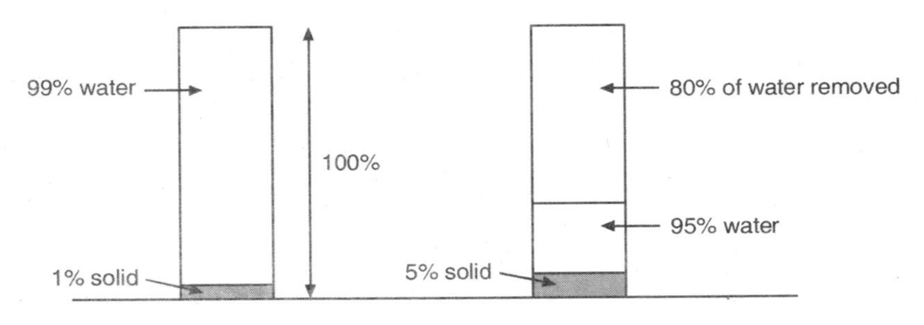

Figure 5.4 Volume reduction owing to sludge thickening.

A concentration of 20 per cent solids which might be accomplished by dewatering results in a 95 per cent reduction in volume with resulting savings in treatment, handling and disposal costs.

Two types of non-mechanical thickening operations are presently employed; the gravity thickener and the floatation thickener. The latter also uses gravity to separate the solids from the liquid, but for simplicity both descriptive terms are used.

A gravity thickener which is commonly used is shown in Figure 5.5.

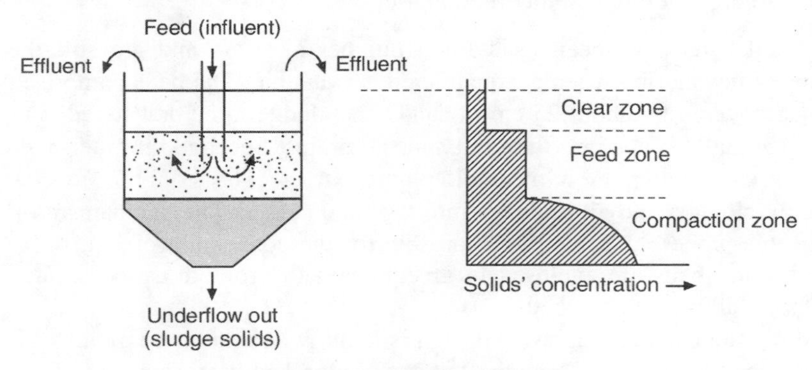

Figure 5.5 Gravity thickener.

The influent or feed enters in the middle and the water moves to the outside, eventually leaving as the clear effluent over the weirs. The sludge solids settle as a blanket and are removed out at the bottom.

A floatation thickener shown in Figure 5.6, operates by forcing air under pressure to dissolve in the return flow and releasing the pressure as the return is mixed with the feed. As the air comes out of the solution, tiny bubbles attach themselves to the solids and carry them upward to be scraped off.

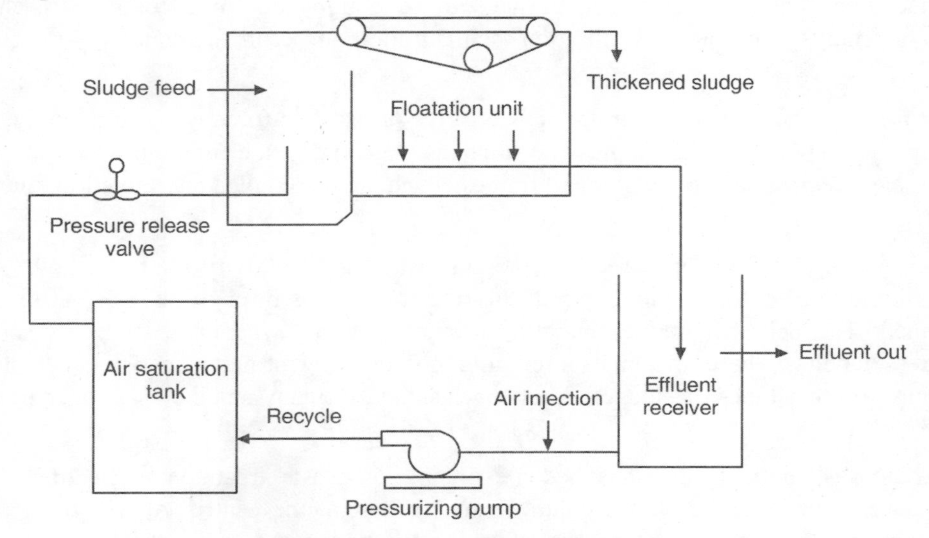

Figure 5.6 Floatation thickener.

Sludge dewatering

Dewatering differs from thickening in that, the sludge should behave as a solid after it is has been dewatered. Dewatering is rarely used as an intermediate process unless the sludge is to be incinerated and most waste water plants use dewatering as a final method of volume reduction before ultimate disposal.

The widely employed dewatering techniques are sand beds, pressure filters, belt filters and centrifuges. These are briefly presented hereinafter.

Sand beds: Sand beds have been used for a number of years and are still the most cost-effective means of dewatering when adequate land is available. The beds comprise tile drains in sand and gravel, covered by about 25 cm of sand. The sludge to be dewatered is poured on the sand. The water initially seeps into the sand and tile drains. Seepage into the sand and through the tile drains, although important in the total volume extracted, lasts only for few days. The sand pores are quickly clogged and all drainage into the sand ceases. The mechanism of evaporation takes over and this process is actually responsible for the conversion of liquid sludge to solid. In some places, sand beds are enclosed in greenhouses to promote evaporation as well as to prevent rain from falling into the beds.

For mixed digested sludge, the usual design is to allow around three months of drying time, making the sand bed to rest for a month after the sludge has been removed. This turns out to be an effective means of increasing the drainage efficiency.

Because raw primary sludge will not drain well on sand beds and will usually have an obnoxious odour, these sludges are rarely dried on beds. Raw secondary sludges are likely to either seep through the sand or clog the pores so quickly that no effective drainage will take place. Aerobically digested sludges may be dried on sand, but usually with some difficulty. In some situations, sludges are intentionally frozen in freezing beds to enhance their dewatering after the spring thaw.

If dewatering by sand beds is considered impractical due to lack of land and high labour costs, mechanical dewatering techniques need to be employed. Three common mechanical dewatering processes are pressure filtration, belt filtration and centrifugation.

Pressure filter: The pressure filter employs positive pressure to force water through a cloth. It is a common practice to build the pressure filters as plate and frame filters in which the sludge solids are captured between the plates and frames which are then pulled apart to allow for sludge cleanout.

Belt filter: The belt filter operates as a pressure filter and a gravity drainage, both. As the sludge is introduced onto the moving belt, the free water drips through the belt but the solids are retained. The belt then moves into the dewatering zone, where the sludge is squeezed between two belts. These machines are quite good in performance and are effective in dewatering different types of sludges. They are installed in many small waste water treatment plants.

Centrifugation: Centrifugation has become popular in waste treatment only after organic polymers were available for sludge conditioning. Although, the centrifuge will work on any sludge, most unconditioned sludges cannot be centrifuged with greater than 60 per cent or

70 per cent solids' recovery. The centrifuge most widely used is the solid bowl decanter, which consists of a bullet-shaped body rotating on its axis. The sludge is placed in the bowl, and the solids settle out under about 500 to 1000 gravities (centrifugally applied) and are scraped out of the bowl by a screw conveyor. Despite the fact that the laboratory tests are of some value in estimating centrifuge applicability, tests with continuous models are considerably better and highly recommended, whenever possible.

The solid concentration of the sludge from sand drying beds can be as high as 90 per cent after evaporation. Mechanical devices, however, will produce sludge ranging from 15 per cent to 35 per cent solids.

Utilization and ultimate disposal

The options for ultimate disposal of sludge are limited to air, water and land. Stringent controls on air pollution complicate incineration eventhough this, certainly, is an option. Disposal of sludges in deep water (such as oceans) is decreasing owing to adverse or unknown detrimental effects on aquatic ecology. Land disposal may be either dumping in a landfill or spreading out over land and allowing natural biodegradation to assimilate the sludge into the soil. Because of environmental and cost considerations, incineration and land disposal are presently very widely employed.

Incineration is actually not a method of disposal at all, but rather a sludge treatment step in which the organics are converted to H_2O and CO_2 and the inorganics are oxidized to non-putrescent ash residue. Two types of incineration units (incinerators) have found applications in sludge treatment; multiple hearth and fluid bed. The multiple hearth incinerator, as the name implies, has several hearths stacked vertically, with rabble arms pushing the sludge progressively downward through the hottest layers and finally into the ash pit. The fluidized bed incinerator is full of hot sand and is suspended by air injection; the sludge is incinerated within the moving sand. Because of the violent motion within the fluid bed, scraper arms are not needed. The sand acts as a *thermal flywheel* allowing intermittent operation.

When sludge is destined for disposal on land and the beneficial aspects of such disposal are emphasized, sludge is often euphemistically referred to as biosolids. The sludge has nutrients (nitrogen and phosphorus), is high in organic contents and as mentioned earlier is full of water. Thus, its potential as a soil additive is often highlighted. However, both high levels of heavy metals, such as cadmium, lead and zinc, as well as contamination by pathogens that may survive the stabilization process, can be troublesome.

Heavy metals entering the waste water treatment plant tend to concentrate on the sludge solids and so far no effective means of removing these metals from the sludge prior to sludge disposal, have been found. Control must, therefore, focus on maintaining stringent regulations (industrial pre-treatment regulations) that prevent the discharge of the metals into waste water collection system.

Reduction in the levels of pathogens is often achieved in the sludge digestion process, but the process is not 100 per cent effective. Sludge that receive the equivalent of 30 days anaerobic digestion is classified by EPA of the United States as class B sludges which can be disposed of only on non-agricultural land like highway median strips; but a 30-day delay in any use of the land is required. Class A sludges are disinfected by other processes, such as composting and quicklime addition in which high temperatures act to kill the pathogens or by non-ionizing radiation. Sludge disposal represents a major headache for many municipal corporations because its composition reflects our style of living, our technological development and our ethical concerns. *Pouring things down the drain* is our way of getting rid of all types of unwanted materials, not recognizing that these materials often become part of the sludge that must be disposed of in the environment. All of us need to become more sensitive to these problems and keep potentially harmful materials out of our sewage system and out of sludge.

5.3.4 Solids Removal

Removal of suspended solids and sometimes dissolved solids may turn out to be necessary in advanced waste water treatment systems. There is a similarity between the solids removal processes employed in the treatment of potable water and those used in the case of advanced waste water treatment. However, the application of the processes is more difficult owing to the overall poor quality of the waste water.

Suspended solids removal

As an advanced treatment process, suspended solids' removal tantamounts to the removal of particles and flocs that are small or too lightweight to be removed in gravity settling operations. There is a possibility of these solids being carried over from the secondary classifier or from tertiary systems in which solids were precipitated.

A variety of methods/techniques are available for removing residual suspended solids from effluents. The successful methods that have been employed include centrifugation, mechanical micro-screening, air floatation and granular media filtration. Basically the same principles that apply to filtration of particles from potable water apply to the removal of residual solids in effluents. There is a likelihood of the differences in operational modes between the two.

Sand filters have been employed to polish effluents from septic tanks and anaerobic treatment units for many years. The process involves essentially a slow sand filter. This type of filter has been applied to the effluent from oxidation ponds effectively. Because they are alternately dosed and allowed to dry, the term intermittent sand filters has been applied to these type of equipments.

Granular media filtration is normally the process of choice in larger secondary systems. Dual or multimedia beds prevent surface plugging problems and allow for longer filter runs. Loading rates have a bearing on both the concentration and nature of solids in the effluent.

Other recent innovations in filtration practices are very effective for advanced waste water treatment. Moving bed filters have been developed in the recent past which are continuously cleaned and the rate of cleaning can be adjusted to match the loading rate of solids. Another improved version termed the pulse bed filter, uses compressed air to periodically break up the surface mat deposited on a thin bed of fine filter media. The filter is backwashed only after a thick suspension of solids has accumulated on the bed which requires frequent pulsing.

Both the moving bed and the pulsed bed filters have the capability of filtering raw waste water. A much higher percentage of solids can be removed by filtration that can be removed in primary settling. The filter effluent, containing lower levels of mostly dissolved organic compounds respond very well to conventional secondary treatment.

Dissolved solid removal

The dissolved organic solid contents of waste water is decreased by the secondary treatment and nutrients removal. However, none of these processes completely removes all the dissolved organic constituents and neither of these removes significant quantities of inorganic dissolved solids. In situations wherein substantial reductions in the total dissolved solids of waste water turn out to be necessary, further treatment is required.

The methods that can be employed to decrease the dissolved solid contents of water include ion-exchange, micro-porous membrane filtration, adsorption and chemical-oxidation. Their use can be adopted to advance waste water requirement if a high level of pre-treatment is provided. Removal of the dissolved organic material (by activated carbon adsorption) is necessary prior to micro-porous membrane filtration to prevent the larger organic molecules from plugging the micro-pores.

It may be mentioned in this context that the advanced waste water treatment for dissolved solids' removal is quite complicated and expensive. Treatment of municipal waste water by these processes can be justified only when reuse of the waste water is contemplated or envisaged.

5.4 ADVANCED WASTE WATER TREATMENT

Advanced waste water treatment also known as tertiary treatment may be found necessary in some instances to safeguard the water quality of the receiving groundwaters and surface waters from added undesirable nutrients, toxic and hazardous chemicals, and pathogenic organisms not removed or inactivated by conventional biological secondary waste water treatment. For instance, nitrogen and phosphorus in plant effluent may promote the growth of plankton and the nitrates may contaminate groundwater; toxic organic and inorganic chemicals may jeopardize fish and shellfish and also endanger the quality of source of water for water supply, recreation and shellfish growing. Pathogens, such as the infectious hepatitis virus, giardia entamoza, ascaris and certain worms not removed or destroyed by the usual sewage treatment including chlorination, place on additional burden on water treatment plants and increase the probability of water-borne disease outbreaks.

Advanced waste water treatment may include combinations of the unit operations/unit processes presented here, following secondary treatment, depending upon the water quality objectives to be met. These are only examples and are not intended to be all inclusive.

For nitrogen removal

For removing or reducing the nitrogen content of waste water, the following techniques are used. Breakpoint chlorination is done to bring about a reduction in ammonia nitrogen level (nitrate and organic compounds are not affected by this process). Ion-exchange which is carried out after the filtration pre-treatment process is to reduce nitrate, nitrogen and ammonia levels. Selective resins are used in each. Phosphate is also reduced by this process. Another technique is nitrification followed by denitrification. By this process, ammonia, if present, is removed or converted to nitrate and then to nitrogen gas. Ammonia stripping (degasifying) is done to remove ammonia nitrogen. Ammonia can be oxidized to nitrate, by the biological activated sludge process. Denitrification can also be achieved by filtration through sand or granular activated carbon or by biological denitrification, usually under anaerobic conditions following activated sludge treatment. Reverse osmosis, following treatment to prevent fouling of membranes, can also be adopted to entail a reduction in total nitrogen level as well as dissolved solids. Electro-dialysis following pre-treatment is yet another technique to reduce ammonia, nitrate nitrogen levels besides dissolved solids. Oxidation pond is also useful for reducing nitrogen levels.

For phosphorus removal

In order to reduce phosphate level, total dissolved solids (TDS), and to remove increased additional nitrogen and heavy metals, coagulation (with lime, alum, ferric chloride and polyelectrolytes) and sedimentation can be employed.

In order to reduce phosphate level further and also to reduce the suspended solids, a combination of coagulation, sedimentation and filtration with mixed media can be resorted to. Here in, additional nitrogen is also removed but the TDS content is increased.

Another technique to remove phosphorus, (pH above 11) is lime treatment (which is carried out after biological treatment) followed by filtration. Suspended solids are also removed by this. Yet another method to reduce phosphate, dissolved solids and nitrogen, is ion-exchange with selected specific resins.

For the removal of dissolved organics

To reduce COD including dissolved organics and chlorine, activated carbon (granular or powdered) adsorption can be used. Dissolved solids can also be removed by reverse osmosis following pre-treatment, electro-dialysis following pre-treatment and distillation following pre-treatment. The level of dissolved organics can be brought down, also by biological waste water treatment. To remove volatile organics, aeration method can be adopted.

For heavy metal removal

The techniques for the removal of heavy metals include lime treatment, coagulation and sedimentation.

For dissolved inorganic solids removal (dimineralization)

To reduce total dissolved solids, the methods that can be used are ion-exchange (with anionic and cationic resins) following pre-treatment, reverse osmosis and electro-dialysis.

For suspended solids removal

For the reduction of suspended solids, nitrogen, ammonia and phosphate filtration (with sand, lime or ferric chloride and possibly polyelectrolytes) and sedimentation can be used. Adding ammonia stripping technique to this will reduce total nitrogen further.

For recarbonation

This is nothing but carbon dioxide addition, which is carried out with an idea of reducing the pH, in situations where the waste water pH has been raised from 10 to 11. This turns out to be necessary to reduce deposition of calcium carbonate in pipelines, equipments or the receiving water course.

For heat removal

To lower the temperature of waste water prior to discharge, open reservoir or evaporative cooler can be employed.

Removal of toxic and hazardous substances

Removal of toxic or hazardous substances must start at the source with in-house process change, if possible, reclamation and reuse, waste control and pre-treatment before discharge to a municipal sewer or water course.

5.4.1 Nutrient Removal

Chemical precipitation of phosphorus

Phosphorus that is found in effluent is principally in the forms of orthophosphate ion, polyphosphates or condensed phosphates and organic phosphorus compounds. Among the above, the predominant phosphorus species is orthophosphate which can occur in varying chemical forms depending on the pH of the effluent. In aqueous solutions the polyphosphates gradually hydrolyze to the ortho form, with the rate of this chemical reversion being accelerated by increasing temperature or by lowering pH or the presence of bacterial enzymes.

During biological waste water treatment organic phosphorus is converted to orthophosphate and the polyphosphates are also hydrolyzed to the ortho form. In the removal of phosphorus from waste water by chemical means, this reversion phenomenon is very beneficial, since orthophosphate is the easiest form of phosphorus to precipitate chemically. Major sources of phosphorus in domestic waste water are human wastes, with food scraps accounting for less than 50 per cent of the total phosphorus while detergents with phosphate builders contribute the remaining 50 per cent.

Many ionic forms are effective in the precipitation of phosphorus from solution. The most important among these owing to their relatively low cost and general availability are aluminium, calcium and iron. In general, the extent of phosphorus removal by chemical precipitation has a bearing on (1) initial phosphorus concentration (2) precipitating cation concentration (3) the concentration of other anions in competition with phosphorus for precipitating cations, and (4) the pH of the effluent. The tendency of aluminium and iron to hydrolyze in aqueous solution gives rise to a competition between the hydroxide and phosphate ions for precipitating metal ion. Thus, the efficiency of phosphorus removal depends on the relative concentrations of these two anions in solution and is consequently pH dependent; a decrease in pH or, more precisely, hydroxide favours the precipitation of phosphates with metallic cations. When calcium is used as a precipitant, the competition for calcium is predominantly between the phosphate and carbonate anions and again the phosphorus removal efficiency depends on the relative concentrations of the anion present and on pH.

Electro-dialysis of phosphorus and nitrogen

Electro-dialysis is a membrane separation process using a voltage impressed across cation-anion membrane pairs to remove dissolved solids from aqueous solutions. Specific ions of phosphorus and nitrogen from waste water can be removed by the electro-dialysis technique. Pre-treatment requirements and ion selectivity are especially important for this application. Development of highly selective, anti-fouling membrane has made nutrient removal possible. The nutrient and other dissolved salts removed from the feed stream are concentrated in a waste stream leaving the electro-dialysis cell. The overall nutrient removal process must provide for treatment or disposal of this concentrated waste stream.

The feed of the electro-dialysis cell passes through pairs of cation-anion permeable membranes across which a DC voltage is imposed. Ions containing phosphorus and nitrogen, as well as other dissolved ions, are removed from the feed stream by migrating under the influence of the applied voltage through the membranes into a concentrated waste stream on the other side. Depending on the membrane porosity and selectivity, approximately one half of the dissolved solids are removed from the feed stream which consists of approximately 10 per cent of the throughput. Suspended solids and free or dissolved organics must be removed from the feed stream prior to introduction into the electro-dialysis stack. Usually the concentrated waste is circulated again to the membrane stack at rates equal to the feed rate in order to equalize the flow and therefore, pressure drops across the membranes. Several membrane pairs are needed and feed may flow through the units in series, parallel or series—parallel depending on the supplier and the extent of removal required.

The electro-dialysis process changes only the concentration of the ions in question. Two product streams are produced, one lower and one higher in concentration of dissolved solids. The overall nutrient removal process must, therefore, include methods for handling the concentrated wastes which are generated in the electro-dialysis cell.

Biological nitrification and denitrification

Certain dissolved forms of nitrogen are oxygen-demanding as well as potential fertilizing elements. Nitrogen can exist in seven oxidation states, consequently numerous nitrogen containing compounds are found in nature. In respect of water pollution, the most significant forms of nitrogen are ammonia, organic nitrogen, nitrite and nitrate. A major advantage of achieving biological nitrogen removal is that all forms of nitrogen ultimately are removed from solution in the non-polluting form of nitrogen gas. The two successive steps involved in the biological removal of nitrogen are (1) nitrification-oxidation; ammonia forms nitrate, and (2) denitrification; the subsequent reduction of nitrate to nitrogen gas.

The oxidation of ammonia to nitrate is a two-step process, is carried to completion. Initially ammonia is oxidized to nitrite by the *Nitrosomonas*, genera of strict aerobic autotrophic bacteria that utilize ammonia as their sole source of energy. The second step, the conversion of nitrite to nitrate, is accomplished by the *Nitrobacter* genera, which is a specific group of autotrophic bacteria that utilize nitrite as their sole energy source.

Biological denitrification is accomplished under anaerobic conditions by heterotrophic micro-organisms that make use of nitrate as a hydrogen acceptor, provided an organic carbon source is available. A variety of common facultative bacteria can accomplish denitrification.

Nitrogen removal by ion-exchange

Phosphate removal can be effected by various chemical precipitation techniques. Processes by which relatively low concentrations of nitrogen can be removed include ammonia stripping with air at high pH, microbial nitrification, biological denitrification, chemical precipitation of ammonia, chlorination, algal harvesting, reverse osmosis and ion-exchange. None of the processes listed has been successfully employed for the removal of such relatively high concentrations of nitrogenous compounds except the ion-exchange process. Ion-exchange can accomplish purification of the effluent to a quality that could comply with zero pollutant discharge criteria or that would permit complete recycle of effluents. The ion-exchange process can also accomplish complete recovery of plant products being lost into the waste stream and can provide for efficient recycle of the recovered products into the plant processes.

Ammonia stripping with air

Ammonia stripping is a mass transfer operation which at high pH can decrease the ammonia concentration of effluent by bringing it into intimate contact with air. The driving force of the mass transfer process is the difference between the partial pressure of ammonia in the air and the equilibrium partial pressure corresponding to the ammonia concentration in the effluent. Intimate contact between air and water is achieved by tower packings with large surface areas.

The effluent is pumped to the top of a packed tower and distributed to cover the full surface of the packing. The water moves down through the packing countercurrent with the air flow. A portion of the water's ammonia content is stripped before it leaves at the bottom and it is necessary to maintain the pH level constant in the influent water.

The size of the ammonia stripping tower depends on the temperature, pH, flow rate and concentration of the waste water and required degree of ammonia removal. The effectiveness of the tower packing also has an influence on the required tower height.

5.5 EFFLUENT DISPOSAL

The effluent after being subjected to treatment is either reused or disposed of in the environment, where it re-enters the hydrologic cycle. Disposal in this context can be viewed as the first step in a very indirect and long-term reuse. The most common means of treated waste water disposal is by discharge and dilution into ambient waters. Another means of disposal is land application, in this case, the water seeps into the ground and recharges underlying groundwater aquifers. A portion of the waste water destined for infiltration also evaporates and in desert locations this evaporated fraction can turn out to be substantial.

A basic element of waste water disposal is the associated environmental impact. The regulatory framework for protecting the environment, affects not only the structures, but also the level of treatment called for. Treatment and disposal are to be considered as mutually dependent and linked. For instance, to accomplish environmental acceptability, a choice may be available between enhanced treatment for one or more effluent constituents or increased dilution of the effluent. Another way out is source reduction, in this case individual dischargers are required to decrease their contribution of specific contaminants to the sewers by effecting process changes or by administering pre-treatment.

The emphasis of environmental impact evaluations of effluent discharges used to be on dissolved oxygen. The assimilative capability of receiving waters, representing the quantity of BOD_5 that can be assimilated without excessively taxing dissolved oxygen levels, was of major concern. This emphasis on dissolved oxygen leads to requirements or necessities for secondary treatment of effluents. The attention has now broadened to a wider range of effluent constituents including nutrients, toxic compounds and a variety of organic compounds. The impacts of these constituents on the environment are diverse and frequently complex. The first aspect in evaluating these impacts is the determination of the distribution and fate of these in water column and bottom sediments. It is of paramount importance to determine the concentrations of the constituents as a function of effluent flow and composition, surrounding water characteristics and discharge structure design. Frequently, environmental criteria or standards exist regulating concentrations directly. In situations where there are large discharges, additional environmental analyses are needed, but the starting point is the distribution of constituents' concentrations. The options available for the effluent disposed are lake and reservoir disposal, river and estuary disposal and ocean disposal.

5.5.1 The Renovation of Industrial Effluents for Reuse

The determination of the treatability of any effluent should go beyond the question of whether it is one that can be affected or degraded by any means of treatment which will remove from the waste, the polluting elements make the water suitable for reuse in the same plant in which the waste originated or in any other plant. The main purpose in treating industrial wastes is to eliminate, abate or prevent stream pollution and actually it makes no difference as to which good method is actually used.

The methods employed for the renovation of industrial effluents are:

1. Adsorption in sludge blanket units, carbon filters, flocculation
2. Electro-dialysis
3. Evaporation
4. Solvent extraction
5. Emulsion breaking
6. Foam separation
7. Freezing
8. Hydration (spray drying)
9. Oxidation
10. Ion-exchange
11. Electro-chemical degradation

The renovated industrial effluents can be used in many ways, some of which are listed here.

Beneficial uses: There are many ways in which water can be used, either directly by people or for their overall benefit. Examples are municipal water supply, agricultural and industrial applications, navigation and water contact recreation.

Direct potable reuse: It is a form of reuse that involves the incorporation of reclaimed waste water directly into a potable water supply system often implying the blending of reclaimed waste water.

Direct reuse: It is the use of reclaimed waste water that has been transported from a waste water reclamation plant to the water reuse site without intervening discharge to a natural body of water. It includes uses such as agricultural and landscape irrigation.

Indirect potable reuse: It is the potable reuse by incorporation of reclaimed waste water into a raw water supply. It allows mixing and assimilation by discharge into natural body of water, such as in domestic water supply reservoir or groundwater.

Potable water reuse: This is a direct or indirect augmentation of drinking water with reclaimed waste water that is normally highly treated to protect public health. Waste water recycling is the use of waste water that is captured and redirected into the same water use scheme. Recycling is practised predominantly in industries, such as manufacturing and it normally involves only one industrial plant or one user.

Waste water reuse: It is the use of treated waste water, for a beneficial use, such as agricultural irrigation or industrial cooling.

The substances found in the effluents of the various industries are summarized in Table 5.3.

Table 5.3 Substances Present in Industrial Effluents

Substance	*Present in Effluent from*
Free chlorine	Laundries, paper mills, textile bleaching
Ammonia	Gas and coke manufacture, chemical manufacture
Fluorides	Scrubbing of flue gases, glass etching, atomic energy plants
Cyanides	Gas manufacture, metal plating, metal cleaning
Sulphides	Sulphite dyeing of textiles tanneries, viscose-rayon manufacture
Sulphites	Wood-pulp processing, viscose film manufacture
Acids	Chemical manufacture, mines, iron and copper pickling, DDT manufacture, brewing, textiles, battery manufacture
Alkalis	Wool scouring, laundries
Chromium	Metal plating, aluminium anodizing, chrome tanning
Lead	Battery manufacture, lead mines, paint manufacture
Nickel	Metal plating
Cadmium	Metal plating
Zinc	Galvanizing, zinc plating, viscose-rayon manufacture, rubber processing
Copper	Copper plating, copper pickling, cuproammonium-rayon manufacture
Arsenic	Sheep dipping
Sugars	Dairies, breweries, glucose and sugar beet factories, chocolate and sweet industries
Starch	Food processing, textile industries, wall paper manufacture.
Fats, oils and grease	Wood scouring, laundries, textile industries, petroleum refineries, engineering works
Phenols	Gas and coke manufacture, synthetic resin manufacture, petroleum refineries, textile industries, tanneries, tar distilleries, chemical plants, dye manufacture.
Formaldehyde	Synthetic resin manufacture, penicillin manufacture.

Major unit processes that are employed in treating the effluents are listed out in Table 5.4.

Table 5.4 Major Unit Processes in Effluent Treatment

S.no.	*Process*	*Brief description*
Unit processes (Physical)		
1.	Balancing	In situations where the flow of effluent generated varies with time, balancing tanks are employed to ensure a constant flow, and consistent quality of effluent is pumped forward for treatment. This entails a reduction in the capacity and cost of treatment.
2.	Screening	Screens remove large particles from waste water/effluent. They are used early in treatment to safeguard other treatment processes. Screens can be stationary, vibrating or rotating drums.
3.	Sedimentation	Special tanks are used to separate organic and inorganic solids from liquids.
4.	Floatation	Small air bubbles introduced at the base of a tank become attached to suspended particles and float. The particles are then skimmed off the surface as a sludge. Employed widely in dairy, paper, meat packing and paint industries.
5.	Hydro-cyclone	Removal of dense particles (such as sand, grit and glass) from effluents is accomplished as it enters a conical tank tangentially. As the effluent spirals through the tank, particles are thrown against the wall by centrifugal forces and fall to the base (point) of the cone from where they can be removed.
6.	Filtration	Treated effluent can be made to pass through a fine media filter (such as sand) in order to further reduce suspended solids' concentration. High performance filters using synthetic fibres to remove particles in the size range of 1 to 500 microns from treated effluent or process streams are employed.
7.	Centrifugation	Separation of solids from liquids by rapid rotation of the mixture in a special tapered vessel. Solids are deposited as a thick sludge (20–25 per cent dry solids) either against the inner wall or at the base. Widely employed in pharmaceutical, pulp, paper, chemical and food industries and also for dewatering sewage sludge.
8.	Reverse osmosis	Under pressure (1500–3000 kPa) water is driven through a semi-permeable membrane with extremely small pores to concentrate ions and other particles in solution, and to purify the water. Employed to remove and recover contaminants from process waters before discharge to sewer.
9.	Ultra-filtration	Similar to reverse osmosis. Particles of 0.005 to 0.1 microns are removed as they are forced through a micro-porous membrane at pressures up to 3000 kPa. Employed for removal and recycling of colloidal material including dyes, oils and even proteins from cheese and whey from effluents. Able to remove the smallest microorganisms including viruses and pyrogenic macro-molecules.

(Contd.)

Table 5.4 Major Unit Processes in Effluent Treatment *(Contd.)*

S.no.	*Process*	*Brief description*
10.	Micro-filtration	Similar to ultra-filtration except used to recover large particles (0.1 to 0.5 microns) at lower pressures (100–400 kPa). Widely employed in food and drink industry. Micro-porous filters can be used for the disinfection of process waters and effluents.
11.	Adsorption	Activated carbon or synthetic resins are used to remove contaminants by adsorption from liquids. Used primarily for the removal of organics from industrial process waters and effluents.
Unit processes (Chemical)		
12.	Neutralization	Non-neutral waste waters are mixed either with an alkali (e.g. NaOH) or an acid (e.g. H_2SO_4) to bring the pH as close to neutral as possible to protect treatment processes. Widely employed in chemical, pharmaceutical and tanning industries.
13.	Precipitation	Dissolved inorganic components can be removed by the addition of an acid or alkali or by changing the temperature, by precipitation as a solid. The precipitate can be removed by sedimentation, floatation or other solids' removal process.
14.	Ion-exchange	Removal of dissolved inorganic ions by exchange with another ion attached to a resin column. For instance, Ca and Mg ions can replace Na ions in a resin, thereby reducing the hardness of the water.
15.	Oxidation-reduction	Inorganic and organic materials in industrial process waters can be made less toxic or less volatile by subtracting or adding electrons between reactants. (e.g. aromatic hydrocarbons, cyanides, etc.).
Unit processes (Biological)		
16.	Activated sludge	Liquid waste water is aerated to allow microorganisms develop and utilize the organic polluting matter (95 per cent reduction). The microbial biomass and treated effluent are separated by sedimentation with a portion of the biomass (sludge) returned to the aeration tank to seed the incoming effluent.
17.	Biological filtration	Effluent is distributed over a bed of inert medium on which micro-organisms develop and utilize the organic matter present. Aeration takes place through natural ventilation and the solids are not returned to the filter.
18.	Stabilization ponds	Large lagoons where waste water is stored for long durations to allow a wide range of microorganisms to break down organic matter. A variety of types and designs of ponds including aerated, non-aerated and anaerobic ponds are available. Some designs rely on algae to provide oxygen for bacterial breakdown of organic matter. Sludge is not returned.
19.	Anaerobic digestion	Used for high strength organic effluents (e.g. pharmaceutical, food and drink industries). Effluent is stored in a sealed tank which excludes oxygen. Anaerobic bacteria break down organic matter into methane, carbon dioxide and organic acids. Final effluent still requires further treatment as it has a high BOD. Also employed for the stabilization of sewage sludge at a concentration of 2–7 per cent solids.

Table 5.5 presents the unit operations and processes used in the advanced waste water treatment.

Table 5.5 Advanced Waste Water Treatment Operations and Processes

Description	*Type of waste water treated*	*Principal or major use*	*Waste for ultimate disposal*
Principal unit operations			
Air stripping or ammonia	EST	Removal of ammonia, nitrogen	None
Filtration-multi median	EST	Removal of suspended solids	Liquid and sludge
Micro-strainers	EBT	Removal of suspended solids	Sludge
Distillation	EST nitrified + filtration	Removal of dissolved solids	Liquid
Electro-dialysis	EST + filtration + carbon adsorption	Removal of dissolved solids	Liquid
Floatation	EPT, EST	Removal of suspended solids	Sludge
Foam fractionation	EST	Removal of refractory organics, surfactants and metals	Liquid
Freezing	EST + filtration	Removal of dissolved solids	Liquid
Gas-phase separation	EST	Removal of ammonia nitrogen	None
Land application	EPT, EST	Nitrification, denitrification, removal of ammonia, nitrogen and phosphorus	None
Reverse osmosis	EST + filtration	Removal of dissolved solids	Liquid
Chemical unit processes			
Breakpoint chlorination	EST/filtration	Removal of ammonia, nitrogen	Liquid
Carbon adsorption	EPT, EST (filtration)	Removal of dissolved organics heavy metals and chlorine	Liquid
Chemical precipitation	EBT	Phosphorus precipitation, removal of heavy metals, removal of colloidal solids	Sludge
Chemical precipitation in activated sludge	EPT	Removal of phosphorus	Sludge
Ion-exchange	EST and filtration	Removal of ammonia and nitrate nitrogen	Sludge
Electrochemical treatment	Untreated	Removal of dissolved solids	Liquid and sludge
Oxidation	EST	Removal of refractory organics	None
Biological unit processes			
Bacterial assimilation	EPT	Removal of ammonia nitrogen	Sludge
Denitrification	Agricultural return water	Nitrate reduction	None
Harvesting of algae	EBT	Removal of ammonia nitrogen	Algae
Nitrification	EPT, EBT	Ammonia oxidation	
Nitrification denitrification	EBT, EPT	Total nitrogen removal	Sludge

EPT - Effluent from Primary Treatment
EBT - Effluent from Biological Treatment
EST - Effluent after Secondary Treatment

The separation processes used in the abatement of pollution are summarized in Table 5.6.

Table 5.6 Separation Processes Employed in Pollution Abatement

S.no.	*Process*	*Feed*	*Separating agent*	*Product*	*Property exploited*	*Application (typical example)*
1.	Filtration surface type	Fluid and solid	Filter media + energy (Pressure reduction)	Fluid and solid cake	Size and adhesive nature of solids	Sludge dewatering, particle removal from stack gas
2.	Filtration in depth type	Liquid and solid	-Do-	Liquid washing, liquid containing solids	-Do-	Water clarification, sewage effluent treatment, treatment of industrial wastes for the purpose of recycling.
3.	Screening	Liquid and large solids or mixed solids	Grid barrier	Liquids and solids	Size difference of solids	Removal of gross solids from sewage, classification of solids in refuse or industrial wastes
4.	Centrifuge	Liquid and solid	Centrifugal force	Liquid and liquid sludge	Density difference	Recovery of insoluble reaction products, sludge dewatering
5.	Settling	Fluid containing solids	Gravity and fluid	Liquid and sludge solids, gas plus solids	Size and density difference	Removal of suspended solids and precipitation from waste waters, separation of particles from gases
6.	Electrostatic precipitation	Gas and fine solids	Electrical fields	Gas and solids	Electrical charge on solids	Dust removal from stack gases
7.	Magnetic separation	Mixed solids or liquids and solids	Magnetic field	Two solids or liquid and solid	Magnetic attraction	Sorting metals from refuse, collection of materials immobilised on magnetic particles, for example, oil
8.	Scrubbers	Gas and solid	Liquid spray or wetted packing	Gas plus liquid	Size and wettability of particles	Air pollution control
9.	Distillation	Liquid or vapour	Heat and cooling	Liquid and vapour	Volatility	By-product recovery, desalination
10.	Stripping	Liquid	Non-condensable gases	Liquid and vapour	Volatility	Ammonia removal, CO_2 removal deaeration
11.	Absorption	Gas	Non-volatile liquid	Liquid and vapour	Preferential solubility	Removal of H_2S from fermentation gas, SO_2 removal from stack gases
12.	Extraction	Liquid	Immiscible liquid	Two liquids	Preferential solubility	By-product recovery
13.	Ion-exchange	Liquid	Solid resin	Liquid and solid resin	Law of mass action applied to available cations and anions	Water softening, metal recovery
14.	Adsorption	Gas or liquid	Solid adsorbent	Fluid and solid	Difference in adsorption potential	Removal of detergents, removal of phenols, odour removal
15.	Leaching or washing	Solids	Solvent	Liquid and solid	Preferential solubility	Leaching metals from refuse
16.	Electro-dialysis	Liquid	Anionic and cationic membranes electric field	Two liquids	Tendency of membranes to pass only ions	Recovery of plating solutions desalting brackish water
17.	Reverse osmosis	Liquid solution	Pressure and membrane	Two liquids	Different combined solubilities and diffusivities in membranes	Desalting sea water
18.	Ultra-filtration	Liquid solution	Pressure gradient and membrane	Two liquid phases	Different permeabilities through membrane determined by molecular size.	Protein recovery and concentration
19.	Drying of solids	Moist solids	Heat	Dry solid and humid vapour	Evaporation of water	Sludge dehydration, by-product recovery
20.	Cyclone air, or liquid	Fluid containing solids	Centrifugal force	Gas and solids, liquid and sludge solids	Density difference	Air pollution control classification of milled refuse

Table 5.7 gives the methods of control of the contaminants of a few selected industries.

Table 5.7 Industrial Process and Control Summary

Industry or process	*Source of emission*	*Particulate matter*	*Method of control*
Iron and steel mills	Blast furnaces, steel making furnaces, sintering machines	Iron oxide, dust, smoke	Cyclones, bag houses, electrostatic precipitators, wet collectors
Grey iron foundries	Cupolas, shake-out making	Iron oxide, smoke oil, dust, metal fumes	Scrubbers, dry centrifuge, collectors
Non-ferrous metallurgy	Smelters and furnaces	Smoke, metal fume, oil grease	Electrostatic precipitator, fabric filters
Petroleum refineries	Catalyst regenerators, sludge incinerators	Catalyst dust ash from sludge	Cyclones electrostatic precipitators, scrubbers, bag houses
Portland cement	Kilns, driers, material handling systems	Alkali and process dusts	Fabric filters, electrostatic precipitators, mechanical collectors
Kraft paper mills	Recovery furnaces, lime kilns, smelt tanks	Chemical dusts	Electrostatic precipitators, venturi scrubbers
Acid manufacture—phosphoric, sulphuric	Thermal processes, rock acidulating, grinding	Acid mist, dust	Electrostatic precipitators, mesh, mist eliminators
Coke manufacture	Oven operation, quenching, materials handling	Coal and coke, dust, coal tars	Meticulous design, operation and maintenance
Glass and fibre glass	Furnaces, forming and caring, handling	Acid mist, alkaline oxides, dust, aerosols	Fabric filters, after burners

The unit operations, unit processes and systems for liquid effluents are summarized in Table 5.8.

Table 5.8 Unit Operations, Unit Processes and Systems for Waste Water Treatment

S.no.	*Contaminant*	*Unit operation, Unit process or treatment system*
1.	Suspended solids	Sedimentation, screening and communition, filtration variations, floatation, chemical polymer addition, coagulation/sedimentation, land treatment systems.
2.	Biodegradable organics	Activated sludge variations, fixed film; tricking filters, fixed film; rotational biological contactors, lagoon and oxidation pond variations, intermittent sand filtration, land treatment systems, physico-chemical systems
3.	Pathogens	Chlorination, hypochlorination, ozonation, land treatment systems
4.	Nutrients—nitrogen	Suspended growth nitrification and denitrification variations, fixed film nitrification and denitrification variations, ammonia stripping, ion-exchange breakpoint chlorination land, land treatment systems
	Phosphorus	Metal-salt addition, lime coagulation/sedimentation, biological-chemical phosphorus removal, land treatment systems
5.	Refractory organics	Carbon adsorption, tertiary ozonation, land treatment systems
6.	Heavy metals	Chemical precipitation, ion-exchange, land treatment systems
7.	Dissolved inorganic solids	Ion-exchange, reverse osmosis, electro-dialysis (membrane separation processes)
8.	Acids and alkalis	Neutralization
9.	Reactive ions and compounds	Chemical oxidation, reduction, encapsulation, fixation, hydrolysis
10.	Solvents and oils	Acid/caustic stripping, distillation, evaporation, filtration, gravimetric, separation incineration, steam stripping

In Table 5.9, the pollution characteristics of different industries are presented.

Table 5.9 Pollution Characteristics of Different Industries

Industry	*Pollution characteristics*	*Suggested treatment methods*
Pulp and paper	Strong colour, high BOD, high COD/BOD ratio, highly alkaline, high sodium content	Chemicals recovery, colour removal by lime treatment, biological treatment
Tannery	Strong colour, high salt content, high BOD, high dissolved solids, presence of sulphides, lime and chromium	Chemical and biological treatment
Textile (cotton)	Highly alkaline, high BOD, high suspended solids	Chemical and biological treatment
Distillery and brewery	Strong colour, high chloride, high sulphate, very high BOD	Biological treatment
Petrochemicals	Oil, high BOD and COD, high total solids	Chemical and biological treatment
Pharmaceuticals	High total solids, high COD, high COD/BOD ratio, either acidic or alkaline	Chemical and biological treatment
Coke oven	High ammonia content, high phenol content, high BOD, low suspended solids, high cyanide	Chemical and biological treatment
Oil refineries	Free and emulsified oil	Oil separation, chemical and biological treatment
Fertilizer	High nitrogen content	Biological treatment
Dairy	High dissolved solids, high suspended solids, high BOD, presence of oil and grease	Biological treatment
Sugar	High BOD, high volatile solids, low pH	Biological treatment

Table 5.10 gives the origin of the various major water contaminants and their characteristics, from a few selected industries, manufacturing food items, industrial chemicals and materials.

Table 5.10 Major Water Contaminants from Some Industrial Sources

Industry	*Origin of contaminants*	*Components and characteristics of contaminants*
Food		
Canning	Fruit and vegetable preparation	Colloidal, dissolved organic matter, suspended solids
Dairy	Whole milk, dilutions, buttermilk	Dissolved organic matter (protein, fat, lactose)
Brewing and distilling	Grain, distillation	Dissolved organics, nitrogen fermented starches
Meat poultry	Slaughtering, rendering of bones and fats, plucking	Dissolved organics, blood, proteins, fats, feathers
Sugar beet	Handling juices condensates	Dissolved sugar and protein
Yeast	Yeast filtration	Solid organics
Pickles	Lime water, seeds syrup	Suspended solids, dissolved organics, variable pH
Coffee	Pulping and fermenting beans	Suspended solids
Fish	Pressed fish, wash water	Organic solids, odour
Rice	Soaking, cooking, washing	Suspended and dissolved carbohydrates
Soft drinks	Cleaning, spillage washing	Suspended and dissolved carbohydrates
Pharmaceuticals		
Antibiotics	Mycelium, filtrate washing	Suspended and dissolved organics
Clothing		
Textiles	De-sizing of fabrics	Suspended solids, dyes, alkaline
Leather	Cleaning, soaking, bathing	Solids, sulphite, chromium, lime sodium chloride
Laundry	Washing fabrics	Turbid, alkaline organic solids
Chemical		
Acids	Wash waters, spillage	Low pH
Detergents	Purifying surfactants	Surfactants
Starch	Evaporation, washing	Starch
Explosives	Purifying and washing TNT, cartridges	TNT, organics, acids, alcohols, oil soaps
Insecticides	Washing, purification	Organics, benzene acid, highly toxic
Phosphate	Washing, condenser wastes	Suspended solids, phosphorus, silica fluoride, clay oils, low pH
Formaldehyde	Residues from synthetic resin production and dyeing synthetic fibres	Formaldehyde

(Contd.)

Table 5.10 Major Water Contaminants from Some Industrial Sources (*Contd.*)

Industry	*Origin of contaminants*	*Components and characteristics of contaminants*
Materials		
Pulp and paper	Refining, washing screening of pulp	High solids, extremes of pH
Photographic products	Spent developer and fixer	Organic and inorganic reducing agents, alkaline
Steel	Cooking, washing blast furnace, flue gases	Acid, cyanogen phenol, coke, oil
Metal plating	Cleaning and plating	Metals' acids
Iron foundry	Various discharges	Sand, clay, coal
Oil	Drilling, refining	Sodium chloride, sulphur, phenol, oil
Rubber	Washing and extracting impurities	Suspended solids, chloride odour, variable pH
Glass	Polishing, cleaning	Suspended solids

EXERCISES

5.1 What are the benefits of primary treatment of effluents?

5.2 Write an explanatory note on primary sedimentation.

5.3 Describe the treatment methods used to remove grit.

5.4 Identify the main components of activated sludge systems.

5.5 How are disease-causing organisms in waste waters destroyed?

5.6 Discuss the advanced waste water treatment for the removal of nitrogen and phosphorus.

6

Air Quality

Air quality is evaluated with respect to the total air pollution in a given area as it interacts with meteorological conditions, such as humidity, temperature and wind to produce overall atmospheric condition. Poor air quality can manifest itself aesthetically and can also result in damage to plants, animals, people and even result damage to objects.

Agencies like Environmental Protection Agency (EPA) of USA directly measure the concentrations of the pollutants in the air, and compare those concentrations with national standards for six major pollutants; ozone, carbon monoxide, nitrogen oxides, lead, particulate matters and sulphur dioxide. When the air we breathe contains quantities of these pollutants in excess of the set standards, it is considered unhealthy, and regulatory action is taken to bring down the pollution levels.

In addition, urban and industrial areas can maintain an air pollution index. This scale, a composite of several pollutant levels recorded from a particular monitoring site or sites, yields an overall air quality value. If the index exceeds certain values, public warnings can be given.

The air pollution index is a value, derived from an air quality scale that employs the measured or predicated concentrations of several criteria pollutants and other air quality indicators, such as Coefficient of Haze (CoH) or visibility. The widely known index of air pollution is the Pollutant Standard Index (PSI).

The PSI has a scale that spans from 0 to 500. The index is indicative of the highest value of several subindices; there is a subindex for each pollutant, or in some situation, for a product of pollutant concentrations and CoH. In case a pollutant is not monitored, its subindex is not used in deriving the PSI. Generally, in the subindex each pollutant can be interpreted as shown in Table 6.1.

Table 6.1 Air Pollution Stage

Index value	*Interpretation*
1	No concentration
2	National ambient, air quality standard
3	Alert
4	Warning
400	Emergency
500	Significant harm

The subindex of each pollutant is derived from a PSI nomogram which matches concentrations with subindex values. The highest subindex value becomes the PSI. The PSI has five health related categories given in Table 6.2.

Table 6.2 The Pollutant Standard Index and the Health Related Categories

PSI range	*Category*
0–50	Good
50–100	Moderate
100–200	Unhealthy
200–300	Very unhealthy
300–500	Hazardous

6.1 AIR QUALITY MANAGEMENT CONCEPTS

Planning for clean air is a comprehensive regional planning problem. The management and uses of our air needs to be planned along the same lines as the use of entire river basin units in the conservation of land and water resources. The major crux of the problems lies in establishing a jurisdictional administrative and financial context for the application of the techniques that already exist.

International mechanisms are needed to (1) take emergency action on imminent and substantial dangers to public health, (2) establish standard international air quality and emission objectives which can be applied on a regional basis, (3) ensure standards inventory techniques which furnish supporting monitoring data to the global net work set-up, World Health Organization (WHO), (4) organise and conduct research and disseminate information on control and prevention, (5) conduct surveys and issue abatement directives requiring remedial and preventive actions.

Five fundamental and urgent areas of action stand out, (1) metrology, standardizing and adoption of measurement terms to express various pollutant concentrations, (2) the coordination and sharing of research with effective dissemination of the knowledge acquired, (3) declarations of aims and attitudes in regional work, (4) formulation and adoption of standards and procedures for air pollution control, (5) the machinery for enforcement of concluded environmental agreements. Within this overall global picture, action depends on an effective regional approach.

6.1.1 Planning at the Regional Level

Contaminants from non-regulated point sources beyond local jurisdictions render local-control useless. Local enforcement is difficult; based on experience it has been observed that the larger the region legally affected, the more varied and complex is the socio-economic base and the less susceptible the planning and control process is, to individual and group pressures. Planning for clean air is an integral part of regional planning. The solutions lie in the regional land use management and in redesigning transport systems, public power supply methods, industrial and domestic processes and systems.

Three main regional strategies are employed, either singly or in combination; they are (1) treating or reducing the generated pollutants before or after they enter the environments; (2) regulating the location of the casual activity of the recipients or both (3) modifying or eliminating the causal activity.

The major tasks include (1) the control and prevention of photochemical smog, (2) the control and prevention of sulphur dioxide smog, (3) the removal of present major sources, like automobile exhausts public power plants *point* sources of industrial emissions, heavy metals contaminators, airborne pesticides residues.

6.1.2 The Regional Planning Process

The regional planning process can be directed towards these objectives. Regional inventory will identify the particular mix of problems, casual relationships and point sources. A series of computer simulations can generate approximate models of the metabolism of the region and give evidence of systems needing the corrective measures. The planners of land-use transport, services and water-resource systems can generate integrated regional planning alternatives to which models of the simulated metabolism of the region can be applied. This, in turn, leads to the selection of the optimum strategies, standards and specific objectives. It is essential or even mandatory that the regional plan be fully supported by all levels of government and given statutory force and strong financial commitment. The regional plan will then provide a realistic context.

Implementation methods would include (1) enforcement of minimal standards and the removal of offensive point sources by public redevelopment (2) allocating a definite *life* to problem sources, (3) using *clean air zoning* for progressive upgrading of problem areas, (4) legal supports, such as an increased enforcement of stricter laws.

The regional planning can be for a 20-year period, updated and revised after every 5 years, and developed in 5 years stages of implementation.

6.2 SOURCES OF AIR POLLUTION

The sources of air pollution may be man-made, such as the internal combustion engine or natural, such as plants (pollen). The contaminants of air can be in the form of particulates, aerosols and gases or microorganisms. Pesticides, odours and radioactive particles carried in the air are also included.

Particulates range from less than 0.01 to 1000 microns (1/1000 of a millimetre) in size. Generally, they are smaller than 50 microns. Particles of 10 microns and larger in size can be seen with the naked eye. Smoke is generally less than 0.1 micron size—soot or carbon particles. Particulates which are below 10 microns can penetrate the lower respiratory tract. Particles less than 3 microns can reach the tissues in the deep parts of the lung. Dust and inorganic, organic, fibrous and non-fibrous particles fall under this category. Aerosols are usually particles of 50 microns to less than 0.01 micron in size; even though generally they are less than 1 micron in diameter. Gases include organic gases such as hydrocarbons, aldehydes and ketones, and inorganic gases (oxides of nitrogen and sulphur, carbon monoxide, hydrogen sulphide, ammonia and chlorine).

6.2.1 Man-made Sources

Air pollution, in general, is the outcome of industrialization, carbon monoxide is the principal pollutant by weight and motor vehicle is the major contributor followed by industrial processes and stationary fuel combustion. However, in terms of hazard it may mentioned that it is not the tons of pollutant that is important, but the toxicity or harm that can be caused by the particular pollutant released. Lead pollution can be lessened, in case the non-leaded gasoline is used as the automobile fuel.

Agricultural spraying of pesticides, orchard heating devices, exhaust from a variety of commercial processes, rubber from tyres, mists from spray cooling towers, and the use of cleaning solvents and household chemicals add to the pollution load on the environment. Toxic pollutant emissions and their fate in the environment need further investigation.

Particulates, gases and vapours that find their way into the air without being vented through a stack are referred to as **fugitive emissions** in the parlance of environmental pollution study. They include uncontrolled releases from industrial processes, street dust and dust from construction activities, and farm cultivation. These need to be controlled at the source on an individual basis.

Wood stoves also contribute significantly to the contamination of air. This type of pollution is a potential health hazard to children and elderly people with chronic lung problems. These stoves have to be redesigned to keep the air pollution at acceptable levels.

6.2.2 Natural Sources

These include dust, plant and tree pollens, arboreal emissions, bacteria and spores, gases and dusts from forest and grass fires, ocean sprays and fogs, esters and terpenses from vegetation, ozone and nitrogen dioxide from lightening, ash and gases (SO_2, HCl, HF, H_2S) from volcanoes, natural radioactivity and microorganisms like bacteria, spores, moulds or fungi from plant decay. Most of these are beyond control or of limited significance. Ozone is found in the stratosphere at an altitude of more than 7 to 10 miles. The principal natural sources of ozone in the lower atmosphere are lightening discharges and in small amount, reactions involving volatile organic compounds released by forests and other vegetation. Ozone is also formed naturally in the upper atmosphere by a photochemical reaction with UV solar radiation.

6.2.3 Types of Air Pollutants

The type of air pollutants are related to the original material employed for combustion or processing, the nature of impurities it contains, the actual emissions and the reactions that take place in the atmosphere. A primary pollutant is one that is found in the atmosphere in the same form as it exists when emitted from the stack; sulphur dioxide, nitrogen dioxide and hydrocarbons are examples. A secondary pollutant is one that is formed in the atmosphere (which can be construed as a reaction vessel) as a result of reactions, such as hydrolysis, oxidation and photochemistry. Photochemical substance is an example.

Most combustible materials are composed of hydrocarbons. In case the combustion of gasoline, oil or coal for instance is inefficient, unburned hydrocarbons, smoke, carbon monoxide, and to a lesser extent aldehydes and organic acids are released.

The application of automobile catalytic converters to control carbon monoxide and hydrogen emissions cause some enhancement in sulphates and sulphuric acid emissions; but this is considered to be of minor significance. The elimination of lead from gasoline has in some situations, led to the substitution of manganese for antiknock purposes with the consequent release of manganese compounds, which are also potentially toxic.

Impurities in combustible hydrocarbons (coal and oil), such as sulphur combined with oxygen to produce SO_2 when subjected to burning. The SO_2 subsequently may form sulphuric acid to other sulphates in the atmosphere. Oxides of nitrogen from high temperature combustion in electric utility, and industrial boilers and automobiles (above 650°C) are released mostly as NO_2 and NO. The source of nitrogen is principally the air used in combustion. Some fuels contain substantial quantities of nitrogen and these also react to form nitrogen dioxide and nitric oxide. Fluorides and other fuel impurities may be carried with the hot stack gases.

Photochemical oxidants including ozone, formaldehydes and peroxides are generated in the lower atmosphere (troposphere) as a result of the reaction of oxides of nitrogen and volatile organics in the presence of solar radiation. Ozone may contribute to smog, problems in respiratory systems, and damage to crops and forests.

Of these sources, industrial processes are the principal source of volatile organics (hydrocarbons) with transportation, the next large contributor. Stationary fuel combustion plants and automobiles are the major sources of nitrogen oxides. Ozone and other photochemical products formed are usually found at some distance from the source of the precursor compounds. Ozone, the principal component of modern smog, is the photochemical oxidant actually measured, which is around 90 per cent of the total (see Table 6.3).

Table 6.3 List of Industrial Sources of Air Pollution and Some of their Emissions

Sources	*Pollutants emitted*
Chemicals	
Acids	
Hydrochloric	HCl, chlorine, organic vapours, catalyst dust
Hydrofluoric	HF, F, oxides of sulphur
Nitric	Nitric acid mist, ammonia, NO_2, NO
Phosphoric (P_2O_5)	Gypsum dust, F, HF, SO_2
Sulphuric	H_2SO_4 mist, SO_2, SO_3, S
Alkalis	
Potassium hydroxide	KOH, chlorine
Sodium hydroxide	Chlorine
Chlorine	Cl_2, H_2, CO, Hg vapour, NaOH mist
Inorganic fertilizer materials	
Ammonia	NH_3, H_2, CH_4, CO
Ammonium nitrate	NH_4NO_3, NH_3, HNO_3
Ammonium sulphate	$(NH_4)_2SO_4$, NH_3, H_2SO_4, SO_2
Rubber	
Synthetic	Organic vapours, particulates
Natural	Organic vapours, particulates

(*Contd.*)

Table 6.3 List of Industrial Sources of Air Pollution and Some of their Emissions (Contd.)

Sources	*Pollutants emitted*
Synthetic detergents	
Soap	Organic vapours, particulates
Synthetic detergents	Particulates, organic vapours
Petroleum and coal	
Petroleum (crude oil) catalytic cracking units	Particulates, SO_x, CO, organic vapours, NO_X, NH_3, H_2S
Metals	
Ore roasting	Metallic dusts, gases (examples F, SO_x)
Ferro alloy electric arc furnace	Particulates, CO
Iron blast furnace	Particulates, CO
Steel furnaces (electric arc, open hearth)	Particulates, CO
Non-ferrous furnaces aluminium, copper, lead, zinc, magnesium	Particulates, SO_x
Re-melting furnaces grey iron cupolas	Particulates, CO
Secondary lead smelting	Particulates, SO_x
Secondary zinc smelting	Particulates
Plating	Acid and metal fumes
Minerals	
Asphalt concrete	Particulates, organic vapours
Cement plants	Lime, limestone and sulphate dusts
Ceramic and brick manufacture	Particulates
Grinding and polishing	Abrasives, metal dusts
Gypsum plants	Particulates, SO_x
Miscellaneous farming	
Fertilizers	Particulates
Crushing, screening and shelling	Particulates
Soil tilling	Particulates
Plant and animal products	
Wood working	Particulates
Grain polishing	Particulates
Sugar refining	Particulates
Kraft wood pulping	Particulates
Surface finishing	
Paint, varnish and lacquer	Organic vapours, particulates
Waste disposal	
Incineration	Particulates, SO_x, CO, NO_x
Decay (aerobic and anaerobic)	Particulates, CO, organic vapours

6.3 AIR POLLUTANTS

6.3.1 Sulphur Oxides

Sulphur dioxide, sulphur trioxide and the corresponding acids and salts (sulphites and sulphates) are air pollution vapours of very high concern to the public. These compounds result from the combustion of solid and liquid fossil fuels containing sulphur in some form. These are naturally occurring oxides of sulphur in the atmosphere, but those of concern to the society or the general public are the result of modern man's industrial technology.

Sulphur dioxide is a non-flammable, non-explosive, colourless gas. Most people can detect it by taste if its concentrations are above 0.3 ppm to 1 ppm in air; in concentration greater than 3 ppm, it has a pungent irritating odour. The threshold concentration of sulphur dioxide should not be permitted to exceed 5 ppm in any industrial activity, and it is recommended that an eight-hour exposure be limited to 0.19 ppm and that continuous exposure be limited to 5.6 ppm (0.15 mg/m^3). The gas is highly soluble in water; 11.3 g/100 ml. Compared to 0.70 g/100 ml for oxygen, nitric oxide, carbon monoxide, and carbon dioxide.

Sulphur dioxide is a gas under ambient atmospheric conditions and can act as a reducing agent or as an oxidizing agent. It has the ability to react either photochemically or catalytically with materials in the atmosphere to give rise to sulphur trioxide, sulphuric acid and salts of sulphuric acid. Sulphur trioxide is immediately converted to sulphuric acid in the presence of moisture.

The oxidation of atmospheric sulphur dioxide results in the formation of sulphuric acid and other sulphates that account for around 5 per cent to 20 per cent of the total suspended matter in urban air.

The corrosion properties of sulphur oxides on materials is of more concern at higher humidity, higher temperature and in the presence of particulate material. Sulphur oxides generally accelerate corrosion by first being converted to sulphuric acid, either in the atmosphere itself or on metal surfaces. The corrosion products are mainly sulphate salts of exposed metals. Most of the building materials are subject to corrosive effect of sulphur oxides including limestone, marble, roofing slate and mortar. Any carbonate containing stone is damaged by having the carbonate converted to relatively soluble sulphates which are then leached away by rainwater.

Cellulose vegetable fibres, such as cotton, linen, hemp, jute as well as rayons and synthetic nylons suffer loss of fibre tensile strength on exposure to sulphurous and sulphuric acid components. Certain types of dyes in dyed fabrics are subject to reduction of colour or destruction by acid compounds present in polluted air.

Sulphur oxides present in the atmosphere and their derived products are also damaging to vegetation. The degree of damage varies with different species, the extent and duration of exposure, climatic conditions and other factors.

A number of different experiments have been conducted using sulphur dioxide and acid mists on animals and insects. The mortality or indication of damage varies with different species, the concentration of the pollutant and the duration of exposure; sulphuric acid is more toxic than sulphur dioxide; the smaller sulphuric acid particles cause more irritation than larger particles.

Humans have been the subject of a number of experiments to determine the effect of various concentrations of sulphur oxides and acid mists. It has been found out that most individuals will not respond to the presence of sulphur dioxide in actual normal living conditions until a concentration of about 5 ppm is reached.

Epidemiological, clinical, physiological and animal experiments have demonstrated the hazards of sulphur dioxides and their associated products. A concentration of 10 ppm for 1 hour is considered as an emergency situation, even the threshold limit of 5 ppm for 1 hour is a serious level of exposure.

It is suggested that in design work, in the absence of stringent regulatory standards, sulphur dioxide be below 0.03 ppm/24 hrs in residential sections, 0.04 ppm/24 hr in commercial sections and 0.06 ppm/24 hr in industrial sections. It is desirable to keep average exposure to below 0.02 ppm.

Sulphuric acid is particulate. A serious level would be 0.01 mg/m^3/30 days.

6.3.2 Hydrides

The hydrides of sulphur, selenium, tellurium are all colourless gases with offensive odours. Hydrogen selenide and hydrogen sulphide are almost as toxic as hydrogen cyanides, which is used in prison gas chambers. Small concentrations of hydrogen sulphide will produce headaches and larger amounts will cause paralysis in the nerve centres of the heart and lungs resulting in fainting and death.

Concentrations of hydrogen sulphite of 1000 to 3000 ppm are fatal after few minutes exposure; concentrations in the range of 100 to 150 ppm cause discomfort after several hours exposure. In industrial situations the threshold limit of hydrogen sulphide should be below 10 ppm at all times. In engineering design work, an exposure of 0.005 ppm/24 hours is suggested as a safe limit. An exposure of 0.1 ppm/1 hr on a one-time basis is permissible, but it is higher than desirable; concentrations as high as 0.9 ppm have been recorded in many urban areas.

6.3.3 Halides

Hydrogen fluoride, hydrogen chloride, hydrogen bromide and hydrogen iodide are colourless gases having sharp penetrating odours. They are very soluble in water and they will fume in moisture-laden air.

Hydrogen fluoride in gas form is readily absorbed by the stomata of plants. Its degree of toxic effect on plants varies much and has a bearing on a number of factors, but depending on concentration and duration of exposure, hydrogen fluoride can bring about serious damage to a number of plant species. Particulate fluoride material of equal concentration deposited on the leaves most likely would not cause injury.

Hydrogen chloride is less toxic to plant life than sulphur dioxide, but it probably causes damage to some plant species that are exposed for sufficient periods of time to concentrations above 10 ppm. In industrial situations, hydrogen chloride should be kept below 5 ppm at all times and in general air pollution design, a level of 0.025 ppm/24 hrs is recommended. Hydrogen bromide should not exceed a concentration of 5 ppm and again a design level of 0.025 ppm/24 hr is recommended. Hydrogen fluoride should not be allowed to exceed 1.5 ppm and an industrial design of 0.008 ppm/24 hrs is recommended.

The chlorides will probably be found in greater concentrations than the fluorides in many industrial cities; the chlorides are produced by a number of industries.

Chlorine is of course quite toxic to humans and animals if present in sufficient concentration; however this magnitude of exposure is usually the outcome of an accidental release of chlorine. Such an accident might take place upon derailment of chlorine tank during shipment; an accidental release might occur at water treatment plants, swimming pools, sewage treatment plants, or other facilities using chlorine. A design figure of 0.01 ppm/24 hr is recommended. Where intermittent exposure does not exceed 8 hrs, values upto 0.5 ppm/8 hrs are acceptable.

6.3.4 Aliphatic Hydrocarbons of Paraffin Series

The hydrocarbons include a large number of compounds. The aliphatic hydrocarbons are known as the **paraffin series** and they occur in mineral oils. The paraffin series start with methane and progresses through gases, gasolenes (light and heavy), lubricating oils and into progressively heavier molecules, such as paraffin wax. Methane, ethane, propane and butane are gases. Isobutene, pentane, isopentane, 2,3-dimethyl propane, hexane, heptane, 2,3-dimethyl pentane, octane, nonane and decanes may occur as vapours. The alkanes (saturated paraffins) are relatively inert chemically, except for their ability to enter into the chemical reactions of combustion, dehydrogenation and halogenation. Their combustion properties make them useful as fuels. The products of this combustion are of major concern in air pollution.

Gasoline is a mixture of hexanes, heptanes, octanes and other hydrocarbon compounds. In the internal combustion engine without adequate air pollution control devices, a portion of the gasoline vapour gets out by the pistons without being burnt. The engine also exhausts a lubricating oil mist. This emission will amount to significant quantity of hydrocarbons per unit mass of gasoline burnt in the engine. The benzyls are the most common aromatics in gasoline, but a number of other hydrocarbons are involved. These hydrocarbon products from internal combustion engines are the chief offenders in the atmospheric interactions that contribute to smog production.

It should be mentioned in this context that carbon monoxide concentration would ordinarily reach a dangerous level before the hydrocarbons. As an example, a concentration of 0.4 ppm of hydrocarbons may produce poor visibility and cause eye irritation. For gasoline, the maximum tolerable level for 8 hours may be as high as 500 ppm. Naphtha would show about the same tolerance level.

Mineral oil mist as an interaction product is more of a nuisance than an outright toxic product in air pollution.

6.3.5 Aromatic Hydrocarbons

The aromatic hydrocarbons, such as benzene, toluene, xylene and solvent naphtha are of more concern in industrial hygiene than in air pollution.

Benzene at room temperature is a colourless and flammable liquid. In vapour form, the flammability limits are 1.4 to 8.0 per cent. Repeated exposure can produce damage to the bone marrow that may not be evident for sometime after discontinuing the exposure. Various toxic effects are indicated in the literature. The threshold limit of benzene is 25 ppm. In air pollution design work exposure should not exceed 0.5 ppm for 1 hour or 0.26 ppm for 24 hrs. Values up to 25 ppm have been indicated in industrial hygiene practice for 8 hrs exposure. However, there are a number of special conditions involved and it is the type of thing that involves the possibility of hazard to workers. The engineer should only consider levels higher than recommended here, when working in close liaison with industrial hygiene personnel and medical specialists.

Toluene is a simple derivative of benzene. Its threshold value is 100 ppm and in air pollution design, an exposure of no more than 0.15 ppm/24 hours is recommended. Levels as high as 200 ppm are considered tolerable by some sources, but the acceptability of such exposure levels should be determined only by specially trained personnel.

Xylene has a threshold limit of 100 ppm and design value of 0.05 ppm/24 hrs is recommended.

Benzene, toluene and xylene are used as solvents in dyes and paints, fumigants, chemical synthesis, phenol, rubber manufacturing, detergents and in other industrial applications.

Textile dyes are mostly made up of aromatic hydrocarbons. A number of these compounds are synthesized from aniline which has a threshold limit of 5 ppm. A design value of 0.015 ppm/hr or 0.008 ppm/24 hrs is the recommended exposure limits.

Phenol has a threshold limit of 5 ppm and is also a skin contact hazard. An air pollution design value of 0.003 ppm/24 hrs is recommended. Phenol is also known as **carbolic acid** and as **benzenol**.

6.3.6 Oxides of Nitrogen

Nitrogen oxide is liberated at a rate of approximately 0.02 lb per lb of gasolene burned in an internal combustion engine lacking adequate air pollution control devices. This is comparable to 0.03 lb of nitric oxide for a gas turbine. Nitrogen oxides are injurious to plant life and are involved in atmospheric interactions. A small concentration (less than 0.5 ppm) can produce considerable reduction in visibility. It is unlikely that lethal concentrations will be encountered in residential areas. Under localized industrial conditions the concentration of both nitric oxide and nitric dioxide should be kept below a 25 ppm threshold limit and a design level of no more than 0.1 ppm/24 hrs is recommended.

Nitrogen dioxide is extremely dangerous in industry because a worker may breathe lethal concentrations of the gas without being physically aware of exposure of any kind.

6.3.7 Carbon Monoxide

Carbon monoxide is a colourless, odourless, and tasteless gas that is only slightly soluble in water. Flammability limits are between 12.5 and 75 per cent. An exposure of 30 ppm for 8 hrs is of epidemiological significance. A concentration of 100 ppm for 9 hrs or 900 ppm for 1 hr will produce discomfort. An exposure of 100 ppm for 15 hrs produces severe distress and a concentration of 4000 ppm is lethal in less than 1 hr.

In air pollution design work, a limit of 3 ppm for a 1 hr exposure and 0.5 ppm for 24 hrs is recommended. Carbon monoxide in cities have been found to be as high as 45 ppm.

6.3.8 Arsenic

Arsenic ignites when heated and a white cloud of As_4O_6 results. Arsenic vapour has a garlic odour and yellow cover. As an air pollution hazard, it can result from impurities in ores and coal. Hence, industrial processes, smelters or coal burning sources may have stack emissions with arsenic in the flue gases. Arsenic has been detected up to 10 kilometres from the point of stack emissions.

The threshold limit for arsenic concentrations should be below 0.25 mg/m^3 and in design it is desirable to strive for 0.03 mg/cubic metre.

6.3.9 Aldehydes

Formaldehyde and acetaldehyde are examples of these alcohols, which represent the first stage oxidation of hydrocarbons. They are produced by the combustion of gasoline fuel oil, diesel oil, and natural gas. If motor fuel oil or lubricating oils are incompletely oxidized, the result can be the formation of aldehydes and organic acids.

Formaldehyde is a colourless gas possessing a pungent and irritating odour and is readily soluble in water. In air pollution, its most important effect is irritation to the eyes. A concentration of 5 ppm or greater poses a grave health problem. A design value of no more than 0.015 ppm/24 hrs is recommended.

Acetaldehyde is also colourless and ethylene gas is also a component of auto exhaust. It is capable of producing abnormal development in some plants at low concentrations. Ethylene and other olefin compounds are known to react with ozone.

6.3.10 Ozone

Ozone is of prime importance in general air pollution, as a component of the total oxidant mixture which is used as an index of photochemical smog. The odour is detectable instantaneously at less than 0.02 to 0.05 ppm.

6.3.11 Peroxyacetylenitrate (PAN)

Principal compounds of peroxyacetylenitrates (PAN compounds), organic nitrogen periodic compounds are identified along with ozone as a part of the total oxidant mixture present in photochemical smog. A concentration of greater than 0.30 ppm significantly increased pulmonary function during a 5-minute exposure of volunteers. A concentration of 0.3 ppm is estimated to be the concentration of heavy atmospheric smog.

6.3.12 Nitrogen Compounds

Ammonia is one of the oxidation states of nitrogen. It is a colourless gas possessing an irritating odour. It is a heart stimulant but at high enough concentrations it has killed people who inhaled it. As an air pollutant it can be injurious to plants exposed beyond a certain concentration and length of time. It is a chemically active compound, highly polar and lighter than air. Gaseous ammonia and gaseous chloride will react to form a cloud of very small crystals of ammonium chloride.

Ammonia is more likely to be of concern in localized areas of a plant and its immediate vicinity than over a widespread area, such as the average city. Concentrations must be kept below 50 ppm as a threshold limit. A design limit of 0.3 ppm/24 hr is recommended. Cyanogens is a colourless gas, possesses a distinctive odour and is extremely poisonous. It smells like green apples. Concentrations approaching 200 ppm would be dangerous, but this is not likely to occur except in localized industrial situations, such as the manufacture of aniline dyes and synthetic rubber. A design limit of 5 ppm/24 hrs is recommended.

6.3.13 Lead

If lead is heated in a stream of air it will burn; lead tetraethyl is a liquid at ordinary temperature and is used as an 'antiknock' compound in gasoline. Any soluble lead salt is dangerous. These salts are cumulative poisons because their elimination from the body by the kidneys is an extremely slow process. Lead poisoning can be caused by breathing in fumes or by ingestion.

The threshold of lead tetraethyl should be less than 0.1 mg/cubic metre and a design value of not more than 0.001 mg/cubic metre is desirable; lead as a particulate in the ambient air quality criteria is given as 6 micro-grams/m^3/day.

6.3.14 Olefin Hydrocarbons of Ethylene Series

The olefins or unsaturated alkenes are members of ethylene series. They are more reactive than the saturated paraffin series and are of greater industrial concern. They include such materials as propylene, ethylene (ethane), butadiene and butane.

High temperature (550°–600°C) combined with an appropriate catalyst results in the dehydrogenation of the saturated paraffins and can produce olefins. Ethylene accompanied by propylene, butylenes and other olefins result. These substances are extremely important as industrial solvents. Olefins are also combined with acids in the preparation of alcohols and ethers.

Ethylene has flammability limits of 32 to 34 per cent of volume of gas in air and propylene 2.2 to 9.8 per cent. Ethylene can be used in general anaesthesia. The ambient air quality criterion is 0.5 ppm for 1 hr or 0.1 ppm for 8 hrs. It will decompose in sunlight forming ammonium oxalate ammonium formate and urea. It is also known as **oxalic acid dinitrile**. In air pollution design work a limit of 0.0225 ppm per 24 hrs is recommended.

6.3.15 Alcohols

The simplest alcohol is methyl alcohol also known as methanol or wood alcohol. It may be used in the manufacture of formaldehyde, antifreeze solvent for shellac or other organic compounds. Ethyl alcohol is also called **grain alcohol**, ethanol or plain alcohol. Ethyl alcohol is widely used in manufacture of many products. It is used as a solvent in many medicines and in medical work. The threshold limit as well as some of the recommended design limits for some of the alcohols is given in Table 6.4.

Table 6.4 Recommended Design Limits for Alcohols

Substance	*Threshold limit (ppm)*
Methanol	200
Ethanol	1000
Allyl alcohol	2
n-Propyl alcohol	400
Isopropanol	400
n-Butanol	100
n-Amyl alcohol	100
Isoamyl alcohol	100
Diacetone alcohol	50
Furfural alcohol	50

6.3.16 Esters

Esters result from the reaction of an acid with an alcohol. Some of the esters and their associated threshold limits are methyl formate (100 ppm), ethyl formate (100 ppm), methyl formate (100 ppm), ethyle formate (100 ppm), methyl acetate (200 ppm), ethyl acetate (400 ppm), propyl acetate (200 ppm), butyl acetate (200 ppm) and amyl acetate (200 ppm).

6.3.17 Ketones

A ketone is essentially an aldehyde in which the hydrogen in the aldehyde group is replaced by a hydrocarbon radical. Acetone, the simplest and most important of the series, is a fermentation product of either corn or molasses. Some other ketones are methylethyl ketones, methyl propyl ketones and methylbutylketones. The threshold limit of acetone should be 500 ppm with a design value of 25 ppm/24 hrs.

6.3.18 Photochemical Oxidants

The two terms employed to describe the levels of photochemical oxidants viz. oxidants and total oxidants are generally indicative of the net oxidizing ability of the ambient air; ozone (O_3), the major photochemical oxidant makes up around 90 per cent of the oxidant pool. Other photochemical oxidants of relevance in air pollution monitoring include nascent oxygen (O), Peroxyacetylnitrate (PAN), Peroxylbutylnitrate (PBN), nitrogen dioxide (NO_2), hydrogen peroxide (H_2O_2) and alkyl nitrates.

6.3.19 Particulates

Particulates may be liquids or solids. Particulates may be identified as any dispersed matter in the solid or liquid form in which the individual aggregates are larger than a single small molecule (about 0.002 m) in diameter, but smaller than about 500 microns.

It is a common practice to classify particulates in accordance with their physical, chemical or biological characteristics. Physical characteristics include size, mode of formation, settling properties and optical qualities. Chemical characteristics include organic or inorganic composition and biological characteristics pertain or relate to their classification as bacteria, viruses, spores, pollens, etc.

Typical gaseous contaminants and their origin are summarized in Table 6.5.

Table 6.5 Typical Gaseous Pollutants and their Sources

Key element	*Pollutant (gaseous)*	*Source*
S	SO_2	Boiler flue gas
	SO_3	Sulphuric acid manufacturing
	H_2SO_4 vapour	H_2SO_4 manufacturing, pickling operations
	H_2S	Natural gas processing, pulp and paper mills, sewage treatment
	R-SH (mercaptans)	Petroleum refining, pulp and paper mills
N	NO, NO_2	Nitric acid manufacturing boiler flue gas
	HNO_3 vapours	Nitric acid manufacturing pickling operations
	NH_3	Ammonia manufacturing
	Other N Compounds (i.e., amines, pyridines)	Sewage, rendering solvent process
C	Inorganic, CO, organics, Volatile Organic Compounds (VOC)	In complete combustion
	VOC, hydrocarbons, paraffins	Solvent uses, gasoline marketing
	Olefins, aromatics, VOC oxygenated	Petrochemical plant
	hydrocarbons, aldehydes, ketones, alcohols, phenols	Surface-coating operations, petroleum processing, plastics manufacturing
	VOC, chlorinated solvents	Dry cleaning, degreasing
Halogen F	HF	Phosphate fertilizer plant, aluminium plant
	SiF_4	ceramics, fertilizer plant
Halogen Cl	HCl	HCl manufacturing, PVC combustion
	Cl_2	Chlorine manufacturing

Table 6.6 presents some of hazardous air contaminants.

Table 6.6 Hazardous Air Pollutants

Hazardous air contaminants	
Arcylonitrile	Hexachlorocyclopentadiene
Asbestos	Inorganic arsenic
Benzene	Manganese
Beryllium	Methyl chloroform
1-3 Butadiene	Methyene chloride
Cadmium	Mercury
Carbon tetrachloride	Nickel
Chlorinated benzenes	Perchloroethylene
Chloroflurocarbon	Phenol
Chloroform	Polycyclic organic matter
Chloroprene	Radionuclides
Chromium	Toluene
Coke oven emissions	Trichloroethylene
Copper	Vinyl chloride
Epichlorohydrin	Vinylidine chloride
Ethylene dichloride	Zinc
Ethylene oxide	Zinc oxide

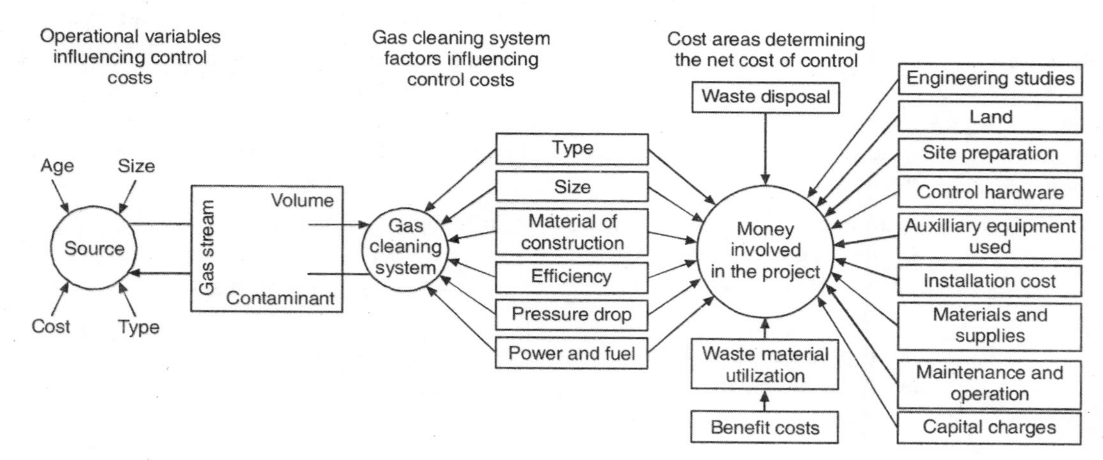

Figure 6.1 Cost evaluation scheme for a gaseous pollutant control system.

Source: Arthur, C. Stern, *Engineering Control of Air Pollution*, 3rd ed., Vol. IV, Academic Press, New York, 1977.

EXERCISES

6.1 Discuss about the typical gaseous pollutants and their sources.

6.2 Describe the methodology of air quality management.

6.3 How are the air pollutants classified?

7

Treatment Systems for Air Pollution Control

The most positive way of dealing with the air pollution problem is to prevent the formation of the contaminants or minimize their emission at the source itself.

In the case of the emission of pollutants from the industrial sources, this can be accomplished by exploring various possibilities at an initial stage of process design and development and choosing those methods which do not give rise to air pollution or have the minimum air pollution potential. They are termed as source correction methods. There are likely to be difficulties in applying these methods to existing plants, but still the possibilities can be explored without severely upsetting the economy of the operation. Pollution control at source can be achieved in many ways, such as going in for raw material changes, variations in operating conditions, modification or replacement of process equipment by more effective utilization of existing equipment and modifications of processes. To cite an example, a modification of the process, such as switching to natural gas in lieu of coal in a power generation plant, will eliminate the immediate air pollution problem; likewise decommissioning a sludge incinerator and placing the sludge on dedicated land for instance, may turn out to be more economical than installing air cleaning equipment for the incinerator.

When source correction methods fail to accomplish the desired objective of air pollution control, one can make use of effluent gas cleaning techniques. Many of the chemical engineering unit operations are involved in this technique.

7.1 CONTROL DEVICES FOR PARTICULATE CONTAMINANTS

The composition of clean air is 78.09 per cent nitrogen and 20.04 per cent oxygen by volume. The balance of 0.97 per cent of the gaseous constituents of air includes small quantities of carbon dioxide, helium, oxygen, krypton and xenon as well as small quantities of other organic and inorganic gases. Water vapour's concentration in air is normally 1 to 3 per cent. Besides the aforementioned constituents, air also contains aerosols, dispersed solid or liquid particles ranging in size from clusters of few molecules to diameters of a few tens of microns. Due to industrial developments all over the world, air also contains pollutants today. The five most common primary air pollutants are carbon monoxide, sulphur oxides, hydrocarbons, nitrogen oxides and particulate matters. The major sources of these contaminants are automobiles, industry, power plants and refuse disposal. The sizes of common atmospheric particles, their significant characteristics and presently available methods of control and abatement are illustrated in Figure 7.1.

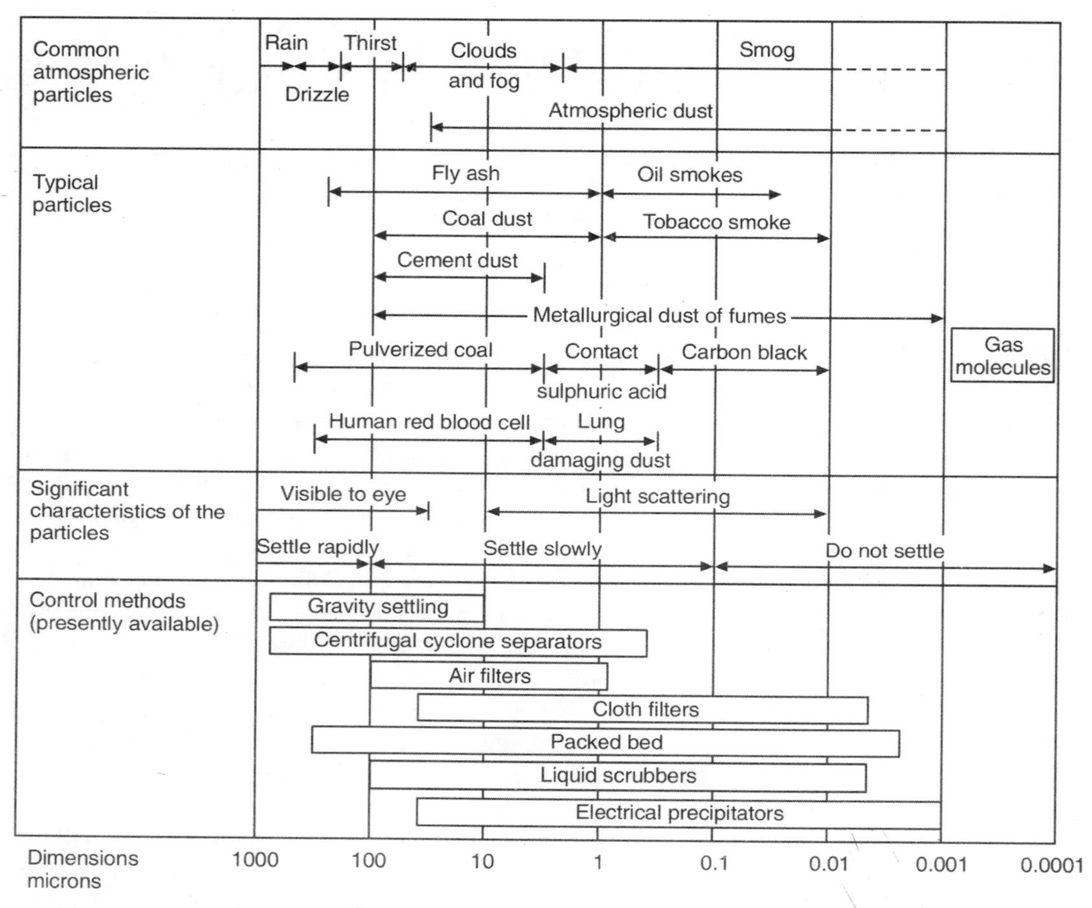

Figure 7.1 Common atmospheric particles—sizes, characteristics and control methods.
Adapted from Frank Krieth, University of Colorado, Boulder, Colorado, USA.

Particulates are the most widespread of all the substances usually termed pollutants of the air. The sources of these contaminants are many; those larger than 10 μ are mostly the outcome of processes such as erosion and grinding; those in the size range of 1 to 10 μ, the most numerous in the atmosphere, emanate from mechanical process but, include also industrial dusts and fly ash, particles in the size range of 0.1 to 1.0 μ, tend to contain products of condensation, products of combustion, aerosols formed by photochemical reactions in the air and ammonium sulphate. The chemical nature of particles in the size range below 0.1 μ is relatively little known, but their likely sources are combustion processes and their concentration is highest over large industrial belts. An important problem in gas-particle reactions in air, is the role of water vapour, but only for the reaction occurring between sulphuric acid and gaseous ammonia, the effect of humidity has been studied.

Industrial sources generate around one-fifth of the total natural tonnage of the five most common contaminants, but they are the largest producer of particulate matter. The increase in regulatory efforts, not withstanding, the situation is continually worsening owing to the industrial growth, failure to apply existing technological and scientific knowledge and in a few cases, because of the non-availability of economically feasible technology.

Gravitation is perhaps the main mechanism by which particles are removed from the atmosphere, but intervening processes by which particles whose size is less than one micron coagulate to form larger particles, are also significant. Particles which are larger than one micron in diameter begin to develop appreciable settling velocity and rain can also remove particles larger than two microns from the air. The oldest method of pollution abatement, although today no longer considered adequate, is to disperse particles into the atmosphere through stacks of sufficient height to carry the pollutants away from the source. Although by itself not really an adequate air pollution control device exhaust stack, if properly designed is capable of preventing localised concentration of contaminants near the source. it should be noted in the context that the chemistry of air pollution reactions is very important for the development of methods to eliminate air pollution.

Devices that collect particulate matter from air streams can be categorized as follows:

1. Gravity setting chambers
2. Cyclones
3. Scrubbers
4. Filters
5. Electrostatic precipitators

The most important drawback of most of the available equipments is that only a few types can handle particles whose diameter is less than 2 μ. Of these, filters cannot operate at high temperatures and electrostatic preciptators suffer from unpredictable loss in efficiency when there is a variation in operating conditions.

To bring about an improvement in the design, operation and effectiveness of air pollution control equipment, a need exists for more fundamental knowledge of their operating principles and of the effects of the contaminants on the efficiency of the control devices. This assumes a special significance in the case of electrostatic precipitators which are used extensively to control emission of small particles form conventional electric power plants.

7.1.1 Settling Chambers

Settling chambers are simple to fabricate, produce only little pressure loss, and need small capital investment. The major disadvantage of these devices is the low efficiency for collecting particles whose diameter is less than 50 microns. The principal design parameters are the chamber length and the gas, which should be less than 30 m/sec to prevent re-entrainment from occurring.

7.1.2 Cyclones

Cyclones are the most widely employed particulate collection device owing to the fact that they are relatively easy to construct, have no moving parts and need little or no maintenance. As illustrated in Figure 7.2, the gas stream with the entrained particulate matter moves in a continuous spiral and the heavier particles are thrown radially outwards by virtue of the centrifugal force field and eventually hit the wall and fall into the hopper by gravity.

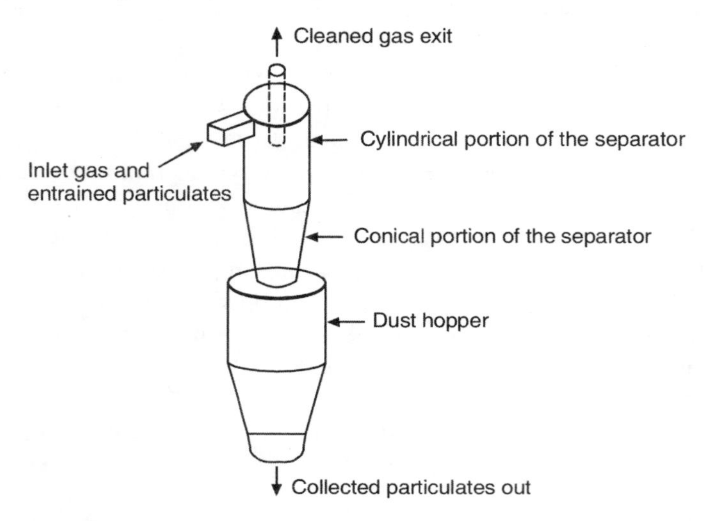

Figure 7.2 Schematic representation of a cyclone separator.

The collection efficiency and the pressure drop have a bearing on the cyclone dimensions, the gas velocity and the particle characteristics. Design criteria for cyclone separators are available in the literature.

7.1.3 Scrubbers

A wet collector or scrubber is essentially a gas-liquid contacting device in which airborne particulates are transferred from suspension in the gaseous phase to a liquid phase suspension or solution. The liquid is usually water; to maximize the gas liquid contact area (interfacial area) and hence, the mass transfer rate, the liquid is usually broken up either into droplets as in spray or into bubbles as in foam.

The merits of wet collectors include particulate re-entrainment, capability to handle high humidity gases and small space requirements.

These devices can also collect particulate and gaseous contaminants simultaneously, and can cool high temperature gases. The main demerits are that the particulates must be removed from the polluted scrubbing liquid and that the washed air has high relative humidity which may bring in steam plume problems. To avoid water pollution, the disposal of the waste scrubbing liquid must be considered in any design using a scrubber.

There is a wide variety of wet collectors including spray chambers, cyclone scrubbers, floating beds, packed beds, venturi scrubbers, induced draft scrubbers, foam scrubbers and wetted scrubbers.

Despite a substantial difference in the appearance of wet collectors, they all have the design parameters given here.

1. Gas-liquid contacting geometry
2. Liquid-gas flow rate ratio
3. Gas pressure drop
4. Gas velocity

For the removal of small particles, venturi scrubber is perhaps the most important wet collector in industrial practice. The venturi scrubber is represented schematically in Figure 7.3.

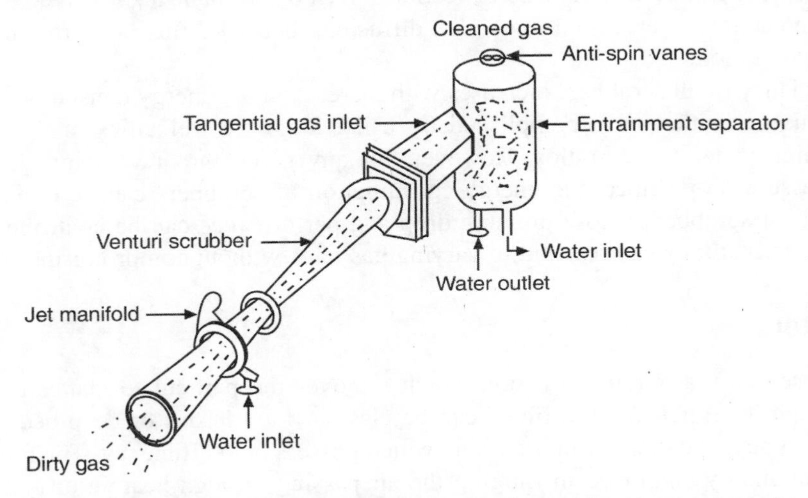

Figure 7.3 Schematic diagram of a venturi scrubber (wet collector).

The additional factors that need to be considered while choosing a wet collector for a specific particulate collection problem are given here.

- Particulate size distribution
- Required removal efficiency
- Heat transfer (gas cooling)
- Gas adsorption requirements
- Particulate solubility in scrubbing liquid
- Gas stream properties
- Flow rate
- Temperature

A few approximate theories of wet collection have been developed. Three basic zones of scrubber operation are considered in the theoretical analysis.

1. Gas humidification
2. Gas-liquid contacting
3. Gas-liquid separation

The gas humidification zone extends from the gas inlet to the point where the gas is saturated with water vapour. The fraction of the scrubber which is required for gas humidification is significant because the evaporation of the water in this zone interferes and hinders particulate collection. The scrubber liquid droplets are removed from the gas in the gas liquid separation zone. The major part of the particulate collection occurs in the gas-liquid contacting zone. The collection of particles by liquid scrubbers involves the capture of the particles by liquid droplets. Both filtration and liquid scrubbing use aerodynamic particle capture mechanism. The gas stream passing through the wet collector takes the particle close to the collecting bodies (droplets or filter fibre), where a number of different collection forces actually accomplish the particle collection. The collection forces or mechanisms involved in the process are inertial impaction, interception, brownian diffusion, thermal diffusion, diffussophoresis and electrostatic migration.

The efficiency of all scrubbers increases with increase in the energy consumption. In venturi scrubbers, this can either be accomplished by using large gas velocities in the throat or by employing high water to gas ratios. However, for any given throat velocity, there exists an optimum pressure drop. Since, the energy consumption of scrubbers can be quite large, there exists a need for scrubbers whose pressure drop and performance can be controlled while they are operating so that they can respond to varying gas rates without compromising on efficiency.

7.1.4 Filters

A filter is essentially a porous structure, which removes the particulate matter from the fluid passing through it. In general, the filters can be classified as fibrous or deep-bed filters which have low efficiency, and cloth or fabric filters which possess high efficiency. The former (fibrous filters) are commonly employed in filtering the air passing through heating or cooling units in air-conditioning and have pressure drops in the range of 0.5 to 6 inches of water. The latter category (cloth or fabric filters) employed in air pollution control are usually arranged as bags. These filter bags are cleaned periodically employing systems, such as shaking, air jets or reverse air flow. The fabric filters are composed of various materials, which include cotton, wool, glass, asbestos and synthetics.

In the design of filter systems, it is customary to specify the gas to cloth ratio, which is defined as the cubic feet per minute of gas flow filtered per unit filter area and is expressed as feet per min. The gas to cloth ratio ranges from 1 to 25 ft/min. A typical gas to cloth ratio is 3 ft/min for intermittently cleaned bags and 12 ft/min for reverse set cleaning.

The fluid mechanics and the entrainment process by which filters made of fibrous woven cotton, wool or glass fabric remove particles from an air stream flowing them, are quite complex. More of basic research will be required to explain quantitatively how a filter removes particles from an air stream, but quantitatively the operational characteristics of most filters follow several steps. The primary particle capture mechanisms in fibrous filters are:

1. Inertial interception
2. Diffusion
3. Electrostatic attraction

The effectiveness of capture due to the first of these mechanisms increases with particle size and velocity while that due to the latter two is reduced. Thus, as particle size is decreased, the major forces in filtration pass from inertial to diffusive and electrostatic. At some critical operating conditions all the three forces will be small and the efficiency of the filter would be expected to pass through a minimum; point of such minimum efficiency has been observed by investigators. Extensive studies have been made of each of the capture mechanisms, but much of the work has centred around the behaviour of a single fibre and extended the results to multiple fibres empirically. This approach does not take into account the mechanism by which a deposit gradually builds up in a filter, causes a decrease in penetration, the formation of an external layer and an increase in the resistance that is an enhancement in pressure drop or decrease in gas velocity through the filter.

Initially when a dirty air stream passes through clean fibrous filter material, the filter does not remove many particles from the air. Even though large particles are trapped at the surface, most of the smaller particles pass through the tiny passages with the air. The pressure drop across the filter material during this phase is small. But with passage of time as more and more particles become trapped in the tortuous passages of the filter, the pressure drop through the filter increases and more and more of the particles are removed from the air stream. This phase of the operation of a filter is termed as **seasoning**. After some time the pressure drop through the filter becomes so large that the flow begins to decrease. At this point it turns out to be necessary to clean the filter by some means. After the cleaning of the filter is carried out, the dust passes through the seasoned filter than did initially through the new material. The cycle has to be repeated several times before the filter material becomes fully seasoned and achieves the desired effectiveness. Periodic cleaning is of course still necessary to maintain the pressure loss at or below the design value. Some preliminary attempts to explain the complex mechanism of filtration and predict the pressure drop have been made analytically and experimentally. The interaction between given filter media and various particles, as well as the transient characteristics of the flow and effectiveness of various cleaning mechanisms, are potential challenging areas for future research.

7.1.5 Electrostatic Precipitators

Electrostatic precipitation is today considered by many experts in this field as the most promising method for getting rid of small particles from exhaust streams of large power generation plants. The process involves the application of an electric field to remove electrically charged particles from their gaseous carrying media. An electrostatic precipitator has a discharge electrode (negative) of small area, such as wire and a collection electrode (positive) of large surface area such as a plate or a tube. A wire in-tube electrostatic precipitator and a wire and plate electrostatic precipitator are represented schematically in Figure 7.4.
There are essentially four steps in the electrostatic precipitation of particles:

1. Placing an electrical charge on the particles
2. Migration of the charged particles in the electrical field to the collecting surface
3. Neutralization of the electrical charge of the particle after impacting on the collection electrode
4. Removal of the precipitated particles from the collection electrode

The particles get charged by the process of colliding with air molecules ionized at the discharge electrode. The neutralized particles are removed from the collecting surface by rapping, scraping, brushing or washing.

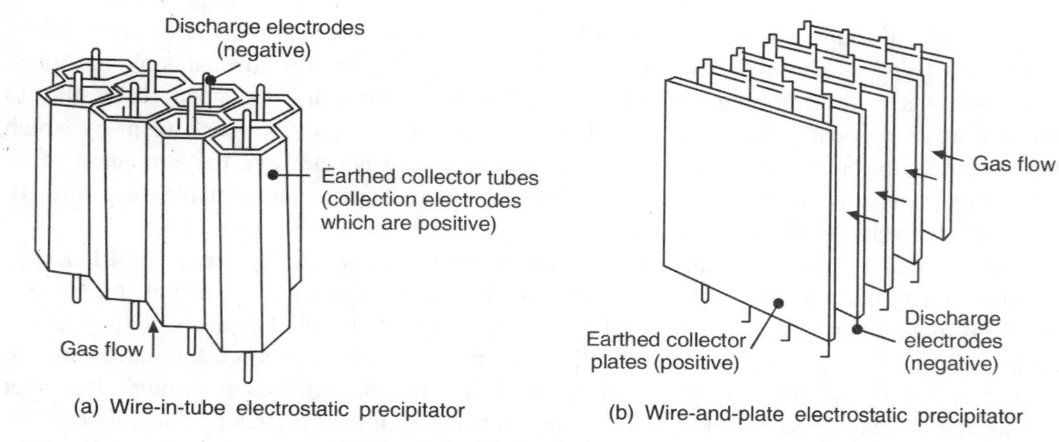

Figure 7.4 Electrostatic precipitators.

There are two general types of electrostatic precipitators, low voltage (two stage) and high voltage (single stage). The low voltage precipitator comprises an ionizing stage followed by a collecting stage. Two stages of electrostatic precipitators are generally employed for removing oil smokes or in air-conditioning. High voltage or single stage electrostatic precipitators combine ionization and collection and are used in removing particulates from highly concentrated emissions. The pressure drop across a precipitator is quite low, 0.1 to 0.5 inches of water being typical magnitudes. The electrical resistivity of the airborne particles has a significant effect on the removal efficiency. Electrical resistivites range from 10^{-3} to 10^{+14} ohm.cm. For high efficiency particle removal, the electrical resistivity should be in the range of 10^4 to 10^{10} ohm.cm. Low resistivity particles are rapidly neutralized and recharged upon impaction on the collection electrode, resulting in the particles being repelled into the gas stream. High resistivity particles build up a layer on the collection electrode resulting in an insulating layer which may eventually break down under the condition of the high potential gradient.

The particle collection efficiency E of an electrostatic precipitator can be described approximately by the relation

$$E = 1 - e^{(-\Delta V/Q)}$$

where

A = area of collecting electrode (ft^2)

V = particle migration velocity (ft/sec)

Q = gas flow rate (ft^3/sec)

The particle migration velocity is a function of the electric field, particle resistivity and the particle diameter and in general its magnitude ranges from 0.1 to 1.0 ft/sec.

From a basic point of view, it will be necessary to study just what the properties of particles are, which must be collected by the electrostatic shield in the electrostatic precipitator. From a physical viewpoint it would appear that the particles of low sulphur and high sulphur coals are not sufficiently different to bring about a major change in the overall performance. This may be a field in which particle physicists could make a contribution which would aid materially in the elimination of particulate pollution. The advantages and disadvantages of electrostatic precipitators are summarized at the end of this section.

7.1.6 Economics of Particulate Control

The economics of pollution control depends on many factors and the cost figures may not be quite accurate. To estimate particulate control costs, the normal assumption is that all particulates would be removed from the exhaust by stack cleaning devices such as electrostatic precipitators, cyclones, wet scrubbers or fabric filters. The value of recovered particulate matter may influence the form in which it is collected. In most cases, the recovered material has no value or the value is so little that reusing it is not economical. But in some cases, for example, detergents, plastics, kaolin or cement, the value of the product may in itself make recovery economical, even if air pollution was not a factor.

Besides the economic factors, the physical characteristics of the gas and dust must be considered. The gas may be corrosive, hot, explosive, dry or moist. Moisture can cause bags in fabric filters to plug with electrostatic precipitators; the chemical composition of the gas can heavily influence corona discharge characteristics. Some gases have a conditioning effect in precipitators and greatly reduce dust resistivities. In collecting fly ash, for example, a precipitator is sensitive to the quantity of sulphur trioxide in the gas stream because its presence significantly enhances collection efficiency.

Particle shape, as well as size distribution can influence collection efficiency. With the mechanical collectors and fabric filters, it is possible to make a fairly good prediction of efficiency from particle size. The efficiency of wet collections, and electrostatic precipitators depends, however, also on wettability for the former and resistivity for the latter. These properties vary from material to material.

Once the various economic and physical factors have been resolved, they can be matched against the advantages, disadvantages and capabilities of different types of equipments or combinations. One such consideration is the energy consumption of the collector. Fabric filters typically operate with a pressure drop of 3 to 6 inches of water, corresponding to a power consumption of 0.6 to 1.2 kW per 1000 cft/min. To obtain high collection efficiency in the sub-micron particles, most high energy scrubbers generally operate at pressure drops of 28 to 50 inches of water although for medium efficiency, it may be possible to operate with pressure drops as low as 5 inches. At higher pressure drops, power consumption can reach 10 kW per 1000 cft/min. Electrostatic precipitators have one advantage in that the energy for separation is applied only to the particles and not to the entire gas stream. As a consequence, power consumption is relatively low, running about 0.2 to 0.4 kW per 1000 cft/min.

Figure 7.5 shows a comparison of collection efficiency for precipitators, scrubbers and cyclones.

Regardless of the type of equipment employed, efficiency is related to cost. The higher the efficiency, the higher will be the energy consumption, in one form or the other and the higher will be the cost. Higher efficiencies are being demanded today and they will have to be paid for. It should be borne in mind, however, that it was not too long ago that the efficiency of modern particulate collection methods could not even be considered. In the overall picture, it appears that particulate pollution control is technologically and economically feasible although additional research could help materially in the development of more effective and less expensive particulate control devices.

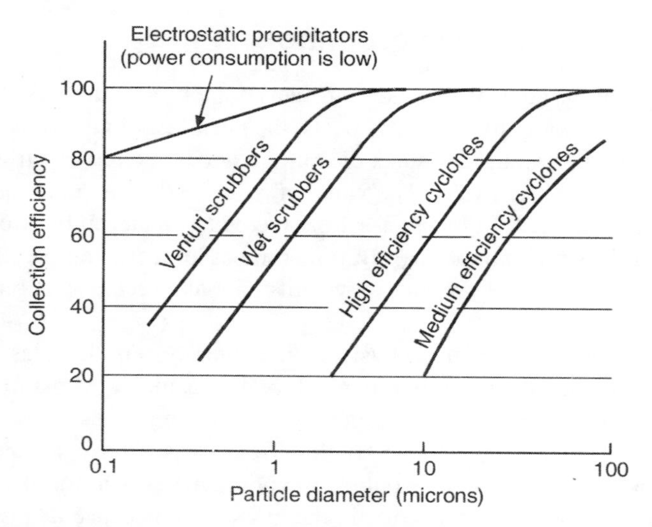

Figure 7.5 Particulate collection efficiencies—typical particulate collection efficiency versus particle diameter for precipitators, scrubbers, and cyclones (particle shape and size distribution can have an impact on collection efficiency).

The advantages and disadvantages of electrostatic precipitators are listed here.

Advantages

- High collection efficiency is possible even in the case of particles as small as 0.01 microns range and collection efficiency is 80 to 99.9 per cent.
- Operating costs are low.
- Low pressure drops of 0.1 to 0.5 inch water are typical.
- Gas flcws as high as 4 million cft/min can be handled effectively.
- Gas pressure and vacuum operating conditions can be used.
- There is essentially no limit to usage of solids, liquids or corrosive chemicals.
- Particulate concentration from 0.0001 to 100 grains/cubic foot can be handled.
- Gas temperatures can be as high as 1200° F.
- The units handle a wide range of gas velocities operation for convenience in cleaning.
- Units of the precipitator can be removed from operation for convenience in cleaning.

Disadvantages

- Installation costs are high.
- Space requirements are high for cold precipitators and even greater for hot precipitators.
- Explosions can occur when the precipitators are collecting combustible gases or particulates.
- Ozone (a poisonous gas) is produced by the negatively charged discharged electrodes during ionization.

- Operating procedures can be complicated. Great precautions must be exercised to maintain safety and proper gas flow distribution, dust resistivity, particulate conductivity and corona spark over rate.

7.1.7 Evaluation of the Removal Methods for Particulate Emissions

This section peruses a variety of air pollution control devices and discusses how each device performs in removing the sub-micrometre sized particles. Such particles are present everywhere in exhaust streams from combustion and other high temperature operations involving boilers and smelters. Since toxic vapours and volatilized heavy metals, generated during high temperature operations, condense on the surface of particles, sub-micrometre sized particles can be deadly carriers of toxic substances.

Once coated with heavy metals, tiny particles pose a greater toxicity threat than larger particles owing to their increased surface area. For instance one gram of 0.1 μm particles has 10 times the surface area as a gram of 1.0 μm particles. The threat posed by minute particulates calls for a concern about worker safety and public health and hence, it turns out to be necessary to look for an effective control strategy for sub-micrometre airborne particles.

Many of the present day's air pollution control devices are not suited for the efficient and effective removal of sub-micrometre particles while settling chambers, cyclone separators, venturi scrubbers, conventional electrostatic precipitators and packed towers remove a variety of gaseous and particulate contaminants, with some performing better than others when it comes to the removal of sub-micrometre sized particles.

When choosing a pollution control system, one must start with a thorough analysis of the gas stream. In many situations, efficiency and cost-effectiveness are often compromised when system specifications are not matched with the parameters of a given gas stream.

Important parameters that will impact or influence equipment selection and performance are:

1. Volumetric flow rate
2. Temperature of the exhaust gas
3. Moisture content
4. Particulate concentration, and
5. Particle size distribution

Particle size is the most important factor. Each process relies on the laws of physics that govern the separation of particles from a gas flow. In general, gas-solid separation requires

1. A differential force that acts on the particle.
2. A differential force that crosses the gas streamline perpendicular to the motion of the gas.
3. A retention zone to collect the separated particles.

Today's devices for gas solid separation rely on a variety of forces depending on particle size, gravity, centrifugal force, electrical impulse and brownian motion. Particles of varying size are displaced as a result of each of these forces. For example, since particles measuring 10 μm or less generally do not respond well to gravitational force, settling chambers (which are based on gravitational force) are not typically effective for the removal of particles less than 50 μm.

In contrast, cyclonic separators rely on centrifugal force, which increases as radius decreases. This makes cyclonic separators more effective than gravity chambers at separating small diameter particles from a gas stream. As particles follow a circular path in a cyclonic separator, centrifugal force is directly proportional to both particle mass and tangential velocity to the second power, but is inversely proportional to the radius of the turn. However, the inlet velocity required for separating small particles (i.e., less than 5 μm) may result in excessive energy consumption and wear on the equipment. This makes centrifugal separators impractical for the removal of tiny particles.

Since small particles typically respond well to electrical forces, electrostatic precipitators (ESP) are particularly effective in removing particles less than 1 μm in size. For example, the electrical forces exerted on an electrically charged particle of 0.1 μm can be more than million times the force of gravity. Particles smaller than 0.1 μm are also responsive to the substantial force of brownian motion which causes random movement and collision with other gas molecules.

Using the elementary kinetic theory one can calculate the velocity of particles subject to brownian motion and show that particle velocity is a function of particle size, gas temperature and viscosity. Equipment based on the relationship between gas temperatures and particle velocity can yield impressive results especially when applied with electrostatic forces.

A number of devices are available to capture minute airborne particles. Users must understand the limitations of each type in a given situation

Fabric filters

These devices, especially pulse jet filters, produce effective removal of fine particles. However, their use is limited by the chemical and physical nature of the particles. For example, fabric filters cannot be used to capture hygroscopic particles or particles that can stick to the filter bags.

Likewise, since fabric filters typically operate at temperature of 350–400° F, they are suitable for the removal of solids, fly ash and particulates. But such systems cannot remove gaseous pollutants and uncondensed vapours unless combined with a dry scrubber, which would further complicate temperature control and wastage disposal and hence, enhance the system cost.

Venturi scrubbers

These devices perform very well when treating moist gas or liquid with particles larger than 1 μm. They are most effective for gas pre-treatment as part of a complete system. However, the use of a venturi scrubber for sub-micrometre particle control can turn out to be expensive since the removal of smaller particles requires an increase in horsepower to draw the gas through the scrubber throat. The increased horsepower requirements consequently increase energy use and operating cost. Besides, in a venturi scrubber, energy is required to move the entire body of gas through the restricted venturi throat. This makes it an expensive option for sub-micrometre particulate control.

Electrostatic precipitators

These devices offer high capture efficiency for sub-micrometre particulates by using electrical forces to remove suspended particles from a gas stream. Because an ESP employs electricity only to charge particles in a passive air stream (not to move the entire gas stream through a venturi throat), ESP energy use is much lower than a venturi scrubber. Typically three steps are involved; charging, collection and removal. A corona discharge, electrically charges, the suspended particles which are collected on electrodes. In a dry ESP, the electrodes are periodically shaken to remove accumulated particles for disposal. In spite of that periodic shaking, dry ESPs are prone to build up on the walls of the collection chamber; such buildup acts as an insulator, reducing the overall efficiency of the equipment, while dry ESPs (also operating at 350–400°F) are commonly employed for particulate removal; they must be used in concert with scrubbers or other equipments employed to capture VOCs and acid gases that are typically contained in industrial exhaust streams.

Wet Electrostatic Precipitator (WESP)

In a WESP, a liquid usually water is used to continuously wash away particulate buildup from the collection surface. The WESP is widely employed for applications where the gas to be treated is hot, has a high moisture content, contains sticky particles and has been subjected to pre-treatment in a scrubber, but requires additional removal of particulate and acid droplets or contains sub-micrometre particles which cannot be treated by other methods. The various control methods are compared in Table 7.1.

Table 7.1 Comparison of Control Methods

Process conditions and requirements	*Wet ESP*	*Scrubber*	*Dry ESP*	*Bag house*
Captures particles less than one micrometre in size	Yes	Yes	Yes	Yes
Handles moist gas	Yes	Yes	No	No
High efficiency for sub-micrometre particles with low energy consumption	Yes	No	Yes	No
Particles collected in liquid	Yes	Yes	No	No
Recommended for fire-sensitive processes	Yes	Yes	No	No

In a conventional *upflow* WESP, the gas stream flows parallel to the water. This requires frequent cleaning because many droplets in the bottom water sprays do not reach the top of the chamber, where sub-micrometre particulates accumulate and sprayers are susceptible to plugging. As such, the equipment must be periodically shut down for cleaning, leading to interruption of operations. In a downward flow or *downflow* WESP, the inlet gas and water sprays move concurrently downward, so these limitations can be partially overcome. By incorporating an appropriate variation on the WESP design, it can be made suitable for the removal of sub-micrometre sized particles. The modified version is termed condensing vertical tube WESP.

Air pollution control devices, particularly those for the removal of sub-micrometre particles should be able to operate as part of an integrated system. For instance, by combining a WESP with a relatively inexpensive, low-energy scrubbing section (which acts as the air-distribution device) large and small particles (down to 0.01 microns) can be easily removed from exhaust with minimum maintenance.

7.1.8 Factors to be Considered While Selecting Particulate Control Equipment

Many factors are likely to influence the choice of a control device employed to reduce particulate emissions. The composition of the particulate matter in terms of concentration, size, and chemical and physical characteristics of the particles should be taken into consideration. If emitted material can be used in the process, dry collection should be used. If the pollutant has little economic value, collection should be accomplished and the pollutant disposed of safely and economically. The industrial process and potential control devices must both be carefully reviewed. It should be borne in mind that the conversion of an air pollution problem into a water pollution problem can entail a more difficult disposal problem. It is necessary to consider many factors which are illustrated hereinafter.

Particle concentration or grain loading

One selection factor to be taken into consideration is the concentration or grain loading of particulate pollutants in the process exhaust stream. Pollutant concentration is typically expressed in terms of pounds per cubic foot, grains per cubic foot and grams per cubic metre. Both the level and fluctuation of grain loading are very important. Some control devices are not affected by high levels or great fluctuations in particle concentration, such as in fabric filters. Other control devices such as electrostatic precipitators do not function effectively with a large fluctuating concentration level. Another pertinent problem can be encountered when the exhaust gas velocity changes rapidly. Some control devices are designated to operate at a specific exhaust gas velocity; large variations can drastically affect the collection efficiency of the unit.

Characteristics of the particle

Particle characteristics, such as size, shape and density must be taken into consideration. Particle size is usually expressed in terms of aerodynamic diameter. The aerodynamic diameter describes how the particle moves in a gas stream. The larger particles with size greater than 60 μm aerodynamic diameter can be collected in simple devices, such as settling or baffle chambers. Particles greater than 5 μm can be collected in cyclones or multi-cyclones. Smaller particles with size less than 5 μm must be collected in more sophisticated devices such as scrubbers, bag houses or electrostatic precipitators.

Particle size, thus, plays a large role in the collection efficiency of a specific control device.

Chemical and physical properties

Chemical and physical properties of particulate emissions greatly affect the choice of control devices. Electrical properties of the particle can both be a hindrance and an aid to collection. Static electricity can create solid buildup in both inertial and bag house collectors. Cakes can form on the bag filters, as a result of the static electrical forces and can be difficult to dislodge.

In an electrostatic precipitator, on the other hand, the collection of particulate matter depends on the ability of the precipitators to charge the particle. Particles passing through an electric field are charged and consequently migrate to an oppositely charged collection plate. Although there are many factors that govern the case of charging the particles, the primary factor governing adequate particle collection is particle dust resistivity—a term that describes the resistance of the collected dust layer to the flow of electric current.

If the particles in the gas stream are explosive, electrostatic precipitators cannot be employed. Fabric filters might be used, but only if no static electric effects exist in the bag house. The logical choice of control would be a scrubber in which water is used as the scrubbing liquid. The water has a dampening effect on the explosive dust.

Hygroscopicity

Hygroscopicity is the tendency of a material to absorb water. It is a physical characteristic of some particles which cause variations in the crystal structure as water is added. Some particles, such as sodium sulphate can absorb up to 12 moles of water per mole of anhydrous salt in high humidity conditions. Other particles from processes, such as alfalfa dehydration and pelleting, and cotton ginning are hygroscopic to varying degrees. The hygroscopic nature of the particle affects the performance of mechanical collectors by causing dust deposits to build up on their internal surfaces. This may cause internal plugging and unpredictable dust cake discharges into collection hoppers at various times. Hygroscopic particles also affect the choice of cleaning in a bag house in that they form cakes on the bags that are difficult to remove.

Particle toxicity

Particles toxicity influences the location of a control device and the air moving system (fan). Highly toxic materials require the use of a negative pressure system so that leaks can be contained within the collector. A positive pressure system could cause fugitive emissions, creating an occupational health and safety problem. In a negative pressure system, the fan is located downstream of the air pollution control devices. The volume of gas to be handled may increase slightly by air leakage into the collector, but little or no contaminant leakage from the collector should take place.

Carrier gas stream

The behaviour of the carrier gas is also of importance in the design phases of air pollution control systems. Gas stream temperature affects a number of variables in the design stages of the control device. The size and thus, the cost of the unit has a bearing on the temperature of the exhaust gas stream being treated. The volume of gas to be cleaned would be larger at high temperatures than that at correspondingly lower gas temperatures. Reducing the temperature reduces the volume of exhaust gas to be handled; however, this could create some additional problems. The gas stream temperature must be maintained above the dew point of the gas to prevent water and acid from condensing in the collector. Water and acid mists could cause corrosion and complete deterioration of the structural material of the collector.

It is also likely that high gas stream temperatures cause equipment failures to components of a fabric filtration system. At exhaust gas temperatures greater than 300°C, most fabric materials deteriorate. Gas temperature can also affect conditions, such as particle resistivity. By bringing about a change in the temperature of the exhaust stream in an electrostatic precipitator one can also change in the resistivity of the particles and thus collection efficiency of the unit. Some of the industrial processes and typical control methods are shown in Table 7.2.

Table 7.2 Industrial Processes and Typical Particulate Control Methods

S.no.	*Industrial process*	*Particulates in the form of*	*Typical control method*
1.	Abrasive blasting	Dust	Bag house or centrifugal scrubber
2.	Aggregate plants	Dust	Spray and/or bag house
3.	Asphalt (hot) paving, patch plants	Dust	Cyclones and centrifugal spray chamber or baffle spray tower
4.	Carbon black production	Dust	Bag house or cyclone and electrostatic precipitator
5.	Cement batching plants	Dust	Bag house, panel filter
6.	Chemical milling	Mists	Wet collector, spray and baffle type
7.	Coffee roasters	Dust, chaff, mists, smoke	Afterburners and/or cyclones
8.	Dryers	Dust, smoke	Scrubber and/or bag house and/or cyclones, spray scrubbers
9.	Electroplating	Acid mists	Spray scrubbers
10.	Fish cannery	Dust, smoke	Cyclones and contact condenser scrubber
11.	Food processing		
	(i) Deep fat frying	Smoke, food particles	Incineration and two stage electrostatic precipitator
	(ii) Smoke houses	Smoke, organic matter	Incineration or two stage electrostatic precipitator with centrifugal scrubber
12.	Glass manufacturing		
	(i) Raw material	Dust	Bag house and panel filter
	(ii) Melting furnaces	Various dusts	Cyclone scrubber or bag house
	(iii) Fanning machine	Smoke	Process control
13.	Heat treating	Salt fumes	Bag house
14.	Insecticide manufacturing		
	(i) Dry	Toxic dusts	Bag house
	(ii) Liquid	Aerosols	Packed tower scrubber

(*Contd.*)

Table 7.2 Industrial Processes and Typical Particulate Control Methods (Contd.)

S.no.	*Industrial process*	*Particulates in the form of*	*Typical control method*
15.	Metallurgical processes	Smoke organic	After burning
	(i) Foundries (a) Coke ovens (b) Sand equipment	acid, dust, smoke, organic vapours	Bag house or scrubber
	(ii) Metal separation processes	Smoke, dust, fumes	Bag house and/or after burning
	(iii) Lead refining	Smoke, fumes, oil vapour and dust	After burning and bag house
	(iv) Aluminium melting	Smoke, fumes	Horizontal wet cyclone or packed column water scrubber
	(v) Brass and bronze melting	Dust, metallic fumes	Bag house
	(vi) Iron casting	Metallic fumes, smoke, oil vapour	Bag house or electrostatic precipitator with auxiliary devices
	(vii) Steel manufacturing	Fumes	Bag house or ESP with condition gas stream.
16.	Paper mills kraft	Chemical dusts	Electrostatic precipitators (ESP), Venturi scrubbers
17.	Petroleum equipment		
	(i) Air blown asphalt	Aerosols	Scrubber and/or incineration
	(ii) Catalytic regeneration	Dusts, oil mists, aerosols	Cyclones wet/or dry and ESP
	(iii) Storage facilities	Aerosols	Scrubbers, for water-soluble products
18.	Phosphoric acid	Fumes, acid, mists	ESP for venturi scrubber and packed tower or packed tower and glass fibre packed filter
19.	Pipe coating	Fumes	Baffle water spray chamber
20.	Pneumatic conveyors	Dust	Bag house
21.	Portland cement	Alkali and dusts	Bag house, ESP, mechanical collectors
22.	Rubber compounding equipment	Dusts, fumes, oil mists	Bag house
23.	Sulphuric acid	Acid mists	ESP, (tube type/or packed bed separators or wire mesh eliminator)
24.	Surfaces coating	Paint particles	Filter pads, water spray, baffle plates
25.	Wood working equipments	Dust, chips	Cyclone or bag house
26.	Zinc galvanizing	Oil mist, fumes	Bag house, ESP

The merits and demerits of the particulate collection devices are presented in Table 7.3.

Table 7.3 Merits and Demerits of the Particulate Collection Devices

Collector	*Merits*	*Demerits*
Gravitational settler, minimum particle size >50 μm	Low pressure loss: easiness in design and maintenance	Much space in required low collection efficiency
Cyclone 5–25 μm	Simplicity of design and easy maintenance; requirement of little floor space; dry continuous disposal of collected dusts; pressure loss is low to moderate; handles dust loadings; temperature independent	Requirement of much headroom; low collection efficiency in the case of small particles; sensitivity to variable dust loadings and flow rates; tends to clog up
Wet collectors 1–15 μm	Simultaneous particle removal and gas absorption; ability to cool and clean hot, moisture laden gases; corrosive gases and mists can be recovered and neutralized; less dust explosion risk; variable efficiency	Corrosion and erosion problems; additional cost of waste water treatment and reclamation; reduced efficiency on sub-micron particles; contamination of effluent stream by liquid entrainment; freezing problems in cold weather; water vapours contribute to visible plume under some atmospheric conditions
Electrostatic precipitator greater than 1 micron	Highly efficient (greater than 99%); very small particles can be collected; particles may be collected wet or dry; pressure drop and power requirements are small compared to other high efficiency collectors; easy maintenance; except in the case of corrosive or adhesive materials; few moving parts; can be operated at high temperature— 600–850 degree Fahrenheit	Relatively high initial cost; sensitive to variations in dust loadings or flow rates; resistivity causes some material to be economically uncollectable; hazard due to high voltage; collection efficiency can come down gradually and imperceptibly
Fabric filters <1 μm (Minimum particle size)	Dry collection possible; decrease of performance is noticeable; collection of small particles feasible; high efficiencies possible	Sensitivity to filtering velocity; hot gases must be cooled to 200–500 degree F; influenced by relative humidly (Condensation); susceptibility of fabric to chemical attack
Combustion devices (Incinerators) after burner direct flame	High removal efficiencies of sub-micron odour causing particulate matter; simultaneous disposal of combustible gaseous and particulate matter; direct disposal of non-toxic gases and wastes to the atmosphere after combustion; possible heat recovery; relatively small space requirement; simple construction and easy maintenance	High operational cost; fire hazard; removes only combustible
Afterburner catalytic (Catalytic combustion)	Same as direct flame afterburner; compared to direct flame; reduced fuel requirement; reduced temperature, insulation requirement and fire hazards	High initial cost; catalysts subject to poisoning; catalysts require reactivation

Table 7.4 presents some of the industrial processes and control summaries for particulate matter.

Table 7.4 Industrial Processes and Control Methods for Particulates

Industry or process	*Source of particulate emissions*	*Particulate matter*	*Method of control*
Iron and steel mills	Blast furnaces, steel making furnaces, sintering machines	Iron oxide, dust, smoke	Cyclones, bag houses, electrostatic precipitators, wet collectors like venturi scrubber, spray chambers, etc
Grey iron foundries	Cupolas, shake-out making	Iron oxide, smoke, oil, dust, metal fumes	Scrubbers, dry centrifuge collectors
Non-ferrous metallurgy	Smelters and furnaces	Smoke, metal fumes, oil, grease	Electrostatic precipitators, fabric filters
Petroleum refineries	Catalyst regenerators, sludge incinerators	Catalyst dust, ash from sludge	(Centrigugal collectors such as) cyclones, electrostatic precipitators, scrubbers, bag houses
Portland cement	Kilns, driers, material-handling systems	Alkali and process dusts	Fabric filters, electrostatic precipitators, mechanical collectors
Kraft paper mills	Recovery furnaces, lime kilns, smelt tanks	Chemical dusts	Electrostatic precipitators, wet collectors like venturi scrubbers
Acid manufacture, phosphoric, sulphuric	Thermal processes, rock acidulating, grinding	Acid mist, dust	Electrostatic precipitators, mesh mist eliminators
Coke manufacture	Oven operation, quenching materials handling	Coal and coke dust, coal tars	Scrupulous design, operation and maintenance
Glass and fibre glass	Furnaces, forming and curing, handling	Acid mist, alkaline oxides, dust, aerosols	Fabric filters, after burners

7.2 TREATMENT OF GASEOUS EFFLUENTS

7.2.1 Absorption

The gaseous pollutants from flue gas can be effectively removed employing a suitable absorbent in a mass transfer equipment. The liquor (scrubbing) and gas can be brought into contact with each other while both are flowing in the same direction (Parallel or concurrent), in opposite directions (Countercurrent flow) or while one stream flows perpendicular to the other (Cross flow). The scrubbing liquor employed for the removal of gaseous contaminants can be a slurry or a solution.

Scrubbing techniques

Scrubbing techniques can be categorized into physical and chemical methods. In all scrubbing techniques, the flue gas containing the gaseous contaminant is intimately brought into contact with the scrubbing liquor.

In physical scrubbing, the gaseous contaminant diffuses into the liquid phase and dissolves in the scrubbing liquor. The flue gas and scrubbing liquor temperatures, the concentrations of the contaminant in both the gaseous and liquid phases and the solubility of contaminant in the scrubbing liquor are the main factors that determine the effectiveness of this method. The scrubbing liquor (usually water) is the solvent and the gaseous pollutant is the solute. Some of the gaseous contaminants that can be removed by physical scrubbing techniques are given here.

- Absorption of hydrogen chloride gas in water
- Absorption of SO_3 with sulphuric acid
- Absorption of fluoride with water

It may be mentioned in this context that this technique is not effective for the removal of low concentration of gaseous contaminants from the flue gases.

In the chemical scrubbing technique, the gas contaminant is chemically reacted with a component (a solvent) in the scrubbing liquor. This is the most effective and commonly employed method to remove gaseous contaminants from flue gases. Parameters, such as reaction rates, temperature, pH of liquor and concentrations that influence chemical reactions are also important in evaluating the effectiveness of the method. Some of the examples of the chemical scrubbing technique are the absorption of sulphur dioxide using calcium carbonate, calcium oxide, magnesium oxide, sodium hydroxide and other alkalis; the absorption of nitrogen oxides using urea solution and the absorption of hydrogen sulphide using green liquor and sodium hydroxide.

This technique is successfully used to produce or recover saleable products. For instance, gypsum can be produced by scrubbing sulphur dioxide contaminants from flue gases with calcium-based alkali. Another application of this technique is the production of liquid sulphur dioxide, elemental sulphur and sulphuric acid which can be produced by scrubbing sulphur dioxide with magnesium oxide slurry or with a sodium sulphite-sodium bisulphite solution.

Yet another example, is the hydrogen sulphide removed from process gas streams in petroleum refinery to meet air pollution regulations.

With further processing, this can be converted to elemental sulphur which is a valuable product in chemical industry.

Scrubbing of gaseous pollutants

The gaseous air contaminants most commonly controlled by absorption include sulphur dioxide, nitrogen oxides, hydrogen sulphide, hydrogen chloride, chlorine, ammonia and light hydrocarbons. The commonly employed scrubbing media used for the removal of various gaseous contaminants are presented in Table 7.5.

Table 7.5 Gaseous Contaminants and Commonly Employed Scrubbing Media

Gaseous contaminant	*Scrubbing media*
Sulphur dioxide	NaOH, Na_2CO_3, Na_2SO_3, KOH, K_2CO_3, K_2SO_3, MgO or $Mg(OH)_2$, ZnO, CaO, $Ca(OH)_2$, $CaCO_3$ and NH_3
Nitrogen oxides	$Ca(OH)_2$, $Mg(OH)_2$, Na_2SO_3, $NaHSO_3$ and urea
Hydrogen sulphide	NaOH, Na_2CO_3, KOH, K_2CO_3 and green liquor
Hydrogen chloride	Water, NaOH, NH_3, calcium hydroxide and $Mg(OH)_2$
Chlorine	Water, NaOH and NH_3

In choosing the scrubbing media, one should remember that removal of gaseous contaminant should not result in water pollution or waste disposal problems. The use of calcium based scrubbing liquor for the removal of sulphur dioxide results in a disposal problem because calcium sulphite, calcium sulphate and the soluble magnesium, potassium and sodium compounds (present as impurities with lime or limestone) are potential water contamination sources. The regeneration of the scrubbing alkali and recovery of the gaseous contaminant in the form of saleable product is the most desirable and practical solution.

One can easily and effectively remove the gaseous contaminant from the flue gas using available scrubbing technology, but it is also important to pay attention to the costs and the resulting water contamination problems. The SO_2 regeneration-recovery processes do eliminate water contamination present and reduce the alkali cost, but they increase regeneration expenses in order to produce a saleable product. The selection of scrubbing liquor is of paramount importance, as the alkali that is most effective for the removal of the gaseous contaminant from flue gas may also be the most difficult to regenerate and may entail water contamination problems. The sodium based alkali namely caustic soda (NaOH), soda ash ($NaCO_3$) and sodium sulphite (Na_2SO_3) are very effective in absorbing sulphur dioxide from flue gases; but because of its very high solubility, the regeneration of the alkali is more difficult. The calcium based alkali namely calcium oxide (CaO) and calcium carbonate ($CaCO_3$) are least effective for the removal of sulphur dioxide and are most difficult to regenerate. The magnesium based alkali, namely magnesium oxide (MgO) and magnesium carbonate ($MgCO_3$) can effectively remove sulphur dioxide and can be effectively regenerated, although it is not compatible with all scrubber designs owing to plugging problems.

Scrubbing (absorption) equipment

A variety of wet scrubbing equipments namely packed, plate and spray towers, and venturi scrubbers are employed to remove gaseous contaminants from flue gases.

Packed towers are the most efficient mass transfer devices, using the suitable packings to provide mass transfer (interfacial) surface. Normally the liquor and gas flow countercurrent to each other, with the gas flowing upward and liquor flowing downward. The use of packed towers is limited to clean gases, as any precipitate or slurry will cause plugging of the packing. Thus, packed towers are not suitable for use with calcium and magnesium based slurry scrubbing liquors.

The plate towers also have the same limitations as packed towers. In the spray towers, the flow arrangements are normally counter current and cross flow. The mass transfer surface is provided by droplets created by spray nozzles, when slurry is used as scrubbing liquor. Spray nozzles should be carefully designed and selected to prevent plugging.

Venturi scrubbers are successful in absorbing gaseous pollutants, employing both clear liquor and slurry as scrubbing liquors. In this design, the gas and liquor flow in the same direction. The liquor is introduced either by spray nozzles or tangentially through pipe nozzles and the gas is accelerated through the converging section reaching the maximum velocity at the throat.

The advantages and disadvantages of wet scrubbers, in general, are listed out in Table 7.6.

Table 7.6 Advantages and Disadvantages of Wet Scrubbers

Advantages	*Disadvantages*
• Absorbs gas phase emissions • Compact size • Efficient through wide loading range • Insensitivity to moisture content • Low capital cost • Low operating and maintenance cost • Re-entrainment rare • Versatility for hazardous emissions	• Corrosion • Inefficient with high temperature gases • Requires high power input • Waste scrubber liquid handling required

The major types of wet scrubbers are presented in Table 7.7.

Table 7.7 Major Types of Wet Scrubbers

Category	*Particle capture mechanism*	*Liquid collection mechanism*	*Types of scrubbers*
Spray scrubbers	Inertial impaction	Droplets	Spray towers, cyclonic spray towers, vane-type cyclonic towers, multiple-tube cyclones
Packed bed scrubbers	Inertial impaction	Sheets, droplets	Standard packed bed scrubbers, fibre bed scrubbers moving bed scrubbers, cross flow scrubbers, grid packed scrubbers
Plate scrubbers	Inertial impaction, differential impaction	Droplets, jet and sheets	Perforated plate scrubbers, impingement plate scrubbers, horizontal impingement-plate (baffle) scrubbers
Venturi scrubbers	Inertial impaction, diffusional impaction	Droplets	Standard venturi scrubbers, variable throat venturi scrubbers, flooded discs, plume bob, movable blade, radial flow variable rod
Orifice scrubbers	Initial impaction, diffusional impaction	Droplets	Orifice scrubbers
Mechanical scrubbers	Inertial interception	Droplets and sheets	Wet fans, disintegrator scrubbers

7.2.2 Condensers

Gaseous pollutants can be removed by condensation initiated by the cooling or compression of the total gas stream. In air pollution control, temperature reduction is usually the more economical method.

When pollutants are present in significant concentrations and with dew points above 30°C, condensation can be considered as a method of removal. Condensation is commonly employed in series with other control equipment; the condenser brings about a reduction in the load on a more expensive control device like an adsorber, absorber or catalytic incinerator.

Condenser designs can be contact or surface types. In contact condensers the coolant and the contaminated gas are mixed together. In surface condensers, on the other hand, heat is removed through a heat exchange surface and the coolant does not come in contact with the gas.

Condenser types

Most surface condensers are heat exchangers of the tube and shell type. Normally the coolant is in the tubes if the condenser is installed in the horizontal position and the vapours condense on the outside of the tubes. On the other hand, vapours can be condensed inside or outside of the tubes in the case of vertical installations. Chilled water is the most common coolant. When the mixture of the condensable and non-condensable gas is exposed to a surface colder than the of dew point of the mixture, certain amount of condensation takes place. A liquid film of condensate forms on the cooling surface and a film of gas collects next to the liquid film. The concentration of vapour in the gas film is less than its concentration in the main gas stream. Because of differences in the partial pressures of the vapour in the bulk gas mixture and in the liquid gas interface, the vapour diffuses from the bulk gas through the gas film to liquefy on the cold surface. The rate of condensation is thus governed by the diffusion of the vapour through a film of non-condensable gas.

Direct contact condensers can be the spray scrubber and the jet condenser types. The spray scrubber type is simply a large vessel in which the gas-condensable vapour mixture is brought into contact with the liquid coolant, as the coolant is injected into the vessel through spray nozzles. Baffles are provided to ensure intimate contact between the liquid spray and the gaseous mixture.

In the case of jet condenser type unit, the gas-vapour mixture is entrained into a high speed water jet. This operation involves heat transfer and mass transfer between the two phases, and the condensed pollutant leaves with the liquid stream. The direct contact condenser needs less cooling water, but may result in an unacceptable water contamination. Figure 7.6 illustrates both types of direct contact condensers.

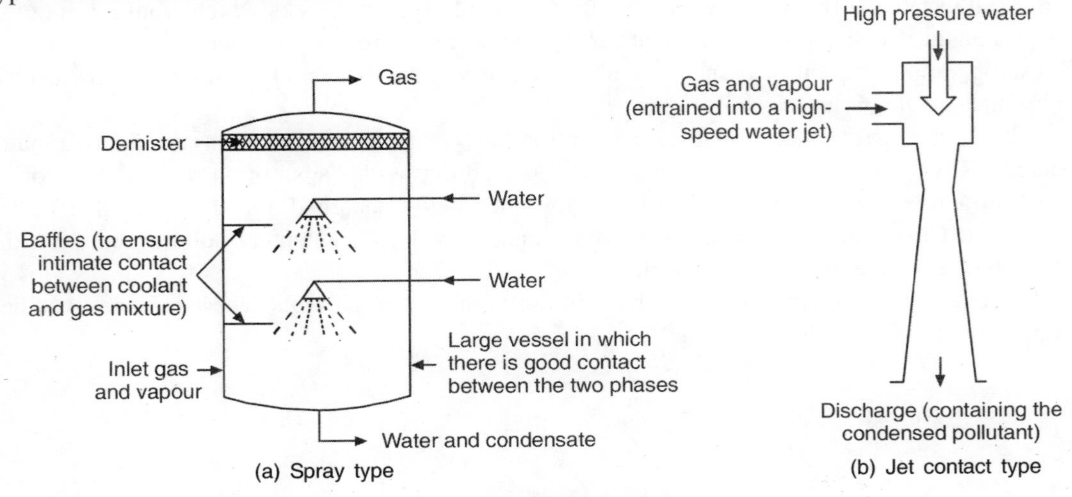

Figure 7.6 Condensers.

Typical installations

Condensers are normally employed in conjunction with other control equipments. They are dependent on the concentration of the contaminant vapour in the gas stream and, therefore, this method turns out to be very inefficient, as the contaminant concentration is lowered by the removal of condensate. It is normally economically attractive to remove the bulk of the contaminant with a relatively inexpensive condenser and then follow it up with a more efficient control device for completing the job. This way, the condenser reduces the load on the more expensive equipment. Figure 7.7 represents a typical scheme for a fat rendering operation.

Condensers are applicable when the vapour present is of relatively high concentrations and the dew point is above 30°C. Applications in the petroleum refining and petrochemical industries include the removal of solvents and organic vapours from air.

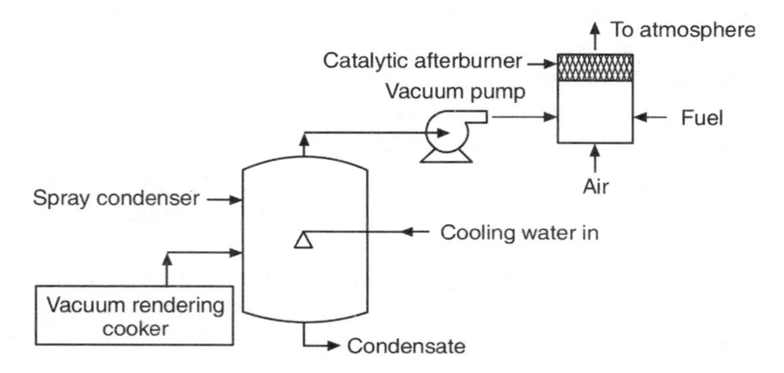

Figure 7.7 Contact condenser and catalytic afterburner in an air pollution control system.

7.2.3 Adsorption

Adsorption is a process whereby molecules of a liquid or gas contact and adhere to the surface of a solid. In air pollution abatement, adsorption is employed to separate the contaminants from industrial gaseous effluent or air. For instance, valuable solvent vapors which would turn out to be pollutants if discharged to ambient air can be so recovered from dilute mixtures with air. These operations exploit or harness the ability of certain solids to preferentially adsorb specific substances to their surfaces.

Physical adsorption and chemical adsorption (also termed chemisorption) are the two major categories of adsorption phenomena. Physical or van der Waals adsorption results from forces to of attraction between molecules of the substance being adsorbed and the solid, that is, the adsorbent. Chemical adsorption, on the other hand, involves a chemical bonding between the adsorbed substance and the solid surface.

Different types of adsorbers have been in use in the industrial units. These include fixed bed systems and fluidized bed systems.

Fixed bed systems

The fixed bed adsorber can be designed as a thin bed of granular solids. The main advantage of this type is the low resistance to gas flow. Flat, cylindrical, conical and pleated shapes of solids have been employed. Figure 7.8 represents an adsorber with a granular bed which is sandwiched between perforated sheets of metal or wire screens.

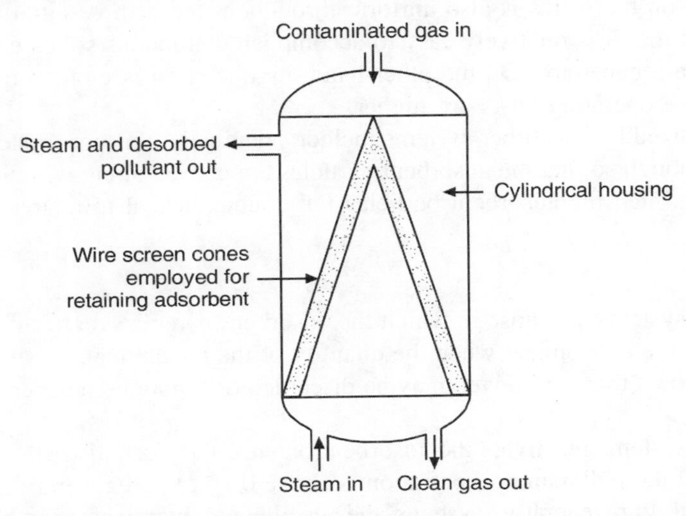

Figure 7.8 Thin bed adsorber (There is low resistance to gas flow).

Thick bed adsorbers use bed depths of 30 to 100 cm. They provide larger adsorbing capability/capacity and sometimes better efficiency. The inlet gas is made to flow down to prevent bed lifting. The bed typically installed in the horizontal, cylindrical vessels. One such adsorbed unit is represented in Figure 7.9.

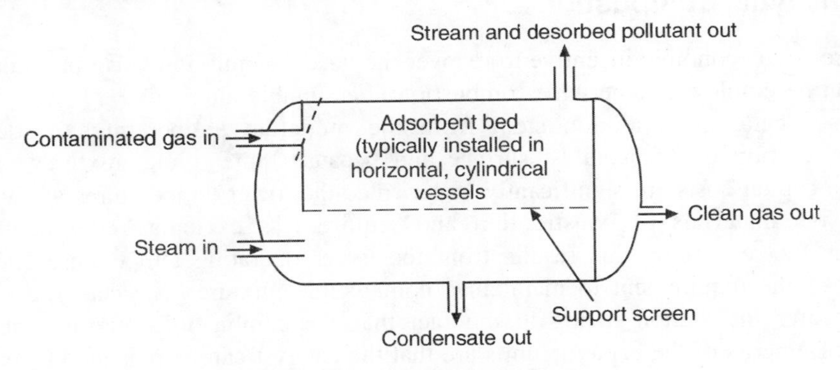

Figure 7.9 Thick bed adsorber.

Fluidized bed systems

Fluidized beds have also been employed in adsorption operations. In this type of unit, the gas passes upflow through a bed of granular solids. The velocity of the gas is such that the pressure drop is equal to the weight of solids in the bed. In this dense bed fluidized system, the solid particles are continually in motion; thus, they are at constant temperature and the amount of adsorbed material on the solids is also uniform throughout the bed. A significant advantage of this type of bed is that it is relatively easy to accomplish continuous solids circulation between the adsorber and a regenerator. On the other hand, its operation is more complex than a fixed bed system and the operating costs are higher.

Types of fluidized bed absorber systems include pneumatic transport systems where the gas velocity is high enough so that the adsorbent particles are carried along with the gas and raining solids systems in which the adsorbent particles fall countercurrent to a stream of rising gas.

Regeneration

Non-regenerative systems are those in which the adsorbent is replaced after its performance has dropped off. They are economical when the quantity of the contaminant is small, such as odour control. The adsorbent that is removed may be discarded or it may be regenerated in an external equipment.

Regenerative systems reactivate the adsorbent on site. They can also allow for the recovery of the adsorbate. If the pollutant concentration is above 0.1%, the regenerative system generally is more economical. In regenerative systems, the adsorbate (contaminant) can be either destroyed or recovered and this choice is a function of the economic value of the adsorbate. Normally, in solvent recovery applications, the value of the solvent is such that recovery is justified. Commercial adsorbents include activated charcoal, alumina, silica gel, other metallic oxides, synthetic zeolites.

7.2.4 Catalytic Combustion

In case there is an economic incentive to recover the gaseous emission, adsorption on activated charcoal can be employed; otherwise combustion is a suitable alternative.

Merits of catalytic oxidation stem from the moderate temperatures needed for the combustion reaction on the catalyst surface and because quite dilute mixtures can also be reacted. The capital costs are significantly lower since the lower temperatures allow the use of less expensive materials of construction and require less exchange equipment. Another advantage is fuel economy; this results from the lower operating temperature and from the elimination of the requirement to maintain a combustible mixture. The catalytic system also provides a safer installation owing to the fact that the combustible mixtures need not be maintained. Demerits of the catalytic units are that the catalyst can be poisoned by atmospheric contaminants, and inadvertent overheating can bring about a loss of catalytic activity.

Platinum and nickel have been the most widely employed catalysts; however, their high cost provides incentive for looking for alternative choices. Other possible catalysts include; copper chromite and oxides of copper, cobalt, chromium and manganese. These have exhibited catalytic properties for the combustion reaction.

Catalytic combustion has been primarily employed to control the emissions of organic vapours. Examples of these vapours are solvents and the organic compounds exhausted from industrial ovens employed in metal decorating and metal coating operations. Emissions from foundry core ovens, fabric coating and fabric backing ovens or from dry cleaning plants can also be controlled by catalytic combustion. When lean mixtures of a combustible vapour and air are present in any industry, catalytic combustion is a possible emission control technique.

The catalytic afterburner comprises a catalytic section preceded by a pre-heat section if the situation warrants as shown in Figure 7.10.

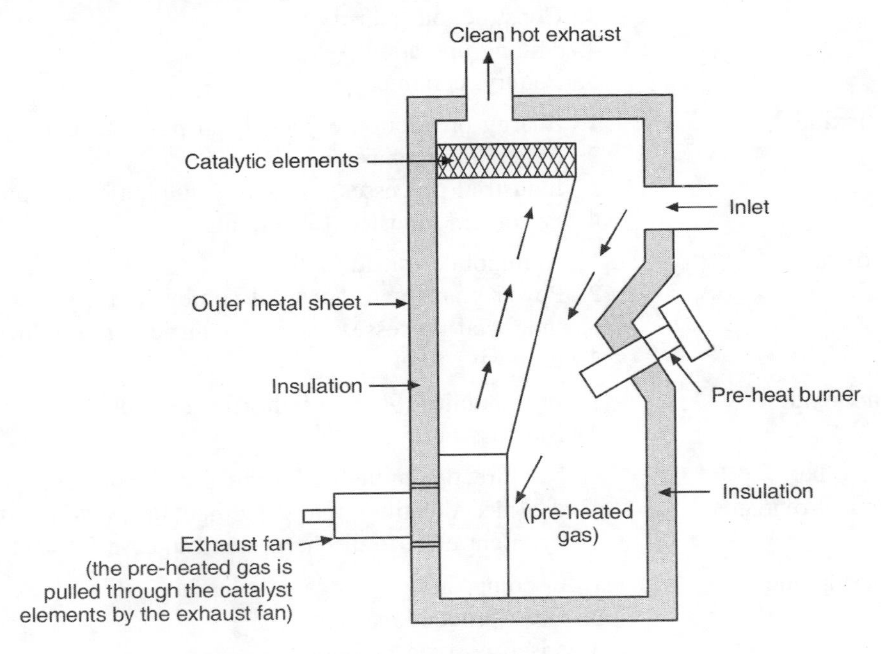

Figure 7.10 Catalytic afterburner chamber.

In the pre-heat zone, the gases are heated by a natural gas burner to the temperature needed to support catalytic combustion. The pre-heated gas is pulled through the catalyst elements by the exhaust fan and combustibles are burned by catalysis.

If the contaminated gas contains a high concentration of combustibles, it is diluted with air drawn in by a suction fan prior to entering the pre-heat zone. When the pre-heat burner is on the discharge side of the fan, the pressure in the afterburner chamber is positive and a pre-mix type of burner is employed. On the other hand, when the fan is between the burner and the catalyst bed, the pre-heat section is under a negative pressure and an atmospheric burner can be used.

All the beginning (start-up) a maximum natural gas fuel consumption is experienced; but as normal operating temperature is reached, the fuel consumption is reduced. This normally is accomplished automatically by temperature control of the discharge gas.

7.2.5 Major Emission Source Categories for WHO's Criteria for Air Pollutants

Substance	*Sources of emission*
Major Pollutants	
Particulate matter	1. Fuel combustion (95% in industrialized countries) 2. Natural and man-made dust erosion 3. Construction industry 4. Mining operations 5. Pottery, ceramics
Sulphur dioxide	1. Energy production especially in power plants 2. Industrial combustion 3. Industrial processes, especially sulphuric acid production 4. Petroleum industry, oil refining
Nitrogen oxides	1. Automobile traffic (50%) 2. Power plants and industrial combustion 3. Industrial processes especially nitric acid production 4. Explosive industry
Carbon monoxide	1. Automobiles (90% in industrialized countries) 2. Biomass burning
Ozone and other photochemical oxidants	1. Not directly emitted, but formed in the presence of nitrogen dioxide, Volatile Organic Compound (VOC) and methane by silent electric discharges and intense UV radiations
Hydrogen sulphide	1. Decomposition of organic material 2. Coke production 3. Viscose rayon production 4. Waste water treatment 5. Wood pulp production with sulphate process 6. Oil refining 7. Tanning industry
Hydrogen fluoride Radon	1. Fertilizer industry, chemical industry, aluminium industry 2. Building materials 3. Natural emissions from ground 4. Well-drilled tap water
Organic Substances	
Benzene	1. Gasoline powered motor vehicles without catalytic converters (80–85 per cent) 2. Chemical and petroleum industries 3. Domestic heating 4. Coke ovens

Substance	*Sources of emission*
Carbon disulphide	1. Production of viscose rayon fibres 2. Coal gasification
1, 2-dichloroethane	1. Synthesis of vinyl chloride 2. Production of ethylene diamines 3. Use as solvent and fumigant
Formaldehyde	1. Catalyst cracking in petroleum refining industry 2. Charcoal production 3. Internal combustion engines 4. Resin and formaldehyde production
Poly-nuclear Aromatic Hydrocarbon (PAH)	1. Residential wood combustion and small combustion sources 2. Aluminium smelting
Vinyl chloride	1. PVC (Poly Vinyl Chloride) production
Volatile Organic Carbon (VOC)	1. Road traffic 2. Use of solvent
Trace Metals	
Arsenic	1. Non-ferrous metal industry 2. Stationary combustion sources 3. Use as catalyst and reagent in inorganic chemical industry 4. Pesticides use
Cadmium	1. Metal production, especially zinc processing 2. Electroplating 3. Production of plasters, pigments and batteries CES 4. Waste incineration 5. Fertilizer processing and application 6. Natural sources, especially volcanoes
Chromium	1. Production of chromium compounds 2. Chromium production from chromite
Lead	1. Road traffic, mainly gasoline powered vehicles 2. Non-ferrous metal industry 3. Iron and steel industry 4. Production of lead-acid batteries
Manganese	1. Use as additive in metallurgical processes 2. Production of dry-cell batteries 3. Chemical industry 4. Glass, leather and textile industries 5. Fertilizer use

Substance	*Sources of emission*
Mercury	1. Coal combustion in stationary sources 2. Electrical apparatus 3. Electrolytic production of caustic soda and chlorine 4. Pharmaceutical industry 5. Production and use of anti-fouling paints
Nickel	1. Burning of residual and fuel oils 2. Nickel mining and refixing 3. Municipal waste incineration
Vanadium	1. Natural sources, especially weathering and volcanoes 2. Metallurgical processes 3. Burning of coal, crude and residue oils

Source: URBAIR, 1993

7.2.6 Atmospheric Sampling and Analysis

There are two approaches to sampling. One is source sampling and the other is atmosphere sampling. In source sampling, pollutant count of a particular source is obtained. Atmospheric sampling deals with the pollutants within the total air mass surrounding the earth.

In the following section atmospheric sampling will be considered briefly.

Atmospheric sampling measures pollution, provides control data and furnishes the background information for application in trend evaluation and source detection. The entire sampling procedure has a bearing on the pollutant sampled, the techniques used in collecting the pollutant, and the device chosen which in turn depends on the technique and the method of analysis.

1 cubic metre (m^3) of air is collected and this is analyzed to determine the quantity of air pollutant collected, which is measured in micrograms (μg). The pollution concentration in the air sample is expressed in $\mu g/m^3$.

Factors to be taken into account in sampling

These factors are (i) the location of the sampling equipment, and (ii) the duration of the sampling period. The duration is established according to the purpose of the test. A 24-hour sampling is recommended for determining the average pollutant concentration.

Peak periods may be ascertained by using a continuous recorder or by collecting periodic samples with a sequential sampler equipped with a timing mechanism, with the sequence of every hour or every two hours, and so on. The sample size has to be considered next. The size of the sample must be large enough to provide statistical accuracy. The rate of sampling is also related to sample size. The rate is based on the contact time required between the pollutant and the reagent (absorbing and adsorbing) used to estimate pollutant concentrations.

Other factors are interference of one pollutant in the sampling of the other, the units employed in expressing results, etc.

Sampling train components: The fundamental components of a sampling device include a sampling train, which embodies an air mover, a flow-measuring device and a sample collection mechanism with a contaminant detector that provides a means for analyzing the sample:

The air mover generates a flow of air that will allow the contaminants in the air to be captured in the sample collection mechanism for analysis. A few examples of air movers are vacuum pumps, ejectors or aspirators, liquid displacers and evacuated flasks. Of the above, the latter two combine the roles of air mover and collection mechanism.

All sampling trains should have a flow-measuring device to determine the quantity of air passing through the train during a given period of time. This device must be calibrated against a very accurate meter accepted as a standard meter.

Particulate matter: Particulates differ in size density, shape and other properties. Oil soot is a suspended particulate as it remains in the atmosphere for prolonged periods and so is pollen. Fly ash and dust, however, are high in density and are medium to large size and, therefore, settle readily.

Collection techniques and sampling devices: A summary of the sampling techniques for the particulates is listed in Table 7.8.

Table 7.8 Sampling Techniques for the Particulates

Collection technique	*Pollutant to be collected*	*Method of analysis*
Gravity (30 days)	Settleable particulars	Gravimetric
Filtration (24 hours)	Suspended particles, organic, inorganic and radioactive compounds	Gravimetric
Filtration 2 hours (paper tape sampler)	Suspended particles (fine soiling matter)	Densitometer
Different sampling devices		
Inertial (impaction)	Pollen, spores	Count
do	Total particulates (these include all measurable particulates)	Gruber particle comparator
do	Bacteria-particulates	Count
do (Anderson sampler)	Total particulates	Microscopic sizing
Different sampling devices		
Inertial (impingement)	Total particulates	Gravimetric
Inertial (Centrifugal separation) (Cyclone sampler)	Particulates of 5 microns in diameter	Gravimetric
Precipitation (Electrostatic precipitator)	Radioactive particulates	Type study, not quantity
Precipitation (Thermal precipitator)	Total particulates	Gravimetric

A similar table of the sampling techniques for gaseous contaminants is given in Table 7.9.

Table 7.9 Sampling Techniques for Gaseous Pollutants

Collection technique	*Pollutant to be collected*	*Method of analysis*
Absorption with fritted gas absorber using 24-hour bubbler or sequential sampler	Inorganic gases, oxidants, NO_2, SO_2, H_2S	Wet chemistry, colorimetry, spectrophotometry
Adsorption by activated charcoal, silica gel, activated alumina, molecular sieves (detector tubes)	Insoluble or non-reactive vapours, NH_3 H_2S, SO_2, CO, NO_2, CO_2	Comparison with colours on a comparative chart
Absorption of SO_2 to form lead sulphate (using lead peroxide candles)	SO_2	Measured in mg SO_2 per day/100 cm^2
Condensation (using sampling train)	Insoluble and non-reactive vapours, HC, radioactive gases	Mass spectroscopy infra-red or gas chromatography
Grab sampling (using evacuated flash or gas displacement collector)	Small quantities of gaseous pollutants-odour measurement	Varied

Table 7.10 given here lists the automatic pollutant analytical methods for gaseous contaminants which also enable one monitor pollutant levels continously.

Table 7.10 Continuous Monitoring Automatic Gaseous Pollutant Analytical Methods

Pollutant	*Analytical method*
Total hydrocarbons	Gas chromatography
Carbon monoxide	Non-dispersive infra-red absorption, gas chromatography
Fluorides	Colorimetric
Nitrogen dioxide	Coulometry, colorimetry, gas chromatography
Sulphur dioxide	Gas chromatography, flame photometry, colorimetry, coulometry
Hydrogen sulphide	Colorimetry
Aldehydes	Colorimetry
Oxidants	Colorimetry, coulometry

The recommended units for expressing the atmospheric pollution data are given in Table 7.11.

Table 7.11 Recommended Units for Expressing Atmospheric Pollution Data

Particle fall out	Milligrams per sq.cm per time interval mg/cm^2/month (or) mg/cm^2/year
Airborne particulates	Microgram per cubic metre $\mu g/m^3$
Particulate counting	Number per cubic metre
Gaseous pollutants	$\mu g/m^3$-normally (for CO mg/m^3)
Gas volumes	Reported as standard; corrected to 25°C and 760 mm Hg pressure
Temperature	Centigrade, °C
Time	Twenty four hundred hour clock
Pressure	mm Hg
Linear velocity	m/sec
Volume emission rate	m^3/min
Sampling rates	m^3/min
Visibility	kilometres

Analytical procedures for particulate and gaseous pollutants

Methods of analysis for interpreting particulate and gaseous pollution data involve sampling, separation, concentration development of a measurable property, recording, calculating and interpreting. Sampling often includes concentrating the sample.

Separation: Separation removes substances that interfere with the development of measurable properties of the pollutant. For gaseous pollutants this can be done by techniques such as chromatography, extraction, ion-exchange and distillation.

Concentration: This step reduces the volume of a given substance. It is realized in practice that by pressurizing a gaseous pollutant to intensify the emission or absorption of radiation or evaporating an inert solvent to produce a more concentrated solution for analysis. In some situations, the concentrating step can be incorporated into the sampling or separating process. The high-volume filter combines sampling with concentration.

Measurable property development: In many determinations, it is necessary to develop some property that is sufficiently marked and distinctive to permit its measurement. One example is a process in which gases are brought into contact with specific chemicals and distinct colour reactions evolve. These colour intensities are easily measured. Gaseous materials that absorb ultraviolet, infrared wavelengths can be determined by spectroscopic procedures.

Measurement of a characteristic property: Quantitative results could be obtained only by measuring the characteristic property or by comparing its intensity with some standard taken as a reference.

The gravimetric method isolates a substance or one of its compounds in the pure state and weighs it on an analytical balance. This process is cumbersome and lacks sensitivity. Some of the instruments using this technique are the high volume filter, a dust bucket impaction and impingement device, a cyclone separator and a lead peroxide candle.

The gasometric method isolates a given substance in the gaseous form or the substance in question is absorbed from gaseous mixture in this method. The volume of the isolated pollutant is determined by comparing the weight of the mixture before and after the gas is removed or gas burette (graduated glass tube used to measure quantities of a gas received or discharged) is used to determine the volume of the gas removed. This method is employed in stack sampling but not in atmospheric sampling because of lack of sensitivity.

The titrimetric method makes use of a liquid burette that releases a titrant (a solution containing a known concentration of reactive chemical) drop by drop into an air sample. This generates a colour reaction, an electrical potential reaction or a photometric reaction or end point. The amount of titrant required to attain an end point is a measure of the amount of the component being measured in the sample.

The absorption of radiation method allows both identification and quantitative estimation of gases. Various light wavelengths including UV and IR rays are made use of in this technique.

Similarly, the emission of radiation method is the basis for several approaches for ascertaining pollutants' concentrations especially of metallic elements.

The methods of measuring the ambient air quality parameters are presented in Table 7.12.

Table 7.12 Measurement Methods for Ambient Air Quality Parameters

Pollutant	*Measurement methods*
SO_2	Ultraviolet pulsed fluorescence, flame photometry, coulometric dilution or permeation tube calibrators
CO_2	Non-dispersive infrared tank gas and dilution calibration gas filter correlation
O_3	Gas-phase chemiluminescence, ultraviolet spectrometry, ozone UV generators and UV spectrometer or Gas-Phase Titration (GPT) calibrators
NO_2	Chemiluminescence; permeation or GPT calibration
Lead	High volume sampler and atomic absorption analysis
PM_{10} (10 μm or small particle)	Tapered element, oscillating micro-balance, automated beta gauge
$PM_{2.5}$	Twenty four hour filter sampling
TSP (Total Suspended Solids) (type, size and composition are important)	High volume sampler and weight determination
Sulphates, nitrates	High volume sampler and chemical analysis deposit dissolved and analysed colorimetrically
Hydrocarbons	Flame ionization and gas chromatography calibration with methane tank gas
Asbestos and other fibrous aerosols	Induced oscillation/optical scattering microscope and electron microscope
Biological aerosols	Impaction incubated 24 hour and microbial colonies counted

EXERCISES

7.1 Discuss the advantages and disadvantages of wet collectors.

7.2 Identify the steps involved in the electrostatic precipitation of particles.

7.3 Discuss the factors to be taken into account in the choice of particulate control equipment.

7.4 Write a short note on

(a) Cyclones
(b) Filters

7.5 Describe the treatment methods used for gaseous effluents.

7.6 What are the sampling techniques used for the particulates and gaseous pollutants?

8

Industrial Pollution and Waste Treatment in a Few Chemical and Processing Industries

The typical pollution problems and nature of the effluents of a few chemical and allied industrial units are outlined in the chapter. At the end of this Chapter, a brief presentation of the methods of deriving useful products from agricultural wastes has been made taking a few agricultural wastes as examples. Also the siting of industries from the perspective of environmental protection is discussed.

8.1 FERTILIZER INDUSTRY

The fertilizer manufacturing industry comprises many basic producers of nitrogen, phosphate and potash materials employed within the industry and several mixers, granulators and blenders that are used in the formulation of the basic products into the final fertilizer products.

Basic phosphate producers, mine and process phosphate rock into wet process phosphoric acid and/or single and triple superphosphate and/or mono and diammonium phosphate. Producers of superphosphate frequently manufacture the sulphuric acid needed. Likewise, the required ammonia for producing ammonium phosphate is frequently manufactured within the establishment.

Basic potash producers merely mine and process the mineral "potash" which is usually in the form of KCl.

Basic nitrogen producers fix atmospheric nitrogen into ammonia. Some of the ammonia is converted into nitric acid which is combined with other ammonia to form ammonium nitrate. Some of the ammonium nitrate is combined with ammonia to produce an ammoniating solution for mix plants. Certain manufacturers make urea from ammonia and from by-product carbon dioxide generated by the ammonia plant. Urea is processed into solid prills and/or added in solution form to ammonium nitrate to produce nitrogen fertilizer solutions. Some ammonium nitrate is usually processed into solid form.

In granulator or mixed fertilizer plants, the three basic plant nutrients are combined to form an NPK (Nitrogen, Phosphorus, Potassium) product, usually in granulated form. Some mix plants manufacture their triple superphosphate by reacting phosphate rock with sulphuric acid which usually is a reprocessed spent acid from another industry. Some plants use phosphoric acid directly in lieu of a superphosphate.

Mix plants ammoniate their product either by the use of ammoniating solutions or by the use of ammonia itself. The three plant food elements are mixed together in such plants usually in a granulating drum. Heat from the acid-ammonia reaction causes the moistened bed of the material to form granules in the rotating drum. The resultant granulated material is dried and screened to size.

In liquid mix plants, which are located in farming areas, liquids such as phosphoric acid and ammonia are combined or high analysis dry fertilizers are dissolved to make a base solution. Other ingredients, such as nitrogen solutions, potash and micro-nutrients are added to produce the desired formulation.

Using dry nitrogen, phosphate and potash fertilizers together with micro-nutrients, the blending plants mix the dry granular fertilizers together in various proportions to make customized formulations.

8.1.1 Pollutants and Their Sources

Sulphur dioxide

Some phosphate fertilizer complexes include sulphuric acid manufacturing plants. Most of these plants are of single absorption type which emit a concentration of around 2000 to 2500 ppm sulphur dioxide. Electrostatic precipitators and mist eliminators and/or water scrubbers could be employed to bring about a reduction in visible emissions. The more modern plants are also employing the double conversion process which reduces the concentration of SO_2 in the emissions to about 500 ppm.

Nitrogen oxides

Essentially all basic nitrogen fertilizer producers possess captive nitric acid plants. Several of such plants are relatively new and include catalytic type tail gas combustors (afterburners) which reduce the orange-coloured nitrogen dioxide (NO_2) in the stack to colourless nitric oxide (NO). Newer units are equipped with improved combustors capable of eliminating virtually all the oxides of nitrogen (NO_x). New emission standards restrict the emission of nitrogen oxides (NO_x) to 1.5 kgs expressed as nitrogen dioxide for each ton of 100 per cent nitric acid manufactured. Nitrogen oxides from all industrial processes, including the fertilizer industry, contribute less than 2 per cent of the total nation wide NO_x emissions.

Ammonia

Ammonia is a gas with a strong odour and is readily absorbed by acidic scrubbing. Recently installed ammonia plants record losses of about 0.2 kgs of nitrogen per ton of ammonia manufactured, which is about 5 per cent of the losses experienced in plants two or three decades ago. It is possible to keep the ammonia loss to a minimum in most urea plants. Once-through urea plant off-gas is frequently channeled through an ammonium nitrate neutralizer for recovery of ammonia. Excess ammonia in the neutralizer gas vent can be effectively scrubbed with a weak acidic ammonium nitrate solution that is recycled into the process. Ammonium losses from mono and diammonium phosphate, granulation and ammoniation plants can be kept to very low levels by web scrubbing.

Fluorine compounds

Emissions of fluorine compounds as gases and in particulates are the most difficult and expensive to control among the air pollutants arising from the production of fertilizers. Fluorine emission from wet-processed phosphoric acid plants, from mono and diammonium phosphate and from granulation plants can be kept to a low level by wet scrubbing of the exit gases. The fluorine compounds produced from mixing single and triple superphosphate are collected in most instances so that emissions to the atmosphere from these operations are low. However, large amount of fluorine compounds are discharged from the curing sheds owing to the fact that only a few sheds have adequate fluorine controls. To adequately control the fluorine emissions from curing superphosphates, a large ventilation system is needed, with wet scrubbing of all exhausted air, and proper treatment and disposal of the resultant scrubber liquor effluent.

Particulates

The main sources of particulate emissions in the fertilizer industry are the phosphate rock drying and grinding operations. Electrostatic precipitators and bag collectors are not able to meet the required removal standards. New techniques have to be identified to reduce such emissions.

Particulate emissions from wet-processed phosphoric acid and ammoniation plants equipped with scrubbers are at an acceptably low level. Dry fertilizer processing and sizing plants generate significant particulate emissions from their stacks and from each piece of material handling equipment. Many of these plants are provided with dust collection systems to control particulate emissions.

In respect of pollution in the ammonium nitrate manufacturing segment of the fertilizer industry, the switch to prilling a high density prill has given rise to severe particulate emissions. To manufacture high density prill, the ammonium nitrate solution is concentrated to 99.8 per cent. Such fortification needs excessive evaporation of the 83 per cent feed stock. The most efficient evaporators for this service are the air-swept falling film type. Use of such evaporators results in significant emissions. Likewise, significant emissions arise due to the spraying of the concentrated solution in the prill tower for the formation of prills. Many of the emissions are ordinary dust size particles while others are sub-micron in size and behave like smoke.

Pollutants and their sources in fertilizer manufacturing are presented in Table 8.1.

Table 8.1 Pollutants and their Sources in Fertilizer Manufacturing

Type of Pollutant	*Sources*
Sulphur dioxide	Sulphuric acid plants
Nitrogen oxides	Nitric acid plants
Ammonia	Ammonia, urea, mono and diammonium phosphate, granulation and ammoniation plants
Fluorine compounds	Wet-process phosphoric acid, mono and ammonium phosphate, single and triple superphosphate and granulation plants
Particulates	Phosphate rock grinding, wet-process phosphoric acid, ammoniation and all dry fertilizer processing plants and ammonium nitrate plants

8.2 THE BREWING INDUSTRY

The brewing industry involves the manufacture of fermented alcoholic beverages, such as beer with cereal grains as the raw material. Two major steps are involved in the overall process. They are malting and brewing.

8.2.1 Malting

The objective of doing the malting is to prepare the grain for brewing. Wheat, corn grits or barley are soaked in water for 2–3 days. The water is aerated, drained and replaced everyday at least once.

When the grain is soft, it is piled in heaps on the 'couching floor', or placed in germination bins, where it is kept at constant temperature and humidity for about a week. During this period, the grain germinates (that is, sprouts small root, shoots) and at the same time, the crude starch in the grain is converted into soluble sugars and starches by enzymatic action. As the germination progresses, the grain is turned intermittently to ensure even sprouting. When the sprouts are about two third the length of the grain, the germination is stopped. The grain is transferred to drying kilns, where it is gently roasted at 150–225° F until it is dark and crisp. Lower temperatures produce a light beer; higher temperature, a dark beer. This dry, roasted grain is termed as **malt**.

8.2.2 Brewing

The dry malt is then crushed with iron rollers and mixed with water to form mash. The mash is thinned with hot water and heated to 150° F while being stirred constantly. The exact temperature is of importance at this stage, for the malt is undergoing chemical change; the remaining starches are being converted to malt sugar. After a while, the temperature is gradually increased to 160° F. Then the cooked mash is filtered, and the liquid, called the **wort** is drained from the grain, which is termed as the **grist**.

The wort which is a clear amber malt extract, is boiled for 4–6 hours with hops, which add flavour and aroma. Hops are the dried flowers, they form the hop vine, and possess a bitter flavour, themselves. For every 100 gallons of wort, 1–12 lbs of hops are added. Besides adding flavour, they also help keep the wort from spoiling. The spent hops are removed and the wort is allowed to cool. As it cools down, the undesirable proteins coagulate and settle out.

The wort is then placed into fermenting vats. About 5 lbs of yeast is added per 100 gallons of wort. Two fermentations take place. Within the primary fermentation tanks, the malt sugars are converted into alcohol and CO_2. Beer undergoes bottom fermentation around 45° F.

The yeast sinks, porter's ale and stout undergo *top fermentation*. The yeast floats on the wort, which is maintained at 60–68° F. After several days, the fermented liquid is poured into settling vats. Secondary fermentation in which all the reactions needed to give the fine distinctive quality to the beverage, then occurs. The yeast is skimmed off or the beer drawn off and the beverage is stored in casks and barrels to age and become clarified. After a short period, the finished beer, ale or stout is ready to go for bottling.

8.2.3 Environmental Problems

Some of the settled yeast from the previous fermentation is used to initiate fermenting in the next batch. Though there is always excess yeast, only part of the yeast is satisfactory for further use. The settling process carries down some materials not desirable in these beverages. This limits the useful quantity. There is always a fairly large quantity of yeast which must be disposed of.

The residual yeast is high in vitamin B complex, proteins and minerals. If the excess yeast is discharged into the sewer, it is objectionable because of the high BOD. It can instead, profitably be dried and used as cattle and poultry feed. The drying deactivates the enzymes, stabilizing the yeast, and also brings about improvement on its digestibility. It has been observed, however, that 10 lakh lbs of yeast solids/year are necessary to economically justify the installation of drying equipment.

The brewing industry consumes much water—about 10 gallons of process water per gallon of product. The BOD levels are quite high, as are the total solids. Typically about half the BOD and over 90 per cent of the suspended solids are generated in the brewing operation. These are solid waste spent grains, hops and sludges which are formed in the brewing and the malting step, which must be disposed of. Packaging operations also generate a large amount of BOD.

A typical treatment system employs an activated sludge process. After mechanical bar screening and grit removal, the wastes are sent to an aerated equalization basin. At that point, various chemicals (such as ammonia and phosphoric acid) are added for pH control and nutrient addition. The liquid wastes are next sent to primary clarifiers, then to the activated sludge basins and the secondary clarifiers. The waste-activated sludge is thickened and then combined with primary sludge, dewatered and disposed of by land-fill procedures.

8.2.4 Sludge Bulking

The growth of a few microorganisms, leads to a phenomenon generally experienced in the brewing industry, i.e., sludge bulking. A bulking sludge is one which settles slowly, leaving a clear supernatant, has a high volume and compacts poorly. It generally leads to a major discharge of solids into the effluent.

Though a number of other microorganisms can contribute to the problem, the majority of the bulking is generally attributable to *Sphaerotilus* whose filaments are composed of chains of rod-like cells with rounded ends that are encapsulated in a tight polysaccharide sheath. The *Sphaerotilus* thrives at temperatures of 15–40° C and at pH values between 6.5 and 8.1.

Within the brewery there are several probable causes of sludge bulking. They include the following:

Low levels of dissolved oxygen: Dissolved oxygen levels in the mixed liquor of less than 1.0 mg/l promote the growth of the filamentous *Sphaerotilus* is retarded less by low dissolved oxygen levels than the growth of other types of organisms, those necessary within the activated sludge, thus allowing it to predominate.

Loading rate: The filamentous organisms will mostly grow if the BOD level is within a certain range.

Nutrients: Inadequate amounts of nitrogen, phosphorus and iron promote bulking. Typically, a BOD-to-nitrogen-to-phosphorus ratio of about 100-to-5-to-1 is desirable.

Waste composition: Wastes high in carbohydrate promote the development of filamentous organisms, possibly due to the occurrence of low dissolved oxygen.

pH: A low pH in the aeration basin (below 6.0) promotes the growth of many types of filamentous fibres.

Regarding the steps to be taken to exercise control over the bulking in activated sludge systems, many techniques have been tried out with varying success. Most approaches involve,

1. Killing the filamentous organisms by addition of chemicals
2. Increasing the sludges settleability by employing chemical coagulants and/or
3. Recirculating the sludge more to prevent build-up

The use of chemical additives is based on the principles that the long, filamentous organisms with their large surface-to-volume ratios are more sensitive to various disinfectants than are the normal activated sludge microorganisms. Typical chemicals used, include chlorine and hydrogen peroxide. It is believed that these can oxidize some of the filaments, destroying the lattice-like framework; hydrogen peroxide can also provide additional oxygen.

Chemical coagulants do not affect the number of filamentous organisms, but simply try to improve the settling in the secondary clarifiers. Frequently, carbonic organic polymers or iron or aluminium salts are used.

Filamentous organism build-up can also be reduced by process modifications. For instance, the rate or return of the activated sludge can be increased to enhance the percentage of high density solids and thus decreases the overall volume of sludge. Likewise, more activated sludge can be wasted to decrease the overall volume. The success of these various methods depends on the nature of the effluent.

Overall, maintenance of adequate dissolved oxygen is the most critical factor in bulking control. The presence of sufficient nitrogen and phosphorus are also essential. Simply maintaining stable operating conditions is very important too. Any changes in recycling rates, sludge wasting rates and so forth most be made very slowly to minimize filamentous organisms growth and to allow appropriate corrective action to be taken when the situation warrants.

8.3 SOAP AND SYNTHETIC DETERGENT INDUSTRY

In the production of soap, vegetable or animal fats, oils or artificial/natural fattty acids are boiled in alkaline solutions. The primary fats employed are mainly those that are not fit for human consumption, e.g. waste fats, fats from waste water, bone fats, horse fats, mutton tallow, fat from rendering plants, etc. In the production of finer toilet soaps, vegetable oils such as olive oil, sesame oil, corn coil, soya oil, etc. are used.

The steps involved in the process in a soap making unit, are essentially as hereinafter:

(i) Purification of the primary fats
(ii) Fat cleavage (or manufacture of fatty acids), and
(iii) Saponification

The following waste waters are generated

(i) Various wastes from the purification of raw fats
(ii) Wastes containing glycerol from fat cleavage and from glycerol refining
(iii) Nigres from the saponification processes
(iv) Washing waters
(v) Condensation, rinsing and cooling waters

8.3.1 Waste Waters from Soap Production with Fatty Acids

To produce soap (alkaline salts of fatty acids), the fats are treated with adsorbent substances, such as charcoal, bone black and activated coal, kieselguhr, fuller's earth or silica gel, at temperatures of 70 to 90° C. Bleaching is carried out by employing sodium peroxides, per carbonates, hypochlorites, sodium hydrogen carbonates, potassium permanganate, sulphuric acid and other reducing agents.

Fat cleavage is carried out in order to separate the fatty acids required for soap making from their bonds with glycerol. The quality of the waste waters has a bearing on the processes employed for fat cleavage.

In all listed fat cleavage processes, a glycerol water containing 10–15 per cent glycerol is got as a by-product. This is processed in the unit to give an 80 per cent concentration of pure glycerol.

Saponification

This is carried out mostly in the form of carbonate saponification, yields a *soap glue* from which the *soap curd* is separated by adding common salt and collected as a half liquid product above the nigre (lower aqueous layer). The waste waters usually have a temperature of 90–100° C and as they cool down, layered soaps and gluey components are separated.

The quantity of effluent is likely to vary according to the type of factory and the process adopted so that generally valid data can not be given.

The figures presented here from a large scale soap factory may be indicative of the quantum of waste water. The total quantity of water in 8 hours could be around 1500 m^3 with the following break-up:

Cooling and condensation water	–	1360 m^3
Discharges from raw fat refining	–	58 m^3
Washing water from fat cleavage facility	–	20 m^3
Nigre and soap washing water	–	20 m^3
Purification water	–	25 m^3

The balance quantity arises from other waste waters, e.g. rinsing water from water purification plants, which are not restricted to soap production units.

Effluents from soap making units have several characteristic qualities that have to be taken into consideration while treating the water. They have an unpleasant odour. They still contain various fats and greasy, soaplike substances, which are detrimental when the water is discharged into the public sewage systems, since they cause greasy deposits on the sides of the sewers.

Discharges from raw fat refining and injection condensers in the glycerol evaporation plant have a strong odour. This can be almost wholly removed by chlorination when treating condensation waters and to a large extent removed in the case of the washing water. The refixing of raw fats by treating with adsorbent agents such as fuller's earth, activated coal, silica gel, etc. actually generates no waste waters, but there are wastes from the purification of the adsorbent agents. This is performed by means of alkaline solutions. The waste water shows a distinct alkaline reaction, contains much suspended matter and small quantities of fat or oil.

In the purification of raw fats by washing, weakly acid waste waters are generated, which contain an amount of oxidizable organic substance which is many times greater than that of normal municipal sewage. If the raw fat refixing process includes bleaching, the waste waters contain residues of excess bleaching agent or its conversion products. Bleaching agents employed in the soap industry include peroxysulphate hypochlorite, sodium peroxide and other oxidizing agents.

8.3.2 Waste Waters from Synthetic Detergent Production

The world wide application of fat, and improvements in manufacturing techniques in oil factories led to a shortage of lower fatty acids, and a rise in their price. Therefore, an attempt was made to synthesize molecules similar to soap molecules that need no natural fats for their production. Such substances are detergents.

The surface active detergents belong to various groups, which can be divided into three different types in accordance with their electrical charge (ionization).

1. Anionic detergents, e.g. alkyl sulphates and alkyl aryl sulphonates, which are most commonly found in household use. Only *soft* alkyl aryl sulphonates should be employed. They have a straight side chain and are easily biodegradable.
2. Non-ionized detergents, which, for instance, are based on polyoxy ethylenes and are employed in combination with fatty acids, and alkyl phenols. These detergents are chiefly used in industry.
3. Cationic detergents, such as pyridine derivatives and quaternary ammonium bases which because of their germicidal properties are mainly used for sanitary purposes as a softening agent in the laundries of restaurants and hospitals, etc.

Heavy-duty detergents have other added ingredients, such as bleaching agents, stabilizers, dirt-absorbing colloids, hydrotropic substances, skin protecting agents, brighteners, etc. In addition, they also contain alkalis and phosphates. They are aptly called 'builders' and are substances that have no surfaces active properties themselves, but which increase the effectiveness of the active substances.

In the past, such builders were ashes, potash and later on it was soda. Most recently, however, certain phosphates for stabilizing the hardness (calcium and magnesium compounds) of the water (tetrasodium phosphate, pyrophosphate, trisodium polyphosphate) have become much more important.

Table 8.2 indicates the composition of the detergents.

Table 8.2

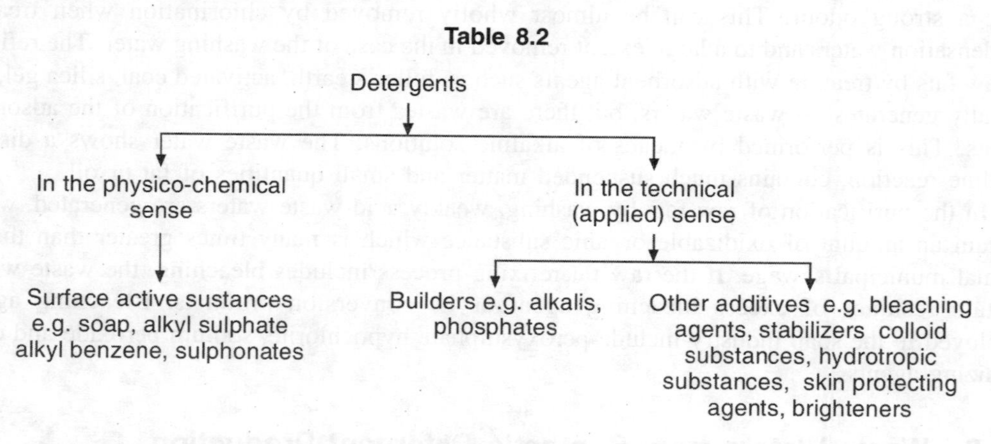

From the manufacturing process, various waste waters are generated with the following qualities:

1. Marked tendency to form foam (due to reduction of the surface tension)
2. Contain emulgated substances, e.g. fats and oils
3. Sudden changes in composition and reactions due to batch discharges of wastes from individual phases of the manufacturing process

Table 8.3 shows the physiological behaviour of some types of detergents.

Table 8.3 Physiological Behaviour of Some Types of Detergents

Chemical structure	*Physiochemical type*	*Physiological behaviour*
Alkyl sulphates	Anion active	Easily biodegradable
Alkyl aryl sulphonates with straight side chain	Anion active	Not easily biodegradable
Alkyl aryl sulphonates (side chain branched with quaternery carbon atom)	Anion active	Hardly biodegradable
Tertiary ammonium or pyridinium salts polyglycol ether	Carbon active non-ionogenous	Toxic, bactericidal not easily biodegradable

8.3.3 Treatment of Waste Waters from Soap and Synthetic Detergent Production

Treatment of waste waters from soap production

The waste waters from plants that are likely to discharge them into the public sewerage system must first be cooled for temperatures below 30° C. Waste waters with a strong odour which result from crude fat refining and injection condensers can be deodorized by chlorination.

Fatty effluents must be principally treated in fat separators, at the place of their origin. The intermediate suds from sponification require a separate treatment. Cooling may effect an additional separation of the substances. The residue generated in the treatment stage might hardly be reusable. The remaining waste waters that arise from fat cracking should also be neutralized by lime. The channels must be provided with airtight covers to avoid odour nuisance.

Before the waste waters are introduced into the receiving water, first the zinc salts must be reclaimed from the acidic water. The glycerine, oil and fat residues contained in the condensation and cooling water can be removed by a fat collector equipped with a coke or activated charcoal filter, before they are discharged into the receiving water.

The introduction of the effluent through an equilibrating and sedimentation system, which should be as large as possible, is recommended. When a predominantly acidic reaction is involved, the addition of lime wash may result in both neutralization and precipitation of some dissolved organic pollutants. Such equalizing sedimentation system should be designed for a capacity corresponding at least to the waste water volume per day.

The biological treatment is considered to be the final purification stage. The effluents from soap production plants, which contain glycerine, protein and fatty acids (except the zinc-rich scrubbing waters from the fat cracking stage) can be decomposed in a crude and diluted condition. The cooling waters are a suitable diluting medium. The nitrogen and phosphoric acid quantities, which are required for biological decomposition may be partly introduced by the inclusion of the plant's toilet and operational waste waters into the biological purification system. If the situation warrants, nutritive salts containing nitrogen or phosphorus must be added.

Treatment of waste waters from synthetic detergent production

While the synthetic detergents are introduced into surface waters and the groundwater through waste waters, they bring about serious damage owing to strong foaming and the high oil and fat concentration. They may also cause disturbance to water processing.

The quantities of cooling and condensation waters are mostly a multiple of the quantity of genuine sewage. They should, therefore, be separated from the contaminated operational discharges. All discharges from operation that are polluted by fat and oil can be treated by good fat and oil separators at their origin. This prevents clogging of the outlet lines and allows for recovery of valuable products. The actual purifications are then best carried out after the combination with other polluted discharges from operation, possibly by means of chemical precipitating agents in a sufficiently dimensioned sewage plant which at the same time, has an equalizing function for the waste water composition. The outcome is a mutual neutralization of the alkali and acidic waste waters, a dissociation of any fat and oil emulsions present and thus a more effective separation of the light-weight substances, an equalization of the waste water surge loads and a more uniform composition of the final discharge. Aluminium salts are suitable precipitating agents, even though only a 50 per cent success of the purification step would result. Such treatment is particularly suited to remove turbidity and any extremely fine oil and fat suspensions.

The suitability of biological methods for final waste water purification must be tested. This is the normal road leading to success, after elimination of the interfering effects of toxic or biologically harmful constituents by preliminary treatment of certain discharges from operation *in situ*.

After a corresponding preliminary treatment, the waste waters from paraffin oxidation are well suited for a biological purification. The low fatty acids, alcohol and ketones, which are contained in these waste waters, are easily accessible to biological decomposition.

When the purified waste water is introduced into the receiving reservoir, the foaming tendency must be considered and correspondingly counteracted (e.g. by floating scum bars, by spray water, with antifoam agents, etc.).

8.4 DAIRY INDUSTRY

The liquid wastes from a large dairy emanate from the following plants: receiving station, bottling plant, cheese plant, butter plant, casein plant, condensed milk plant, dried milk plant and ice cream plant. The other sources of wastes are water softening plant, and bottle and can washing plants.

At the receiving station, the milk is received from the farms and is transported to processing plants in large containers. At the bottling plant the raw milk delivered by the receiving station is stored. The processing includes cooling, clarification, filtration, pasteurization and bottling. In the aforementioned two sections, the liquid wastes originate out of rinsing and washing of bottles, cans and other equipment and hence, contain milk drippings and chemicals employed for cleaning containers and equipments.

In the cheese plant, the milk (whole milk or skimmed milk) is pasteurized and cooled. It is then placed in a vat where a starter (lactic acid producing bacterial culture) and rennet are added. This enables the separation of the casein of the milk in the form of curd. The whey is then withdrawn and the curd compressed to make excess whey drain out. Other ingredients are now added and the cheese blocks are cut and packaged for sale. Waste waters from this plant include the discarded whey and the wash water used for cleaning vats, equipments, floor, etc.

In the creaming process, the whole milk is pre-heated to about 30° C to separate the cream from the milk. In the butter plant the cream is pasteurized and may be ripened using a selected acid and a bacterial culture. This is then churned at a temperature of about 7–10° C to produce butter granules. At the appropriate time the butter milk is drained out of the churn and the butter is washed and after standardization, packaged for sale. Butter milk and wash waters used to clean the churns come out as the waste from butter plants.

The skimmed milk may now be sent for bottling for human consumption or for further processing in the dairy for meeting other products such as non-fat milk powders. Milk powders are made by evaporation followed by drying by either roller process or spray process. The dry milk plant wastes consist of wash waters employed to clean the containers and equipment.

The soured or spoiled milk and sometimes the skimmed milk are processed to produce casein used for preparation of some plastics. The process involves the coagulation and precipitation of the casein by the addition of some mineral acids. The waste from this section includes whey, washings and the chemicals used for precipitation.

The dairy wastes are very often discharged intermittently with the nature and composition of the waste depending on the type of products produced and the size of the plants.

8.4.1 Treatment of the Dairy Wastes

Since the ratio of COD to BOD is low, the dairy wastes can be treated effectively by biological processes. Moreover, these wastes contain sufficient nutrients for bacterial growth. Reduction of volume and strength of the wastes can be accomplished by (i) the prevention of spills, leakages, etc., (ii) by reducing the amount of wash water, (iii) by segregating the uncontaminated cooling water and recycling the same, and (iv) by utilizing the butter milk and whey for the production of dairy by-products of good market value.

Owing to the intermittent nature of the waste discharge, it is desirable to provide equalization tank with or without aeration before the same is for biological treatment. A provision of grease trap is also needed as a pre-treatment to remove fat and other greasy substances from the waste. An aeration for a day prevents not only the formation of lactic acid, but also brings about a reduction in the BOD by about 50 per cent.

Both high rate trickling filters and activated sludge plants can be employed very advantageously for a complete treatment of the dairy waste. But these conventional methods involve much maintenance, skilled personnel and special type of equipment. On the other hand, the low cost of treatment methods like oxidation pond, aerated lagoon, waste stabilization pond, etc. can be employed with simpler type of equipment and less maintenance.

8.5 IDENTIFICATION AND TREATMENT OF LIQUID WASTES OF MAN-MADE FIBRE INDUSTRY

All toxic organics are soluble to some extent in water even though the dissolution limit may be barely measurable. Many a time all toxic organics may be found present in a water medium in concentration far exceeding their solubility which might be the resultant of their solubilization in water by another pollutant or by their presence as absorbed species on suspended solids in the water medium.

In the man-made group of textile fibres there are two broad classifications; cellulosic and non-cellulosic fibres. Rayon and cellulose acetate are the two major cellulosic fibres while nylon, polyester, acrylics and mod-acrylics constitute the major non-cellulosic fibres. No doubt, there are other fibres in both the categories, but these constitute an insignificant percentage of the total man-made fibre consumption. Of the above, the largest volume of man-made fibres consumed by textile mills in India comes from the cellulosic group. Consumption of non-cellulosic fibres, especially polyester has picked up and has become a major textile fibre.

Synthetic fibres can be converted into fabrics in one of the following two ways. Continuous filament yarns can be used to manufacture 100 per cent synthetic fabrics, spun yarns from staple fibre can be used to produce blended and/or mixed fabrics. Blended fabrics are normally processed as per the natural fibre component of the yarn. The first process in which synthetic fibres would be subjected to an aqueous treatment is stock dyeing unless the fabric is to be piece-dyed, printed or used in white. In case, stock dyeing is employed, the liquid waste discharge will vary from about 8 to 15 times the weight of the fibres dyed. Owing to the low moisture regain of the synthetics, static electricity is a problem during processing. To alleviate this, antistatic materials are applied to the yarns which incidentally serve as lubricants and sizing compounds. The compounds generally employed are polyvinyl alcohol, poly acrylic acids, styrene base resins, polyalkylene glycols, gelatin and polyvinyl acetate. The aforementioned compounds turn out to be a source of water pollution when they are removed from the fabrics during scouring. Since the manufacture of synthetic fibres can be well controlled, chemical impurities are relatively absent in these fibres; so only light scouring and little or no bleaching are called for, prior to dyeing and if synthetics are bleached, the process is not normally a source of organic or suspended solid pollution. The process may, however, generate dissolved solids when chlorine bleaches are employed. The outline of the process of manufacture of both the categories of man-made fibres viz., cellulosic and non-cellulosic, the characteristics of their wastes and methods of their treatment are given hereinafter.

8.5.1 The Man-made Fibre Industrial Wastes

As mentioned earlier, synthetic fibres are essentially composed of pure chemical compounds and have no natural impurities. Because of this, only light scouring and bleaching are necessary to prepare the cloth for dyeing. Of the man-made fibres, rayon is chiefly composed of regenerated cellulose; cloth is readily done on the conventional machinery. Acetate rayon is a cellulosic acetate fibre and nylon is the generic term for any long-chain synthetic polymeric amide. Orlon is a trade name for synthetic, orientable fibres from polymers containing a preponderance of acrylic units, the newest of which are the acrylonitriles and ethyl or methyl acrylate. Dacron is a condensation polyester fibre manufactured from ethylene glycol and terephthalic acid. In the following section, the characteristics and treatment of effluents of viscous rayon, acetate rayon, synthetic polyamide fibres and acrylonitrile are presented.

8.5.2 The Wastes of Viscose Rayon

Cellulose in the form of sheets or wet pulp (if a chemical pulp factory is located in the vicinity) is the raw material for the manufacture of viscose rayon. The cellulose as sheets shredded in a machine is first soaked in an 18 per cent solution of caustic soda. The resultant alkali cellulosic is separated from the spent liquor in presses or on rotary vaccum filters. The alkali cellulose is then subjected to *ripening period* during which oxygen is taken up form the air. It is then treated with carbon disulphide from which cellulose xanthate is obtained, simultaneously resulting in the following secondary reaction:

$$3CS_2 + 6NaOH \rightarrow 2Na_2CS_3 + Na_2CO_3 + 3H_2O$$

The xanthate is then dissolved in caustic soda solution giving rise to a spinning fluid known as **viscose**. After repeated filtration through filter presses, ripening and aeration, the viscose is forced through small orifices in acid proof spinnerettes into a coagulation bath containing dilute sulphuric acid, sodium sulphate and zinc sulphate. The fibres are then passed through water baths from the spinning machine where both the carbon disulphide and the residues from the acid bath are got rid of. The viscose rayon is also desulphurized and treated in finishing baths which is termed *lustering*.

Characteristics of wastes

The kinds of effluents coming out of a viscose rayon plant are:

- Alkaline wastes generated in dialyzers as a result of the washing of alkali-cellulose tanks and presses, xanthating equipments, floors, etc. Stronger alkaline wastes are also obtained from the washing of filter cloths and sometimes from the desulphurization process.
- Acid wastes coming out of spinning machines, the acid bath recycle, the washing of filters and the first fibre rinse.
- Neutral wastes from further washing of the fibre on a machine with bar conveyor and from the ancillary water treatment and softening plant.
- Cooling water from the acid bath evaporators and cooling apparatus.

The largest quantities of wastes are generated during the fibre spinning and the finishing processes; they comprise some eighty per cent of the total quantity of concentrated effluents. The remaining wastes usually arise during the preparation of alkali-cellulose and viscose. Sulphuric acid, sodium sulphate, zinc sulphate, waste fibre, hydrogen sulphide and carbon disulphide are present in the wastes from the spinning machines and acid baths. The alkaline wastes also contain hemicellulose and residual viscose. Those from the disulphurization process contain sodium polysulphides and sodium thiosulphate. The total quantity of concentrated waste ranges from 200 to 240 m^3 per ton of staple fibre produced.

Treatment of wastes

In essence, it may be mentioned that the treatment revolves around the recovery of the substances employed in the manufacturing process to the extent possible. Thus, spent caustic soda separated from the alkali cellulose on presses or vacuum filters is purified using dialysers. One method of getting sulphuric acid from sodium sulphate is by precipitating barium sulphate using barium hydroxide or barium chloride with added coal dust. Apart from the removal and utilization of sulphates, a considerable portion of the colloidal suspended organic matter is also got rid of, by this method. The recovery of carbon disulphide is effected by installing a degasifier between the spinning machine and the washer. Further treatment of wastes from viscose rayon plants is usually by physical and chemical methods.

8.5.3 The Wastes of Acetate Rayon

Purified dry cotton linters or enriched pulp forms the fibrogenic raw material in the acetate process; the first step in the manufacture being the purification of the raw material by boiling in a solution of caustic soda after which it is washed with water. Acetic acid or acetic anhydride in the presence of a catalyst like sulphuric acid, perchloric acid, zinc chloride, etc. are employed then to convert it into cellulosic acetate. The syrupy cellulose triacetate is poured into water (in which it is insoluble) and separated. The cellulose acetate is dried after a thorough wash with water. The spinning process is a dry one, employing the cellulose triacetate dissolved in a suitable solvent.

Characteristics and treatment

The wastes could be categorized into alkaline and acid wastes with the later containing sulphuric acid and acetic acid. The mixed wastes may be treated biologically on trickling filters after previous neutralization. The quantity of wastes from triacetate production alone is estimated at 10–15 m^3/ton, the total quantity of waste being dependent on water consumption (may be reduced much by effecting water economy) is around 1300 m^3/ton of the fibre produced. Ion-exchange method has been employed to recover copper, in the case of cuprammonium rayon wastes.

8.5.4 The Wastes of Synthetic Polyamide Fibres

Polyamide fibres are obtained by the polycondensation of dibasic aliphatic acids with aliphatic diamines and the polycondensation of amino acids, e.g. e-amino caproic acid. The wastes generating from their manufacture are of the order of 4–5 m^3/ton of product. One method of treatment of wastes from nylon production is by a process using activated sludge to reduce the chemical oxygen demand. Another method involves pumping of the concentrated nylon plant wastes into the ground (underground disposal) which turns out to be 10 times cheaper, than the usual physical and biological techniques, with regard to operating and capital costs.

8.5.5 Acrylonitrile Production and Waste Streams

Acrylonitrile which is used in the manufacture of acrylic fibres is produced by the ammoxidation of propylene according to the reaction.

$$\underset{\text{(Propylene)}}{2CH_2{=}CHCH_3} + 2NH_3 + 3O_2 \longrightarrow \underset{\text{(Acrylonitrile)}}{2CH_2{=}CHCN} + 6H_2O$$

The raw materials (air, ammonia and propylene) are introduced into a fluid bed catalytic reactor operating at 5–30 psi gauge and 200–260° C.

The reactor effluent is scrubbed in a counter current absorber and the organic materials are recovered from the absorber water by distillation. Hydrogen cyanide, water, light ends and high boiling impurities are removed from the crude acrylonitrite by fractionation to produce an acrylonitrile of specified purity. The major water pollution source of this process is the process waste waters discharged from the steam stripping columns. A survey conducted a few years ago reports that most of the acrylonitrile plant wastes are disposed of by deep well injections.

The principal parameters used to characterize waste effluent are the flow, biochemical oxygen demand, chemical oxygen demand, total suspended solids, oil and grease.

There is also scope for employing the advanced treatment methods such as ultra-filtration, electro-dialysis, activated carbon adsorption and chemical clarification, in the treatment of textile wastes.

In respect of cotton, it may be mentioned that many of the mechanical operations employed in the manufacture of textile fabrics are common to the industry as a whole and the characteristics of the waste waters are similar. Typically the textile fibres are combined into yarns and then the yarns to fabrics. After the fabrics are manufactured, they are subjected to several, wet processes collectively termed as *finishing* and these finishing operations offer scope for the generation of major waste effluents.

8.6 POLLUTION PROBLEMS OF THE PETROCHEMICAL INDUSTRY

The petrochemical industry engaged in producing many items like olefins (ethylene, propylene, etc.), aromatics (benzene, toluene, o-xylene, p-xylene, etc.), synthetic fibres (nylon, polyster, etc.), alcohols (methanol, ethanol, phenol, etc.), ethylene oxide, ethlyene glycol, polypropylene, phthalic anhydride, PVC, synthetic elastomers, thermoplastics, rubber chemicals, etc. is one of the industrial units responsible for large-scale environmental pollution.

Refineries and petrochemical plants deal with petroleum hydrocarbons and hence the pollution problems and hazards are different from those encountered in other chemical industries.

The pollution problems can be broadly classified as follows:

8.6.1 Oily Waste Waters and Oily Sludges

Sources of process waste waters containing heavy hydrocarbons like crude petroleum, lubricating oils and other process products are summarized broadly as under:

1. Drainage from bottom of the storage tanks and vessels containing petroleum crude and allied products.
2. Disposal from cooling water sockets and bearings of rotating equipment.
3. Disposal from quench towers, strippers scrubbers, etc.
4. Disposal from recovery units.
5. Leaks and drips from pump seals, valve glands sampling points, etc.
6. Wash water of floors of processing units.
7. Washings of equipments, etc. before undertaking the maintenance work.

While oily waste waters are handled through pumps, pipelines, tankers, etc. sludges generally need manual or mechanical scrapping and haulage by trucks. In order to prevent accidental fire from the downstream to the upstream side through oily waste sewer upto the source of hydrocarbon emissions, fire brakes and water-seal manholes are provided.

In the processing area, in a naptha-cracker significant quantities of oily and hydrocarbon intermediates get accumulated in small channel/drains which flow to the separating and/or treatment area through oil sewers. Normally the collecting points/drains are open.

8.6.2 Handling of Toxic Waste During Manufacture Treatment and Disposal

Many processes involve toxic materials which need special care and handling. For instance, an acrylonitrile plant manufactures hydrogen cyanide as a by-product which gets mixed with the off gases which are burnt in a separate flare system and in waste waters. It is required to be incinerated since organic cyanides in higher concentrations are difficult to biograde. Special care is to be taken while removing oily sludges, etc. from such plants because of cyanide content. Apart from ensuring personal protective devices while handling/collecting such wastes, sufficient care has to be taken while disposing of the same. It is worth-mentioning here that vinyl chloride can cause stillbirths and miscarriages in women whose husbands come into contact with this material. Vinyl chloride is also known to cause a form of liver cancer.

During the manufacture of boiler feed water in ion-exchange plants, HCl is used to regenerate cation exchange resins. During the regeneration process HCl gas fumes are likely to be released. There is also frequent release of the HCl gas from the breather valve on vents of the acid storage tanks. This gas causes eye irritation and damages the internal organs of the body, besides causing corrosion problems. Similarly, phenol is one of the major constituents of effluents from petrochemical plants. It is a toxic compound capable of destroying bacteria, fish and other aquatic life in rivers and streams. Before phenol bearing effluent is discharged into normal water-courses, it must be reduced to permissible limits so that aquatic life is not endangered and the water is made unsuitable for further use. The primary effluents generated during the process of manufacture of phenol from benzene contain polychlorobenzene, waste HCl, an alkaline solution of phenol from washing of the products and waste pitch compounds. Total concentration of phenol in the effluents formed during manufacture varies between 10–30 mg/litre.

8.6.3 Cleaning of Vessels, Sewers, Pipelines, etc.

Cleaning of vessels/tanks also frequently creates hazards, especially while removing sludges etc., from bottoms. In spite of elaborate purging and washing operations to make the tanks gas-free, pockets of entrapped gases under the sludge may escape to the surface and cause danger. Waste waters are carried to the sewer which too requires cleaning up at times. Sewer gas could be fatal if enough care is not taken for its safe entry. Similarly, when vessels, storage tanks, etc. are emptied according to the process requirements or drained accidentally, resulting in fire and explosion due to pollution.

8.6.4 Generation of Wastes in Petrochemical Plants

Natural gas and petroleum are sources of alcohols, plastics, synthetic rubber solvents, detergents, pesticides, fertilizers and a whole range of products based largely on unsaturated hydrocarbons. Waste liquor is an inevitable result of purification of raw materials and from unit operations and reactions by which the final product is obtained, such as distillation and rectification.

These wastes are characterized by a relatively high concentration of impurities; for example, fractional distillation of ethylene oxide yields about 0.5 m^3 of liquor per ton of pure product. The main pollutant is ethylene glycol at a concentration of 10 gms/litre. The synthesis of ethyl alcohol from ethylene yields about 8 m^3 of wastes per ton of product, which are chiefly polluted with about 500 mg/litre alcohol and with varying amounts of phosphoric acid, which is used as a catalyst.

In the manufacture of acetaldehyde by oxidizing ethylene, the main source of wastes is rectification of the crude aldehyde which gives distillation residues. These residues have a large concentration of organic compounds where biochemical oxygen demand is about 100 gms/litre. The chief pollutants are acetic acid, aldehydes and resins.

In the manufacture of high pressure polyethylene, purification of the gas gives rise to the formation of concentrated alkaline wash liquors containing sulphides and other sulphur compounds—chiefly organics, such as mercaptans.

In the chlorhydrin saponification reaction in the manufacture of propylene oxide from propylene 85 m^3 of slurry calcium oxide wastes are produced per ton of product. The chief pollutants are $CaCl_2$ (approximately 3 per cent) and propylene derivatives, such as propylene glycol. The permanganate valve of these wastes about 20 gms/litre.

In the synthesis of acrylonitrile from HCN and acetylene the reaction products are extracted with water and then stripped from the aqueous solution by distillation. The distillation bottoms (21 m^3 per ton of acrylonitrile) contain acetaldehyde and organic nitriles particularly lactonitrile.

Table 8.4 presents a general idea of the composition of certain petrochemical wastes showing their main constituents that are, hydrocarbons, their oxidation products (alcohol and aldehydes), phenolic compounds, esters, acids, bases and salts. Oil impurities are quite frequently present in the emulsified state.

Table 8.4 Products from Petrochemical Plants and the Resulting Wastes

S.no.	*Products*	*Wastes*
1.	Aldehydes and alcohols	Rectification residues with dissolved hydrocarbons and aldehydes
2.	HCN from natural gas and ammonia	Distillation residues (from the cyanide stripping process) with a small HCN and unreacted hydrocarbon content
3.	Chlorinated derivatives of methane and ethylene	Column effluents containing chlorinated hydrocarbons, salinated to a certain extent with lime sludge
4	Acetylene from hydrocarbon cracking	Wash liquors HCN and hydrocarbons
5.	Ethylene and propylene by thermal cracking	Wash liquors and oily wastes
6.	Polymerization and alkylation	Alkaline wastes (NaOH), hydrocarbons, benzene derivatives, catalysts (H_3PO_4 and aluminium chloride)
7.	Alcohols from olefins by sulphonation and hydrolysis	Large quantities of wastes containing sodium sulphate, polymerized hydrocarbons and butyl or isobutyl alcohol
8.	Acetone by dehydration of alcohols	No wastes
9.	Ethylene and propylene oxides and glycols	Large quantities of wastes containing lime sludge, calcium chloride, dichloro ethylene or propylene and glycol
10.	Aldehydes and alcohols by oxidation of hydrocarbons	Organic acids, formaldehyde acetaldehyde, acetone, methanol and higher alcohols
11.	Butylene from butane and butadiene	Small quantities of wastes with a higher butylene, hydrocarbon content
12.	Aromatic hydrocarbons by reforming processes	Condensates polluted with catalyst, H_2S and ammonia

8.6.5 Aqueous Effluents from Petrochemical Plants

The aqueous effluents from petrochemical plants contain a diverse range of pollutants including oil, phenols, sulphides, dissolved solids, suspended solids, toxic metals and BOD bearing materials. The quantitative amount of pollutants produced in a process plant is a function of the type of source of process feedstock and the type of process involved. The concentration of pollutants in the plant raw wastes depends on the amount of process steam, process water and cooling water used in the plant. The total aqueous wastes from a chemical plant can be minimized by using a minimum stripping steam and a maximum air cooling.

Aqueous effluent sources

Majority of the effluent sources in refineries and petrochemical plants can be broadly categorized as one or more of the following types:

- High or low dissolved solid contents
- Oily or non-oily
- High or low in phenols and/or sulphides
- Chemical or non-chemical

Water withdrawn from process drums containing H_2S and NH_3 is designated as 'Sour water', caustics used to scrub pentane contain essentially sodium sulphide. Such caustic is defined as sulphide spent caustic.

Breaking units called 'steam crackers' are expected to produce foul condensates containing small amounts of the solvent.

Treatment methods

Aeration methods in towers of various types may be used for the removal of many volatile compounds (acetylene, HCN, lower hydrocarbons, etc.). In many cases, pollutants are stripped with steam or waste gases. For example, benzol may be recovered from the wastes by (a) stripping with N_2 which is recirculated in a closed cycle, (b) adsorption on activated carbon, and (c) distillation. Waste water which contains acetaldehyde may be distilled. Butanol, butyl acetate and acetic acid may be removed and recovered by *adsorption* on activated carbon and other adsorbents or on an ion-exchange resin.

Generally speaking, the important economic problem of regeneration and recovery (hence, controlling pollution) leads to the development and application of a number of special methods.

- Wet scrubbing is used for the removal of CO_2 from hydrogen streams in conventional synthesis gas plants. Scrubbing solutions include hot aqueous carbonate solution, ethanol amines and water under pressure.
- Adsorption on molecular sieves or activated carbon is used for separating hydrogen from hydrocarbons.

- Cryogenic scrubbing, removes all the impurities which boil at higher temperatures from the low boiling hydrogen, by condensation. This process is useful when a refrigerant such as liquid air or liquid N_2 is readily available, as in partial oxidation synthesis in gas plants.
- Membrane permeation is based on the selectivity of palladium in adsorption of hydrogen from petrochemical effluents.

Inorganic waste acids from petrochemical plants constitute yet another problem. HCl, for example, may be passed on to the production of vinyl chloride or chlorinated hydrocarbons.

Several other processes are also more frequently used for the recovery of petrochemical wastes, viz.

- Distillation, allowing the recovery of chlorinated hydrocarbons from methane chlorination, oil from the thermal manufacture of ethylene, acetaldehyde from the productin of vinyl acetate and synthetic latex solvents.
- Crystallization, used for the separation of xylene from ethylbenzene.
- Adsorption, permitting the recovery of CS_2, CCl_4 and other solvents.
- Extraction, allowing the recovery of isobutylene, naphthalene, paraffin and phenols from various waste materials.
- Ion-exchange methods, used for the recovery of amines, alkaloids and organic acids.

Mercaptans and other components in alkaline waste liquors form phenol washing, can be oxidized electrochemically. In this process, the mercaptans are precipitated as disulphides and hydrogen sulphide is oxidized to free sulphur.

8.6.6 Treatment Method for Aqueous Wastes

The commonly used waste segregation, collection and treatment facilities for aqueous wastes can be broadly classified as *inplant pre-treatment* and *secondary treatment*. Inplant pre-treatment facilities are those which are upstream of or which precede the API separator. Secondary treatment facilities are those which are downstream of or which follow the API separator. Design procedures are available for the following units

1. Sour water strippers
2. Spent caustic oxidizers
3. Spent caustic neutralizers
4. API oil—water separator
5. Oxidation plants.

Secondary treatment units include

1. Chemical coagulation
2. Activated sludge process
3. Trickle filter processes
4. Oxidation ditches

When secondary treatment is installed, the typical complete waste treatment facilities may include

1. In-plant segregation and pre-treatment
2. Storm water surge pond
3. API oil-water separation
4. Air floatation or chemical coagulation (for further removal of oil and suspended solids)
5. Biological oxidation
6. Sludge dewatering and disposal (probably by incineration)
7. Facilities for special hazardous or toxic wastes.

Treatment of petrochemical effluents includes separation of aqueous effluents from suspended and floating organic material, followed by neutralization. The treated effluents are conveyed to the central waste water treatment plant where after being subjected to carefully controlled further oil removal and separation of solids, they are treated in biological reactors where vigorous aeration is obtained. The effluents are separated from suspended biomass and other solids and the clear oxygenated stream is finally conveyed out after a thorough check of quality. Air quality can be preserved by minimizing the emissions and its simultaneous flaring and dispersal.

8.7 WEALTH FROM AGRICULTURAL WASTE

The exponential increase in our population added with the demand for a higher standard of living has thrown much demand on food, fuel, fertilizers, etc. India has an agro-based economy and about 70 per cent of its population is involved in agricultural work. However, at present, only a quarter of the agricultural output is being used as food and the rest is rejected as agricultural waste. Since minimization of agricultural waste is impossible, utilization of agricultural wastes to produce various products which are in short supply would do a world of good to our economy. Since agriculture is a field of immense renewable potential, we should focus our attention in this particular area and derive maximum utilization of our agricultural output to bridge the gap between demand and supply of various products.

The following section presents the various possibilities of producing valuable products like food, fuel, fertilizer, etc. from agricultural waste. Processes that are commercially possible, technically feasible above all economical and last but not the least, desirable from the point of view of reducing environmental pollution problems are given hereinafter.

8.7.1 Waste Materials

Waste materials available can be classified into agricultural residues like rice straw, wheat straw, corn stalks, etc. and agricultural processed wastes like bagasse, rice bran, fruit waste, etc.

The quantum of various wastes available would give an idea of how much of 'waste' that is not used can be utilized for the manufacture of valuable chemicals, pharmaceuticals, fuel, fertilizer, edible oil, etc.

Table 8.5 gives a rough estimate of agricultural residues and processed wastes available in India per annum.

Table 8.5 Quantity of Agricultural Residues and Process Wastes

Material	*Tons*
Rice bran	3500×10^3
Bagasse	750×10^3
Pineapple skin, core	45×10^3
Grape stem, skin, seed	12×10^3
Banana peel, stem	1150×10^3
Mango peel, stone	5200×10^3
Citrus peel, seed	900×10^3

The various ways in which agricultural wastes can be utilized are given hereinafter.

Ethanol

It has been reported that corn stalks can be hydrolyzed using sulphuric acid and the resulting mixture of glucose and xylose can be converted to ethanol using microorganisms like yeast. A brief description of the process is as follows:

Ground corn stalk is treated with 4.4 per cent sulphuric acid at 100° C for 5 minutes. This mixture is then filtered. The xylose-rich liquid is processed by electro-dialysis for acid recovery. After the solids are dried and impregnated with 85 per cent sulphuric acid, dilution water is mixed with the powder to form 8 per cent sulphuric acid and hydrolysis is carried out at 110° C for 10 minutes. By electro-dialysis the acid is recovered from the glucose-rich stream.

The combined yield of xylose is 94 per cent based on 13.9 per cent hemicellulose content of corn stalk used. The yield of glucose is 89 per cent based on cellulose content of 36.8 per cent. Dilute glucose and xylose streams are then fermented to ethanol using *Saccharomyces cerevisiae*, *Fusarium oxysporum*, respectively.

Fixed film or immobilized cell reactor with high cell density has been suggested. High throughput rates can be achieved in these types of reactors without cell washout. This reactor reportedly converts all the sugar to ethanol.

This can be further distilled to get ethanol of desired purity. As much as 415 lbs of ethyl alcohol can be obtained from 1 ton of corn residue by the aforementioned method based on 90 per cent conversion of the hexasans and pentosans.

Chemical compositions of various agricultural residues are given in Table 8.6. The primary constituents are pentosans, hexasans and lignin.

Table 8.6 Chemical Compositions of Various Agricultural Residues

Material	*Pentosan %*	*Hexasan %*	*Lignin*
Corn stover	15.0	35.0	15.0
Corn cabs	28.0	36.5	10.4
Wheat straw	19.0	39.0	14.0
Rice straw	17.0	39.0	10.0
Bagasse	20.4	41.3	19.9

India loses considerable foreign exchange by way of importing petroleum crude. Production of ethanol from agricultural residues and using it as substitute fuel would help to solve India's fuel problem in a large way.

Rice bran oil

Edible oil is a food commodity which is in short supply in India. Extracting edible oil from sources other than the conventional ones would be a better solution to the problem than importing edible oil to make up the deficit amount. One such material is rice bran.

The process consists of removing a part of the oil from rice bran by pressing in an impeller or by other mechanical means. The remaining oil has to be extracted by using solvents like hexane, alcohol, ether, etc. using solvent extraction technique.

Rice bran is pre-treated and thereafter circulated with the solvent for an optimal time for good extraction. The oil extracted and present in the solvent phase is known as miscella. By distillation, solvent can be recovered and recycled. On further stripping crude rice bran oil is obtained. For good quality, the oil has to be further refined by dewaxing, bleaching and deodourization. The refined oil is fit for human consumption.

The by-products of refining wax and soap stock can be used for polishing and soap manufacture respectively, thereby making the refining step economical. Rice bran oil is high quality edible oil containing mostly unsaturated acids like oleic acid and linolenic acids, small amounts of tocopherol (vitamin E), oryzanol; which accelerates human growth, facilitates blood circulation, and stimulates hormonal secretions and squalene (a substance of pharmaceutical interest). Apart from being consumed as oil it can be used in the manufacture of vanaspati also.

Of the 6.0 lakhs tons that can be recovered, the units in India recover a meagre 0.8 lakhs tons. The reason attributed for this under utilization of rice bran is the time lag that exists between production of bran and its extraction. One other reason is the scattered location and the small size of rice mills, making bran collection difficult and uneconomical.

One way of overcoming the problem is to stabilize the bran to arrest deterioration of bran, thereby preserving all its constituents to ensure production of high quality edible oil. Heat stabilization with steam or hot air has been found to be very easy and economical.

Fertilizer

A process for converting sewage to inexpensive fertilizer, high protein animal food and clean water is being developed by the scientists. The three-step process uses green algae, blue-green algae and plants such as water hyacinth and sunlight. The process in a nutshell comprises decomposition of waste and preparation of fluid medium by the green algae and water hyacinths for the growth of nitrogen-fixing blue-green algae which is then harvested as a fertilizer.

Since the process essentially involves recycling of organic waste, when commercially successful it is expected to largely reduce the petroleum consumption for fertilizer production. The unique features of the process are that it is an attempt to cultivate algae as fertilizer on large scale, apart from simultaneously treating sewage and production of fertilizer without great expense of energy by recycling nitrogen with sunlight as energy source instead of fuel.

Industries processing fruits can use their sewage sludges for producing various algae which can be utilized as fertilizers.

Plastics

A process to manufacture plastics from raw materials based on agricultural wastes like corn stalks, leaves and bagasse has been developed by scientists. This opens up the possibility of producing plastic materials ranging from soft foams to ultra hard substances from non-conventional sources. This process, if commercialized will turn out to be very advantageous to developing countries, which have to depend presently on petroleum for the production of plastics.

Paper

Bamboo has been the main raw material in India for the production of paper. India is required to explore the possibilities of producing paper from non-conventional sources, because bamboo and wood alone as raw material cannot meet our requirements. Deforestation for wood would result in environmental problems. The solution lies in effectively using agricultural fibres since a major structural part of these is the cellulose fibre. Pulp derived from agricultural fibres can be used either alone or in blends because of desirable properties that the pulp can give to the end products.

Banana fibre can be utilized in the preparation of pulp for making paper. It is estimated that about 10 lakh tonnes of banana fibre can be obtained as by product from banana cultivation in our country. The paper made from banana fibre pulp reportedly, has good tensile strength and higher splitting strength.

Cereal straws the like rice straw, wheat straw and corn stalk can be used for the manufacture of paper, but pose problems of collection and storage for a modern plant operating continuously all the year. Though straw cannot compete with wood except when special properties are required, it can, however be used by blending with wood pulp.

Flax straw used for paper making is a by-product of production of flax seed for linseed oil. Flax straw is 12 to 15 inches long and relatively strong. India, being one of the largest flax growing countries in the world can take best advantage of the material for producing paper. It has been reported that special quality papers like cigarette paper, carbonizing tissues, scuff-resistant papers and currency papers can be made from Flax straw.

As regards pulping, alkaline pulping reagents have been preferred for agricultural fibres as mild treatment has been found essential for pulping as well as refining of the pulps.

Conditions for continuous pulping of agricultural fibrous materials are listed in Table 8.7.

Table 8.7 Conditions for Pulping of Agricultural Fibrous Materials

	Process	Na_2O (%)[a]	*Time* (mins)	*Pressure (psi)*	*Result yield %*
Wheat straw[b]	Soda	4.6	8	75	67
Rice straw[b]	Soda	9.8	5.5	100	39
Bagasse[c]	Sulphate	12.0	10	130	52
Reeds	Soda	13.8	20	130	48
Bamboo	Sulphate	18.2	30	130	45
Cotton linters	Soda	5.6	18	100	75

a—chemical applied on dry raw fibre basis, b—uncleaned raw fibre used, c—depithed or cleaned raw fibre used

Boards

Though very good quality paper cannot be economically obtained from the use of agricultural residues alone, boards can be economically made using agricultural residues.

Corn stalk has been suggested for making insulating board, hard board, etc. and bagasse for insulation board, wall board. A method of producing insulating board is briefly indicated here:

In a mechanical process, shredded stalks are passed into a water suspension through two refiners to storage tanks. 3 per cent rosin paper size is added between refiners, to increase the strength of board newspaper pulp to the extent of 10 per cent to 20 per cent or corn stalk, paper pulp made by a caustic process could be added. The pH is carefully maintained at 4.5 by adding alum or sulphuric acid. The sized pulp must have reduced moisture content before being dried in a roller type tunnel drier. To reduce moisture content it is passed between three pairs of smooth rolls.

If a steamed pulp is desired, the shredded stalks are boiled in water at atmospheric pressure for about three hours, before refining. Cooked pulp is made by digesting the stalks in water at pulp 50 psi steam pressure for about 3 hours.

It is reported that board manufactured under such conditions is light weighted and springy, but is however rigid. It is formed from a heterogeneous mixture of fibre bundles upto about 1 inch long together with the pith particles and the added paper pulp. The pith and the hydrated paper pulp cement the fibre bundles together at their points of contact while still leaving an abundance of air films which give the board its thermal insulation properties. The board is suitable for sheathing in buildings as a plaster base and for other purposes where both structural and heat insulation properties are desired.

Oxalic and acetic acids

Strong nitric acid oxidizes sugar and cellulose to oxalic acid. Since corn cobs are composed mainly of pentosans, lignin and cellulose, it should be possible to oxidize them to oxalic acid by using nitric acid.

The process consists of adding slowly to one part by weight of cobs, three parts by weight of cold 90 to 100 per cent nitric acid with constant stirring. The mass is heated until the solids go into solution and then cooled. Then 0.1 per cent by weight of freshly prepared vanadium pentoxide and 3 parts (based on original weight of cobs) of 50 to 55 per cent nitric acid are added. The mixture is allowed to stand for two or three days to complete the oxalic acid formation and crystallization. Further investigation carried out shows that oxalic and acetic acids could be produced by fusing corn cobs with sodium hydroxide. The best results are obtained with 3 : 1 ratio of alkali to corn cobs. Corn stalks and corn cobs give identical yields, but the stalks are bulkier to handle. The optimum condition is heating for 1.20 hrs at 220° C to 225° C. The yields of oxalic acid and acetic acid are 20 per cent and 30 per cent respectively, based on the weight of air dried corn cobs. Since the product will be a mixture of oxalates, it necessitates addition of sulphuric acid for the conversion of oxalates to a mixture of acids which can be separated thereafter. Less drastic conditions are necessary while using corn cobs than using softwood.

Single cell protein (SCP)

Single cell protein can be produced from waste stream of fruit processing industries. The inputs there would be inoculum, the energy is solar energy and outputs include heat, carbon dioxide and cells. Oxygen transfer and heat removal add to the cost of fermentation. In addition to this, unit operations concerned with collecting and utilizing the cells in the fermenter effluent and conversion of waste into a suitable substrate for the organisms would form major operating costs.

Largest biomass productivity for a fermentation plant is achieved when the process is run on a continuous basis, but the fact that fruit processing plants do not run continuously even during seasons, poses the problems of conversion of waste to SCP. If material is processed throughout the year using various raw materials, the concentration of waste streams will vary. Hence, continuous processing becomes difficult, though not impossible. Another problem of consequence in the case of many wastes, is the content of non-fermentable substances which may accumulate in cells or fermenter effluent which may be non-nutritive or even toxic. Examples include lignin, degradation products, heavy metals, agricultural chemical residues, etc. Extra cost of processing for their removal may result in prohibitive cost of the product.

If microbial protein is used as a major source of protein in the diet, its nuclei acid content must be reduced by some suitable treatment which may be required to remove flavour or to build in special textile textural characteristics. Such considerations should not be overlooked. Moreover the product should be of the form which would be readily absorbed into a local social culture. This is an important aspect because even though food produced from waste may have excellent nutritive value it will not benefit man unless it is eaten. The object must, therefore, be to produce palatable cheap food which will be accepted by the consumers. Moreover, foods from biosynthesis must be sent to market only after they pass through stringent chemical and toxicity, feeding tests.

Presently it would be much simpler to use microbial protein for animal food. Proper treatment to render wastes free from pathogens and make it safe for animal feeding is essential.

To avoid significant additional energy expenditure to treat these waste streams, a great deal of attention will have to be paid to more efficient plant and process design. One other criterion is that waste streams from food processing industries should have a high biological oxygen demand (BOD). If they have small BOD, preconcentration is required and small units cannot operate profitably if they choose to process these waste streams.

Single cell oil and fats

With increasing technological development in the fermentation industry, it is now possible to consider the production of Single Cell Oil (SCO) in the same way as a single cell protein production from waste, that is food processing plant sewages. However, the microbial production of fat is much slower than protein production and this necessitates careful selection of fat accumulating species as not all microorganisms have the ability to produce fat. High concentrations of fat and continuous good yields can be obtained by growing appropriate organisms usually yeasts or moulds.

Some of the advances in this direction are described as this method of culture has only recently been recognized as being feasible. It is important to select a good growth substitute and growth conditions. The substrate must be available in large quantities. Most of the substrates which are currently used for SCP production can be successfully and efficiently used for growing fat accumulating yeast and fungi.

Microbial oils and fats might be able to achieve economical parity with conventional sources of oils and fats. If cheaper sources of carbon than molasses were available in suitable quantities, then it is almost certain that single cell oil will be a feasible and economical proposition.

Products from fruit waste

In our country the total fruit production is over 70 million tonnes. There are about 2000 fruit and vegetable processing industries in our country. The solid wastes from small-scale fruit and vegetable processing, have little economic value and are not marketable as the quantity is very little. The best way to dispose them of would be to distribute them as cattle feed. But when they are available in large quantities they can be used to manufacture valuable chemicals. A few such chemicals are indicated here.

Pineapple waste is a potential source for the preparation of oxalic acid. On subjecting pineapple waste to oxidation with fuming nitric acid using vanadium pentoxide as the catalyst, oxalic acid can be obtained. Possibilities of obtaining yields as high as 75 per cent to 80 per cent on the dry basis have been reported. Acetic acid can be prepared from juice extracted from pineapple wastes. Fermentation process is employed.

Grape seeds can be used to prepare fatty oils. By solvent extraction technique, fatty oils to the extent of 10 per cent to 15 per cent on dry basis of seeds can be obtained. Though it cannot be used as edible oil it can be used in soap and paint industries.

Mango kernel flower which contains more fat and calcium compared to cereals, can replace wheat flour to the extent of 10 per cent. It is prepared by crushing and washing mango kernels in running water and thereafter drying them and grinding to flour. Vinegar can be prepared by concentration and fermentation of the water extract from mango peels.

Citrus peels can be used to obtain essential oils and pectin. The inner protein of the peels is rich in pectin and the balloon shaped cells under the surface contain essential oils. They contain cellulose and other polysaccharides. Economically valuable products like citric acid, vitamin C, riboflavin can also be produced. Colours extracted from citrus peels can be utilized for colouring citrus products.

Furfural

Furfural can be manufactured from bagasse, corn stalk, etc. One method of producing furfural is from corn cobs. The corn grain is arranged around the outside of the cob. The outer layer of cob consists of fine and course chaff. Under these, is a hard tough woody ring surrounding a centre of pith. On heating, material like corn cobs with a dilute mineral acid, the pentosans are formed. Further heating results in elimination of water and formation of furfural. This has to be dehydrated further to get pure furfural.

Furfural is a valuable intermediate in the manufacture of range of organic chemicals.

It could thus be seen that there is a potential scope for converting the agricultural wastes into useful products. Turning these wastes to desired valuable products can contribute in a small measure to national economy apart from reducing unemployment problems and environmental pollutions problems.

8.8 INDUSTRIAL SITING BASED ON ENVIRONMENTAL CONSIDERATIONS

The human life support system has two components, one industrial and the other ecological. It is of prime importance to strike a balance between these two factors for the survival of the present social system. Growing concern over the quality of the environment has caused the 'environmental impact' of new plants and many of the existing ones, to be closely scrutinized. Consequent to a perceptible change in the attitude of a pollution conscious public born out of concern for the habitat, environmentalists are expressing their concern over the setting up of new plants and industrial establishments.

8.8.1 The Need for Environmental Considerations

Although a number of measures could be used for size or the magnitude of production, land acreage is normally used as the size indicator. Figure 8.1 illustrates the importance that is attached to different site selection factors which increase with the size of the facility involved.

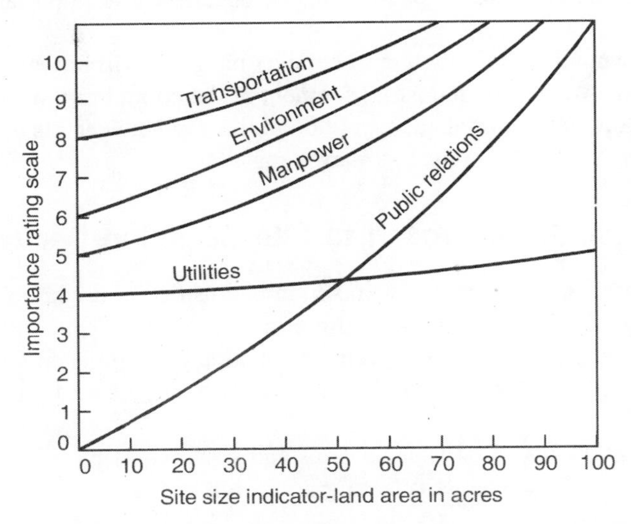

Figure 8.1 Plot of site selection factors against the size of the facility.

As seen from the graph, environmental considerations, start at a high level and rapidly increase on the 10 point importance rating. Utilities raise only slightly, whereas public relations ascend from a rating of 0.5 to a rating of 10 as the industrial size expands to 100 acres. Transportation, which is the most important of the five factors considered, is given a high importance regardless of the size of the industry.

A similar graph of the specific environmental factors involved in the site selection which is given in Figure 8.2 shows that air and water pollution control are the most important industrial site selection factors for large industrial units.

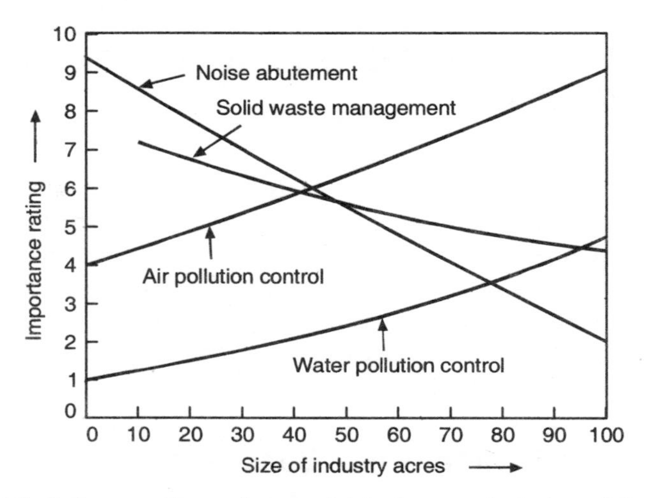

Figure 8.2 A plot of the specific environmental factors against size of the industrial units.

The importance of solid waste management declines with the increase of the industrial size while noise abatement, important to smaller plants, becomes less important as the size of the facility increases.

Environmental planners would incline to believe that the environment is the most important size selection factor. However, the industrial segment of the community would feel that the profit motive is still of paramount importance, even though the gap between clean environment on one hand and industrialization on the other, is narrowing down.

8.8.2 The Systematic Approach to Site Selection Decision-making

A number of environmental factors may be taken into consideration while selecting a site for an industrial unit. They are (1) location of emissions, (2) quality and quantity of pollutants, (3) control of pollutants, (4) impact on the environment and compliance with regulations. Each of these should be analyzed while considering the effects of any of the four major environmental control categories.

Figure 8.3 shows how by a step wise analysis and systematic approach one can arrive at the final decision in favour of or against selecting a site for locating a plant.

8.8.3 Industrial Plant Location Based on Air Pollution Considerations

Control of air pollution assumes major importance when an area is considered as a potential plant site. Proper designing and correct location of industry incorporating safety and control features prevent precipitation of imbalances in the ecology and human life style in the environment.

The environment and the habitat of the surroundings have to be safeguarded against the pollutants of the atmosphere and the emissions of the industry, regardless of whether the locale is industrialized or not. A review of the pollution levels in the area and a knowledge of the effect

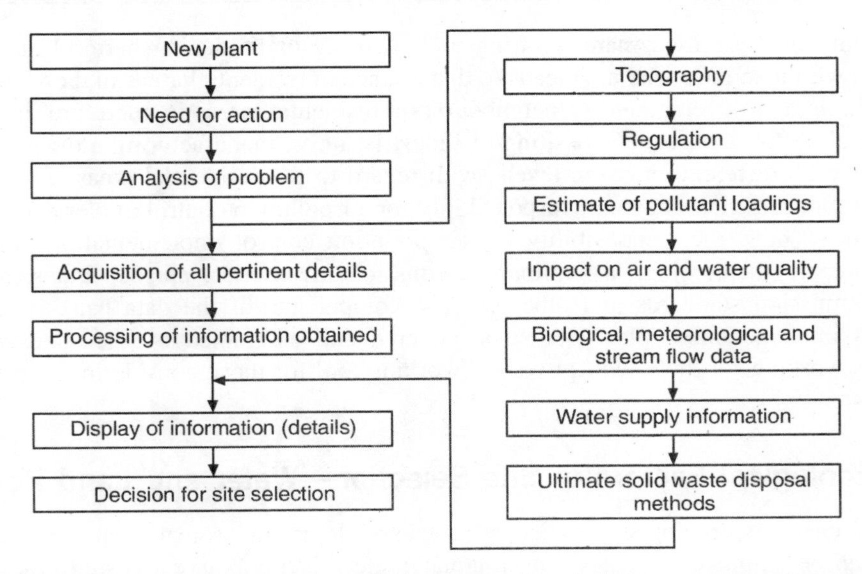

Figure 8.3 Environmental factors in plant site selection decisions—systematic approach to decision-making.

Adapted from Gerard Kiely, Irvin, *Environmental Engineering*, McGraw Hill, New York, 1985.

that the load of emissions from the planned industry would have on pollution, would be of great advantage for the determination of safety levels of emission of contaminants. This necessitates a survey of air pollution with reference to urban and rural areas. The rural dwellers may bear the existing conditions until the activities of the new plant commence.

Fulfillment of applicable pollution regulations, however, is far more important than mere knowledge of pollution level. The pollution put out by a plant, the air quality standards for the area, the background levels of pollution that exist are interrelated by the simple equality.

$$\text{Air quality standard} - \text{Background level} = \text{Maximum plant emission}$$

This forms the basic guideline for the environmental engineer when a decision on the location of industry in the area is to be made.

The ideal site for disposal of airborne wastes should possess the following qualities:

1. It should be comparatively level terrain.
2. The average wind velocity of the region should be about 15 km/hr or more.
3. Deep temperature inversions are to be a rare occurrence.
4. It should not be upwind of valuable farm land or a populated community.
5. The site area must be large enough for maximum effluent concentrations at ground level, downwind to occur will within company premises rather than on surrounding private property.

Valley sites require more pollution control. Stacks must be tall enough to enable the wind to carry the atmospheric wastes out of the valley, rather than to be trapped by the obstructing hills.

Proximity of diversified manufacturing units capable of producing harmful and injurious admixtures within an industrial area leads to disturbance in residential areas in the vicinity of the units. Planning and development of neighbourhood residential areas is, therefore, inevitable in the industrial sector. Consequently, siting of industrial units, manufacturing a diversified range of products with different emission levels, with regard to the residential areas, assumes prime importance. State agencies have the responsibility for air pollution control implementation plans. These agencies have the responsibility for air pollution control implementation plans. These agencies prepare meteorological summaries, emission inventories, episode plans, air quality standards, emission standards and other specific components of the data banks necessary to prepare an implementation plan. Statistics gathered and assembled by the states or by the consultants working for them, will provide a working tool for those considering selection of an industrial site.

8.8.4 Ecological Factors in Site Selection—Water and Land Pollution

A team of chemists, ecologists, engineers, a bacteriologist, a protozoologist and a sanitary engineer can be employed to carry out an aquatic site selection survey to study the chemical, physical and biological water quality, upstream and downstream of the potential site location. Such a study helps not only in determining the waste assimilative capacity, but also determines the pre-existence of man-made or natural loads on the receiving system. These data are particularly useful where receiving systems are already polluted from other waste discharge and it is essential to establish the presence of both man-made and natural waste in the beginning and then avoid blame for pollution damage after the commencement of the plant. Based on this survey the ecological risks and benefits of each site can also be studied.

A new site should be chosen so that the receiving capacity of the water system is adequate, because the effluent standards limit the quantity of waste that can be discharged irrespective of the assimilative capacity of the system. Industrial plants also should not be located near the areas where there is a possibility of aquatic life being affected.

Predictive assays

Predictive assays can provide us with valuable information in determining the possible environmental impact of an industrial plant on the proposed site. Latest developments in laboratory techniques have made it possible to predict upto a certain degree of accuracy, the possible consequences of introducing various waste materials into the receiving system. Simulated wastes should be used to predict their ecological consequence at each site being considered.

Before the final design of the plant is complete, some bioassays should be carried out utilizing a simulated plant waste which would be similar in quality to the anticipated operating waste under the worst possible conditions.

An industry, which is discharging heavy metals as waste should consider location in an area where the hardness of water is high, because a buffer zone which prevents acute or chronic effects caused by its waste discharge is created. No longer can an industry be solely concerned with determining the acute toxicity of its waste products. The increased sophistication of bioassay procedures indicates that acute toxicity determination using lethality as an end sub-lethal end points. These bioassays provide us with information, about the 'biologically safe concentration' of various toxic subtances and these points should be considered during site selection.

Simulation techniques

Scale models can be used to provide valuable ecological information concerning plant site selection. Simulation techniques using scale models become important when more intensive use is made of finite space available for setting up more plants. A protective measure to be used to prevent major ecological and environmental problems is to simulate prospective uses in scale or laboratory models. These models need not be very expensive and can be used to generate data which would be useful in preventing large-scale mistakes. Physical and mathematical tables can also give information in ecologically evaluating a proposed plant site.

Construction surveys, predictive bioassays and simulation techniques help us gather valuable information for a new plant site location. This information is useful in determining waste assimilative capacity, predicting potential toxicity problems and establishing the credibility of an industry's environmental protection policy.

8.8.5 Employing the Services of an Environmentalist and Use of a Check List

Delays and even abandonment of chemical processing construction projects can be avoided when potential environmental problems are identified early. One way to supplement an early identification programme is to use the checklist well before preparation of the environmental impact assessment.

Usually at the site selection phase of the project, it is wise to use an experienced environmentalist's services. But too often due to project momentum, environmental guidance is deferred until *later*. This usually results in loss of credibility, project delays, prolonged negotiations and unnecessary frustration. All of these translate into increased cost.

The important aspects of a project environmental checklist which relate to industrial siting are (1) general site data, (2) climatology, (3) water pollution, (4) air pollution, (5) solid waste, (6) noise, (7) transportation, (8) construction, (9) aesthetics, and (10) regulatory agencies.

The checklist gives an early insight into the questions that should usually be tackled at the construction phase, like the control of turbid rain water run-off during construction and regulation of traffic which carries the construction workers and equipment. It also saves the entrepreneur from the shock and embarrassment of the questions on waste disposal control which the regulatory agencies may pose before him.

Of late, pollution control authorities all over the world are stricter in specifications of waste disposal facilities of new plants than old ones because at the rate at which new plants are being built they will soon outnumber the existing ones. Not only that, it is much easier to incorporate waste disposal systems in new plants right from the inception than to make changes after the plant commences production. Resiting of an existing plant due to waste disposal problems might involve a huge amount of money and resources. Not only the pollution control authorities but also the fear of public outcry have kept the industrialists constantly aware of the waste disposal problem while selecting a site.

The current rule of thumb for industrial locations seems to look for sites in industrial zones where other plants are already established. This suggests that more attention should be given to the establishment of industrial plants where wastes including those related to particulate emissions can be treated through some kind of centralised system much like our existing sewage works. These questions are forcing the industrial site location people, city and regional planners, and air management staff to get together and try to understand different point of views.

EXERCISES

8.1 Identify the pollutants in the fertilizer industry and write a brief note on their sources.

8.2 Discuss the environmental pollution problems of brewing industry.

8.3 Write an explanatory note on the treatment of liquid effluents from soap and synthetic detergent manufacture.

8.4 What are the wastes of a dairy unit? What treatment methods are available to treat their wastes?

8.5 What are the characteristics of wastes of a synthetic fibre industry? How is the waste stream of acetate rayon plant treated to mitigate its impact on the environmental pollution?

8.6 What are the sources of aqueous effluents in refineries and petrochemical plants?

8.7 Discuss briefly some of the ways in which agricultural wastes can be utilized to give rise to new products.

8.8 Write a note on the ecological factors in industrial site selection.

9

Solid Wastes

Solid waste, often considered as third after air and water pollution, is that material which comes up from various human activities and which is normally discarded as useless or unwanted. It normally consists of highly heterogeneous mass of discarded materials from the urban community as well as the more homogeneous accumulation of agricultural, industrial and mining wastes.

9.1 SOURCES OF SOLID WASTES

Sources of solid wastes in a community, in general, have a bearing on the land use and zoning. Although the development of any number of source classifications or categorizations is possible, the following categories have been found to be adequate and useful: (1) residential, (2) commercial, (3) institutional, (4) construction, (5) municipal services, (6) treatment plant sites, (7) industrial, and (8) agricultural. Typical waste generation facilities, activities or locations that go along with each of these sources are presented in Table 9.1.

Table 9.1 Sources of Solid Wastes within a Community

Source	*Typical facilities, activities or locations where wastes are fenerated*	*Types of solid wastes*
Residential buildings	Single family and multifamily dwellings, low, medium and high-rise apartments	Food wastes, paper, card board, plastics, textiles, leather wood glass, tin cans, aluminium and other metals, ashes, street leaves, special wastes including batteries oil and tyres
Commercial buildings	Stores, restaurants, markets, office buildings, hotels, motels, print shops service station and auto repair shops	Paper, card board, plastics, wood, food wastes, glass, metal wastes, ashes hazardous wastes
Institutional buildings	Schools, hospitals, prisons, governmental offices	As above for commercial (sources)
Industrial (non-process wastes) establishments	Construction, fabrication light and heavy manufacturing refineries, chemical plants, power plants demolition	Paper, card board, plastics, wood, food wastes, glass metal wastes, ashes, hazardous wastes

(*Contd.*)

Table 9.1 Sources of Solid Wastes within a Community (Contd.)

Source	*Typical facilities, activities or locations where wastes are generated*	*Types of solid wastes*
Municipal solid waste (This term normally is assumed to include all of the wastes generated in the community with the exception of waste generated from municipal services treatment plants, industrial processes and agriculture)	All of the aforementioned facilities, activities and locations	All of the aforementioned types of solid wastes
Construction and demolition activities	New construction sites, road repair renovation sites, razing of buildings broken pavements	Wood, steel, concrete and dirt
Municipal services (excluding treatment facilities)	Street cleaning landscaping, catch basin cleaning, parks and beaches other recreational areas	Special wastes, rubbish street sweepings, landscape and tree trimmings, catch-basin debris, general wastes from parks, beaches and recreational areas
Treatment plant sites (effluents)	Water, waste water and industrial treatment processes	Treatment plant wastes mainly composed of residual sludges and other residual materials
Industrial establishments (process wastes)	Construction, fabrication light and heavy, manufacturing refineries, chemical plants, power plants demolition	Industrial process wastes scrap materials, non-industrial wastes including food wastes rubbish, ashes demolition and construction wastes, special wastes, hazardous wastes
Agricultural activities	Field and row crops, orchards, vineyards dairies, feedlots, farms	Special food wastes, agricultural wastes, rubbish, hazardous wastes

As noted in Table 9.1, municipal solid waste is normally assumed to include all community wastes with the exception of wastes generated from municipal services, water and waste water treatment plants, industrial processes and agricultural operations. It is important to be alive to the fact that the definitions of solid waste terms and the categorization of solid waste differ greatly in the literature and in the profession.

9.2 CHARACTERISTICS OF SOLID WASTE

Important characteristics of solid waste include the composition, quantities and specific weight.

Composition

Typical data on the general characteristics of municipal solid waste are given in Table 9.2.

Table 9.2 Approximate Composition of Residential Solid Wastes

Component	*Percentage by weight*
Food waste	6–10
Paper products	30–40
Rubber, leather	4–10
Textiles, wood	8–15
Plastics	6–8
Glass and ceramics	7–9
Yard wastes	16–18
Rock, dirt, miscellaneous	1–3

Averages are subject to adjustment depending on many factors, time of the year, habits; education and economic status of the people; number and type of commercial and industrial operations, whether from urban or rural area, and location. Each community should be studied and actual weighing made to get representative information for design purposes.

Quantities

Various estimates have been made for the quantity of solid waste generated and collected per person per day. The quantity of municipal solid waste collected is estimated to be 6 lb/capita/day, of which about 3.5 lb is residential. Community wastes are not expected to exceed 1 ton/per capita year, with the emphasis being placed on source reduction (such as, less packaging) and waste recovery, and recycling (such as of paper, metals, cans and glass), the amount of solid waste requiring disposal is reduced. Recovery and recycling of hazardous wastes and toxicity reduction by substitution of less hazardous or non-hazardous wastes and materials should also be emphasized.

Specific weight

The volume occupied by solid waste under a given set of conditions is of importance as are the number and size or type of solid waste containers, collection vehicles and transfer stations. Transportation systems and land requirements for disposal are also affected. For example, the specific weight of loose solid waste will vary from about 100 to 175 lb/yd^3. Specific weights of various solid waste materials are given in Table 9.3.

Table 9.3 Weight of Solid Waste for Given Conditions

Condition of solid waste	*Weight* lb/yd^3
Loose solid waste at curb	125–240
As received from compactor truck at sanitary landfill	300–700
Normal compacted solid waste in a sanitary landfill	750–850
Well-compacted solid waste in a sanitary landfill	1000–1250
In a compactor truck	300–600
Shredded solid waste, uncompacted	500–600
Shredded solid waste, compacted	1400–1600
Compacted and baled	1600–3200
In incinerator pit	300–550
Brush, dry leaves, loose and dry	80–120
Leaves loose and dry	200–260
Leaves, shredded and dry	250–450
Green grass, compacted	500–1100
Green grass, loose and moist	350–500
Yard waste, as collected	350–930
Yard waste, shredded	450–600

The variabilities in the aforementioned data are due to variations in moisture content.

9.2.1 Commercial and Household Hazardous Waste

The contamination of ordinary municipal waste by commercial and household hazardous wastes has intensified the potential problems associated with the disposal of municipal waste by landfill, incineration and composting. Based on a number of past investigations, the quantity of hazardous waste typically represents less than about 0.5 per cent of the total waste generated by households. Typically, batteries and electrical items and certain cosmetics accounted for the largest amount.

From a practical standpoint, it would seem that more can be accomplished by identifying and prohibiting disposal of commercial hazardous waste with the municipal solid waste. The minimal household hazardous waste could, with education and municipal cooperation, be disposed of by voluntary actions. These could include periodic community collections and provisions of central guarded depositories. Many communities have established ongoing programmes for the collection of household hazardous waste. The quantity of waste can be expected to decrease as old stockpiles are discarded. Restricting sales and promoting development and substitution of non-hazardous household products would also be indicated. It should be noted that a large fraction of the household hazardous waste ends up in the sewer.

9.2.2 Construction and Demolition Debris

Construction and demolition debris comprises uncontaminated solid waste resulting from the construction, remodelling, repair and demolition of structures and roads and uncontaminated solid waste consisting of vegetation from land cleaning and grubbing, utility line maintenance, and seasonal and storm related cleanup. Such wastes include, but are not limited to, bricks; concrete and other masonry materials, soil; wood; wall coverings; plaster; drywall; plumping fixtures; non-asbestos insulation; roofing shingles; asphaltive pavement; glass; plastics—that are not sealed in a manner that conceals other wastes; electrical wirings, and components containing no hazardous liquids and metals that are incidental to any of the above.

Solid wastes are not only construction and demolition debris even if resulting from the construction remodelling, repair and demolition of structures, roads and land clearing, but they also include asbestos waste, garbage, corrugated container board, electrical fixtures containing hazardous liquids such as fluorescent light ballasts or transformers, carpeting, furniture, appliances, tyres, drums and containers, and fuel tanks. Specifically excluded from the definition of construction and demolition debris is solid waste (including what otherwise would be construction and demolition debris) resulting from any processing technique, other than that employed at a construction and demolition processing facility, that makes individual waste components unrecognizable, such as shredding and pulverizing.

Some of these materials, such as bricks, rocks, wood and plumbing fixtures can be recycled. However, care must be exercised to ensure (by monitoring each load) that hazardous materials, such as those mentioned above are excluded and that fire, odour and groundwater pollution is prevented. Engineering plans and reports, hydrogeologic report, operation and maintenance reports, and permits from regulatory agency are usually needed.

9.2.3 Collection of Special Wastes

In every community a number of wastes materials are collected separately from residential and commercial solid wastes. Special wastes include medical wastes, animal wastes, waste oil, and old used tyres.

Medical wastes—infectious and pathological

Medical waste is defined as *any solid waste which is generated in the diagnosis, treatment or immunization of human beings or animals, in research pertaining thereto, or in the production or testing of biologicals.*

Infectious waste usually comes from a medical care or related facility. It includes all waste materials resulting from the treatment of a patient on isolation (other than patients on reverse or protective isolation), renal dialysis, discarded serums and vaccines, pathogen contaminated laboratory wastes, animal carcasses used in research, and other articles that are potentially infectious such as hypodermic and intravenous needles.

Regulated medical waste, includes the following waste categories: cultures and stock of infectious agents and associated biologicals, human blood and blood products, pathological wastes, used needles, syringes, surgical blades, pointed and broken glasses and contaminated animal carcasses. The potential hazards associated with the handling of infectious waste necessitate certain precautions. Infectious waste needs to be segregated at the source, and clearly colour coded and marked. The packaging is expected to maintain its integrity during handling, storage and transportation with consideration of the types of materials packaged. The storage time should be minimal (treated within 24 hours), the packaging should be moisture proof, puncture resistant, rodent and insect proof, and the storage places and containers are transported in closed leak-proof trucks or dumpsters. It must be kept separate from regular trash and other solid wastes everytime. Health care workers and solid waste handlers must be cautious.

Most infectious wastes can be treated for disposal by incineration or auto claving. The residue can be disposed of in an approved landfill. Liquids may be chemically disinfected; pathological wastes may also be buried if permitted or cremated; and blood wastes, may under controlled conditions be discharged to a municipal sanitary sewer, provided secondary treatment is employed. Infectious waste may also be rendered innocuous by shredding, disinfection (by sodium hypochlorite), thermal inactivation and gas-vapour treatment. Infectious waste is only one component of medical waste.

In all cases, local, state and central regulations should be followed closely. Public concerns and fears associated with the possible detrimental effects need to be taken into consideration.

Animal wastes

It is likely that animal wastes contain disease organisms causing different types of illnesses. Manure contaminated with foot-and-mouth disease virus should be buried in a controlled manner or otherwise properly treated. The excretion from sick animals should be stored 7 to 100 days or as long as is necessary to ensure destruction of the pathogen, depending on temperature and moisture. Dead animals are best disposed of at an incinerator or rendering plant or in a separate area of a sanitary landfill. Large numbers may be buried in a special trench with due consideration to protection of groundwaters and surface waters if approved by the regulatory authority.

Waste oils

Large quantities of waste motor and industrial oils find their way into the environment as a result of accidental spills, oiled roads, oil dumped in sewers and on the land, and oil deposited by motor vehicles. Used oils contain many toxic metals and additives that add to the contamination received by sources of drinking water, aquatic life and terrestrial organisms. The lead content of oils is of particular concern. In the United States for example, the industrial facilities, service stations and motorists produce around 120 crores gallons of used oil. Approximately 60 per cent is reprocessed and used for fuel, but air pollution controls are required. Around 25 to 30 per cent is re-refined and reused as a lubricant. The remainder is used for road oil, road dust control and stabilization and other unacceptable uses. Refined oil is so classified when it has physical and chemical impurities are removed, and when by itself or blended with new oil or additives, is substantially equivalent or superior to new oil intended for the same purposes.

Old used tyres

The dumps can cause major fires and release many hazardous chemicals, including oils contributing to air and ground water contamination. Tyres collect rainwater in which mosquitoes breed, and provide harborage for rats and other vermin. Tyres are not suitable for disposal by landfill, but may be acceptable of shredded or split, although recycling is preferred and may be needed.

9.3 DESCRIPTION OF THE FUNCTIONAL ELEMENTS OF A SOLID WASTE MANAGEMENT SYSTEM

A short description of the functional elements of a solid waste management system is given in Table 9.4.

Table 9.4 Functional Elements of a Solid Waste Management System

Functional element	*Description*
Generation of waste	Those activities that go along with the ones in which materials are identified as no longer being of value and are either thrown away or gathered together for disposal.
Onsite handling, storage and processing	Those activities that are involved in handling, storage and processing of solid wastes at or near the point of generation.
Collection	Those activities associated with the gathering of solid wastes and the hauling of wastes after collection to the location where the collection vehicle is emptied.
Transfer and transport	Those activities with (1) the transfer of wastes from the smaller collection vehicle to the larger transport equipment, and (2) the subsequent transport of the wastes, usually over long distance, to the place of disposal.
Processing and recovery	Those techniques, equipment and facilities employed both to improve the efficiency of the other functional element and to recover usable materials; conversion products or energy from solid wastes.
Disposal	Those activities that are engaged in ultimate disposal of solid wastes, including those wastes collected and transported directly to a landfill site; semi-solid wastes (sludge) from effluent treatment plants, incinerator residue, compost or other materials from the various solid waste processing plants that are not of any further use.

Adapted from H.S. Peavey, D.R. Rowe and G. Tchobanoglous, *Environmental Engineering*, McGraw Hill, New York, 1985.

9.4 DISPOSAL OF SOLID WASTE

When we think of solid waste, often the problems and processes which immediately come to our mind are the ones associated with municipal disposal. These substances are, no doubt, the products of various industries, but their disposal is not directly the responsibility of the industry which generated them. The industries have their own problems—their own types of solid waste which must be disposed of.

The ideal solution to this problem, economically, energetically and environmentally would be to recover and reuse many of the solid wastes. Attempts have been and are being made by many industries with varying degrees of success. As with most industry related issues, pollution control and economics are inseparable. The primary responsibility of an industry official is to safeguard his organization's interest and financial position. If he does not do this, the company's prosperity and growth will be affected. The financial incentives may be to avoid fines, court cases or expensive enforcement squabbles or it may be by-product recovery, but unless the incentive is made available, little progress will be made.

There are several types of solid wastes an industry may have to deal with. There are of course, the sludges that originate from water treatment. In addition to these, there are process solids, like collected particulates and slags. Many of these are composed of various minerals, though their form and actual chemical composition may vary significantly, depending upon the source.

A very large quantity of solid wastes are generated annually by the minerals processing industries alone. These ores contain only a small percentage of the desired metallic elements such as copper, iron, gold or silver; thus the spent ores or tailings accumulate very rapidly. Tailings are typically composed of silica (sand) and various silicates and carbonates of calcium, magnesium and possibly aluminium. These tailings, often consisting of very fine particles are piled near the processing plant, creating a nuisance because of their size and physical instability; plant growth often must be encouraged to stabilize the piles. Few recovery methods have been found to be economically feasible for many of these wastes.

Many industries also generate fly ash and the coal ash which result from power generation besides a few other sources. Fly ash is one substance on which a lot of research has been carried out, looking for more and better ways for its utlization. Fly ash has been used fairly widely as an additive for cement. The fine fly ash can be added to the ground cement clinker, increasing for some purposes the desirable cement characteristics. Unfortunately, this is not the solution for fly ash disposal. Many cement companies have found the cement not to be marketable, primarily due to its dark colour.

Fly ash is also employed as one of the raw materials for the production of sintered lightweight aggregate, such as the one used in concrete blocks and other precast forms. It can be employed as a filter in asphalt pavings, as a soil stabilizer for embankments, as raw material for bricks and in the bases for road beds. Recent investigations have shown its feasibility as plastic filler.

9.4.1 Sanitary Landfills

In case the various solid wastes need to be land filled, they must usually be disposed of in sanitary landfills. A sanitary landfill is one in which the wastes are deposited, spread into layers, compacted and covered daily with earth. There are two basic methods by which this can be done. They are area and trench. Area landfills are suitable where land depressions exist, such as canyons, ravines or valley. The waste is simply spread, compacted, covered and then compacted again until the area is filled, after which a thicker layer of earth is spread over the whole area. In the case of trench landfill, the trench is first cut into the ground and then the spreading-compacting sequence is carried on, as before. This method is suitable for flat or slightly rolling land if the water table is low enough and the soil is deep enough for trenching.

Similar landfills are particularly important in disposing of biodegradable wastes where harmful gases and microbes may form; often they are not used for mineral wastes, such as cement dust. The anaerobic decomposition of organic materials leads to the production of gases, such as methane (CH_4) and hydrogen sulphide (H_2S) as well as CO_2 and N_2. CH_4 can be an explosion hazard if high concentrations accumulate in enclosed areas. H_2S smells similar to rotten eggs, and is toxic in higher concentrations. CO_2 can dissolve in ground water, forming carbonic acid, which will dissolve the rocks and mineralize the water. To minimize the effects, the gases can be vented from deep in the landfill to the atmosphere. This will bring down the possible hazards of CH_4 and H_2S by not allowing any build-up of the gases.

Leaching is another problem that is associated with landfill sites. Groundwater or surface water which infiltrates the solid waste can dissolve some of the waste substances or form suspensions of solid matter and microbial waste products. The leachate can then percolate through the soil to either the underground water systems or to nearby lakes and streams. This problem is not confined to biodegradable wastes; mineral wastes can also have major leaching problems; the leachate often being either acidic or basic, depending on the source.

Leaching can be minimized by lining the site with fine grained soils, clays or synthetic liners, locating the site in areas far from natural water systems, and covering the site with properly graded, nearly impervious soils such as clays. Other areas must have sumps to collect the leachate or special drainage systems.

There is yet another major problem associated with landfill sites, that is simply the scarcity of suitable sites. In many areas, suitable sites are not located at reasonable distances from the various manufacturing units or plants. As more industries dispose of their wastes by landfill, the number of suitable sites gets smaller and smaller and the costs for disposal rise. Some types of waste materials will eventually decompose, providing areas which can be reclaimed for recreational areas; other types of wastes, particularly mineral wastes do no generally produce such amenable areas. Many such areas are difficult to reclaim, particularly if future construction is desired. Recovery of the waste materials is currently the most economically favourable method of solid waste handling for many industries, such as cement industry. The probability is that the feasibility will increase as shortages loom in non-fuel mineral supplies, such as chromium, cobalt, manganese, titanium, aluminium, tin, strontium, sheet mica and similar minerals.

9.4.2 Disposal of Toxic Substances

Recently much interest has been shown in 'hazardous' waste disposal. Hazardous wastes are those which are ignitable, corrosive, chemically reactive or toxic. Many wastes from plants are toxic, and once collected, the plant is faced with the question of final disposal. In the past, such substances were often simply drummed, then disposed of in a landfill area. This was generally unsatisfactory and inadequate. Several approaches are presently adopted towards appropriate disposal techniques.

Secured landfills, those which can produce no toxic leachate are available in some locations and can sometimes be employed to dispose of toxic wastes. These landfills are located in thick, natural clay deposits, or employ specially designed liners and leachate collection systems. Often, however, appropriate sites are too far from the plant which generates the waste, to make them an economically viable disposal method.

Another method involves the use of a liquid waste incinerator. This type of incinerator generally consists of a waste-feed system to deliver the liquids, slurries, sludges and/or solids to thermal destruction chambers. There are usually two thermal chambers, one for oxidation and a separate one for gasification. If particulate matter is a product of the combustion, venturi scrubbers or baghouses are also employed. Typical incinerators operate at 1000–2400° F with residence times of the waste of 1–3 seconds or perhaps longer. Successful test *burns* have been made on ethylene dichloride, chlorotoluene and PCB's. Inclusions of heat recovery units can make the systems economically more feasible.

Another potentially useful disposal technique for hazardous wastes comprises combined stabilization and solidification of these wastes. These techniques are widely employed in Europe and Japan currently. If properly carried out, the hazardous waste can then be disposed of in any well-designed landfill. Though more expensive, the encapsulating costs can often be offset by lesser transportation costs.

There are three steps involved in the process (1) analysis, (2) stabilization, and (3) solidification.

The analysis is to determine the toxic components, whether the expected ones are trace impurities or by-products of the process. Biological screening to test for potential mutagenicity is also useful.

Stabilization converts the toxic components to stable forms that will resist leaching. Adjustment of the pH is often critical, for solidification processes are frequently pH-sensitive. Stabilization of heavy metals typically involves the addition of sulphur-containing chemicals.

A wide variety of solidification techniques are available. These include portland cements, asphalt, pozzalanic (natural) cements, silicates soil-binding agents and ion-exchange resins. Two different processes are actually possible; encapsulation, where the solidifying agent physically surrounds the waste particles or chemical fixation, where a chemical reaction takes place between the waste and a solid matrix. Asphalt and portland cement can be used for encapsulation and ion-exchange resins are among the materials employed in chemical fixation.

Much interest is being generated currently in ocean incineration of toxic chemicals. This was originally developed in Europe, but the EPA is now very interested in possibly having many of the toxic wastes disposed of, in this manner in the near future.

For incineration, the wastes must have a minimum calorific value of 5800 BTU/lb high-grade coal, for comparison, is in the range of around 12000–14000 BTU/lb.

The technique to be employed is governed by the nature of waste. In the years to come, it is very likely that processes will be changed to generate fewer toxic wastes and more *wastes* will be converted to useful by-products. But with the growing stockpiles of indispensable wastes and residues and increased restrictions on transportation of hazardous substances, new disposal techniques will be increasingly important.

9.5 ON-SITE HANDLING AND STORAGE OF SOLID WASTE

Where solid waste is temporarily stored on the premises between collections, an adequate number of suitable containers should be provided.

9.5.1 Low-rise Residential Area

To a large extent, the type of container used for the collection of residential solid waste will depend on the type of collection service provided and whether source separation of wastes is employed.

Low and medium-rise apartments

Large containers located in enclosed areas are used most commonly for low and medium rise apartments. In most applications, separate containers are provided for recyclable and commingled non-recyclable materials.

Curbside collection service is common for most low- and medium-rise apartments. Typically, the maintenance staff is responsible for transporting the containers to the street for curbside collection by manual or mechanical means. In many communities, the collector is responsible for transporting containers from a storage location to the collection vehicle, where large containers are employed, the contents of the containers are emptied mechanically using collection vehicles equipped with unloading mechanisms.

9.5.2 High-rise Apartments

In the case of high-rise buildings (higher than seven stories), the most common methods of handling commingled wastes involve one or more of the following: (1) wastes are picked up by building maintenance personnel from the various floors and taken to the basement or service area, (2) wastes are taken to the basement or service area by tenants, (3) wastes, usually bagged, are placed by the occupants in a waste chute system employed for the collection of commingled waste of a centralized service location. Typically, large storage containers are located in the basements of high rise apartments. In some locations, enclosed ground level storage facilities have been provided. In some of the more recent apartment building developments, especially in Europe, underground pneumatic transport systems have been used in conjunction with the individual apartment chutes.

9.5.3 Commercial and Institutional

Bulk containers or solid waste bins are recommended in situation where large volumes of solid waste are generated, such as hotels, restaurants, motels, apartment houses, shopping centres and commercial places. They can be combined to advantage with compactors in many instances. Containers should be placed on a level, hard cleanable surface in a lighted open area. The container and surrounding area must be kept clean, for the reasons mentioned earlier. A concrete platform provided with a drain to an approved sewer with a hot-water faucet at the site to facilitate cleaning, is generally satisfactory.

9.5.4 Solid Waste Collection

Collection costs have been estimated to represent about 50 to 70 per cent of the total cost of solid waste management, depending upon the disposal method. In view of the fact that the cost of collection represents such a large percentage of the total cost, the design of collection systems must be considered carefully.

9.5.5 Type of Service

The type of collection service provided will depend on the community solid waste management programme. Typical examples of the types of collection service provided for the collection of commingled, and source separated and commingled wastes are presented in Table 9.5.

Table 9.5 Typical Collection Services for Commingled, and Source Separated Solid Waste (The Method of Waste Preparation for Collection is Often Selected for Convenience and Efficiency of Collection Services and Subsequent Materials Processing Activities)

Preparation method for waste collected	*Type of service*
Commingled wastes	Single collection service of large container for commingled household and yard waste
	Separate collection service for 1. commingled household waste, and 2. containerized yard waste.
	Separate collection service for 1. non-containerized yard waste.
Source-separated and commingled waste	Single collection service for a single container with source-separated waste put in plastic bag and commingled household and yard wastes.
	Separate collection service for 1. source-separated waste put in a plastic bag and commingled household waste in the same container, and 2. non-containerized yard wastes.
	Single collection service for source separated and commingled household and yard wastes making use of a two-compartment container; separate collection service for 1. source-separated and commingled household wastes utilizing a two-compartment container, and 2. containerized or non-containerized yard waste.
	Separate collection service for 1. source separated waste, and 2. containerized commingled household and yard waste.
	Separate collection service for 1. source separated waste 2. commingled household waste, and 3. containerized yard wastes.
	Separate collection service for 1. source separated waste 2. commingled household waste, and 3. non-containerized yard wastes.

It should be mentioned in this context that numerous other variations in the service provided have been developed to meet local conditions. Besides, the routine collection services, presented in Table 9.5, annual or semiannual special collections for appliances, tyres, batteries, paints, oils, pesticides, yard wastes, glass and plastic bottles and *spring cleaning* have proven to be an appreciated community service while at the same time providing and ensuring environmental protection.

9.5.6 Collection Frequency

The frequency of collection will have a bearing on the quantity of solid waste, time of year, socio-economic status of the area served, and municipal or contractor responsibility. In residential areas, the frequency is twice a week solid waste collection during the most part of the year. In business areas, solid waste including garbage from hotels and restaurants, should be collected daily. Depending on the type of collection system, the containers used for the on-site storage of solid waste should be either emptied directly in the collection vehicle or hauled away, emptied and returned or replaced with a clean container. Solid waste transferred from on-site storage containers will invariably cause spilling, with resultant pollution of the ground and attraction of flies. If other than curb pickup is provided, such as backyard service, the cost of collection will be quite high. Nevertheless, some property owners are willing to pay for this extra service. Bulky wastes should be collected every two to three months.

9.5.7 Types of Collection Systems

Solid waste collection systems may be classified from several perspectives, such as the mode of operation, the equipment employed and the types of wastes collected. Collection systems can be categorized, according to their mode of operation into the groups (1) hauled container systems, and (2) stationary container systems. The principal operational features of these two systems are given here.

Hauled container systems

These are collection systems in which the containers employed for the storage of wastes are hauled to a materials recovery facility, transfer station or disposal site, emptied and returned to either their original location or some other location. There are three main types of vehicles employed in hauled container systems (1) hoist truck, (2) tilt-frame container, and (3) truck tractor trash-trailer.

Hauled container systems are ideally suited for the removal of wastes from sources where the rate of generation is high owing to the fact that relatively large containers are employed. The use of large containers eliminates handling time as well as the unsightly accumulations and unsanitary conditions associated with the use of numerous smaller containers. Another advantage of hauled container systems is their flexibility. Containers of different sizes and shapes are available for the collection of all types of wastes.

Stationary container systems

In the stationary container system, the containers employed for the storage of waste remain at the point of generation, except when they are moved to the curb or other location to be emptied. Stationary container systems may be used for the collection of all types of wastes. The systems vary according to the type and quantity of wastes to be handled as well as the number of generation points. There are two main types: (1) systems in which manually loaded collection vehicles are used, and (2) systems in which mechanically loaded collection vehicles are employed.

The major application of manual loading collection vehicles is in the collection of residential source-separated and commingled wastes, and litter. Manual loading is used in residential areas where the quantity picked up at each location is small and the loading time is short. In addition, manual methods are used for residential collection because many individual pickup points are inaccessible to mechanized mechanically loaded collection vehicles. Special attention must be given to the design of the collection vehicle intended for use with a single collector. At present, it appears that a side-loaded compactor, equipped with a standup right-hand drive, is best suited for curb and alley collection.

9.5.8 Personnel Requirements

In most hauled container systems, a single collection driver is used. The collector driver is responsible for driving the vehicle, loading full containers on to the collection vehicle, emptying the contents of the containers at the disposal site (or transfer point) and re-depositing (unloading) the empty containers. In some situations, for safety reasons, both a driver and helper are employed The helper is responsible for attaching and detaching any chains or cables employed in loading and unloading containers on and off the collection vehicle; the driver's job is to operate the vehicle. A driver and helper should always be employed where hazardous wastes are to be handled.

Labour requirements for mechanically loaded stationary container systems are basically the same as for hauled container systems. In situations where a helper is used, the driver often assists the helper in bringing loaded containers mounted on rollers to the collection vehicle and returning the empty containers.

In situations, wherein the containers to be emptied must be rolled (transferred) to the collection vehicle, from inaccessible locations like in congested cities, commercial areas, driver and two helpers are used. In stationary container systems, where the collection vehicle is loaded manually, the number of collectors varies from one to four, in most cases, depending on the type of service and the collection equipment. While the aforementioned crew sizes are representative of current practices, there are many exceptions. In many metropolitan areas, multiperson crews are employed for curb service as well as for backyard carry service.

9.5.9 Collection Routes

To decide on the most economical route for collecting solid wastes and hauling to disposal points can be difficult. Interrelated variables like labour costs, crew size, union restrictions, frequency of collection, distance (travel time) to disposal and the performance and annual costs of various types of waste-handling, will govern the choice. Alley or curb pick-up are the two basic types of local collection. Alley pickup, where possible has certain merits, a homeowner does not have to set out cans, scheduled service is not needed, there is no interference with street traffic and both sides of the alley can be served at the same time with minimum walking. In some municipalities setting out and/or returning the cans to the front or backyard is not provided, but most residents will forego these benefits owing to the additional cost. System analysis can be applied to waste routing problems, but for local collection a trial and error approach is commonly employed.

In situations where wastes from different collection districts can be directed to several possible points, it may be difficult to decide which wastes should go to which location for the most economical solution. Waste allocation problems of this type are normally solved by linear programming. Where a programme or a computer is not available, approximate methods can be employed for ascertaining a solution close to optimal.

9.5.10 Transfer and Transport

The urban localities around cities have been spreading, leaving fewer nearby acceptable solid waste disposal sites. The shortage of acceptable sites has led to the construction of incinerators, resource recovery facilities or processing facilities in cities and their outskirts or the transportation of wastes to longer distances to new landfill disposal sites. However, as the distance from the centres of solid waste generation increases, the cost of direct haul to a site increases. A *distance* is reached (in terms of cost and time) when it becomes less expensive to construct a transfer station or incinerator at or near the centre of solid waste generation where wastes from collection vehicles can be transferred to large tractor-trailers for haul to more distant disposal sites. The ideal situation would be the location of the transfer station at the central point of the collection service area.

The cost factor analysis of transfer operations

A comparison of direct haul versus the use of a transfer station and haul for various distances is useful in making an economic analysis of potential landfill sites. The transfer station site development, transportation system and social factors involved in the selection of site, should also be taken into account while making the comparison.

If the cost of disposal by sanitary landfill is added to the cost comparison, the total relative cost of solid waste transfer, transportation and disposal by sanitary landfill can be compared to the corresponding cost for incineration, incase that is on option. The relative cost of incineration with the cost of landfill for various haul distances and a given population has to be weighed. Based on past experience, a direct haul distance (one-way) of 40 to 50 kilometres is about the maximum economical distance, although longer distances are not uncommon, in situations where other options are unacceptable or cannot be implemented for a number of reasons, including cost. A cost comparison between incineration and transfer, and haul to land fill needs to be carried out.

Types of transfer stations

Transfer stations are employed to accomplish transfer of solid wastes from collection and other small vehicles to larger transport equipment. Depending on the method used to load the transport vehicles, transfer stations can be categorized into two general types (1) direct-load, and (2) storage load. Combined direct-load and discharge load transfer stations have also been developed. Transfer stations may also be categorized with respect to throughput capacity (the quantity of material that can be transferred and hauled; small, less than 100 tons/day; medium, between 100 and 500 tons/day and large, more than 500 tons/day).

Direct-load transfer stations

When direct-load transfer stations are used, the wastes in the collection vehicles are emptied directly into the vehicle to be used to transport them to a place of final disposition or into facilities employed to compact the wastes into transport vehicles or into waste bales that are transported to the disposal site. In some solutions, the waste may be emptied onto an unloading platform and then pushed into the transfer vehicles after recyclable materials have been removed. The volume of waste that can be stored temporarily on the unloading platform is often defined as the surge capacity or the emergency storage capacity of the station.

Storage-load transfer station

In the case of the storage-load transfer station, wastes are emptied directly into a storage pit from which they are loaded into transport vehicles by different types of auxiliary equipments. The difference between a direct load and a storage load transfer station is that the latter is designed with a capacity to store waste, say for one to three days.

The various types of transfer stations employed for municipal solid wastes are presented in Table 9.6.

Table 9.6 Types of Transfer Stations Used for Municipal Solid Waste

Type	*Description*
Direct load transfer stations	
Large and medium capacity direct-load transfer station without compaction	Wastes to be transported to landfill are loaded directly into large open-top transfer-trailers for transport to landfill site.
Large and medium capacity direct-load transfer stations with compactors	Wastes to be transported to landfill are loaded directly into large compactors and compacted into specially designed transport trailers or into bales, which are then transported to landfill site.
Small capacity direct-load transfer stations	Small capacity transfer stations are used in remote and rural areas. Small capacity transfer stations are also employed at landfills as a convenience for residents who wish to haul wastes directly to landfill site.
Storage-load transfer station	
Large capacity storage-load transfer stations without compaction	Wastes to be transported to a landfill are discharged into a storage pit where they are pulverized before being loaded into open trailers. Waste is pulverized to to bring about a reduction in the size of the individual waste constituents to accomplish more effective utilization of the transfer trailers.
Medium capacity storage-load transfer station with processing and compaction facilities	Wastes to be transported to a landfill are discharged into a pit where they are further pulverized before being baled for transport to a landfill site.
Other types of transfer stations	
Combined discharge load and direct-load transfer stations	Wastes to be transported to a landfill can either be discharged on a platform or discharged directly into a transfer trailer. Wastes discharged onto a platform are typically sorted to recover recyclable materials.
Transfer and transport operations at Materials Recovery Facilities (MRFs)	Depending on the type of collection service provided, materials recovering and transfer operations are frequently combined in one facility. Depending upon the operation of the Materials Recovery Facility (MRF), wastes to be landfilled can be discharged directly into open trailors or into a storage pit to be loaded later into open-top trailers or baled for transport to a landfill site.

Adapted from G. Tchobanoglous, H. Theisen and S. Vigil (1993), *Integrated Solid Waste Management; Engineering Principles and Management Issues*, McGraw Hill, New York.

9.5.11 Vehicles for Uncompacted Wastes

Motor vehicles, rail rods and ocean going vessels are the principal means now employed to transport solid wastes. In addition to the above means, pneumatic and hydraulic systems have also been employed. In recent years, owing to their simplicity and dependability, open top semi-trailers have also been used for the hauling of uncompacted wastes from direct-load transfer stations. Another combination that has proven to be very effective and useful for uncompacted wastes is the track-trailer combination. In a few places, transport trailers employed for hauling solid waste over long distances are all of monocoque construction, where the bed of the trailer also serves as the frame of the trailer. Using monocoque construction allows greater waste volumes and weights to be hauled.

9.5.12 Criteria for Siting Transfer Station

A transfer station, resource recovery facility or processing facility should be located and designed with the same care as needed for an incinerator. Drainage of paved areas and adequate water hydrants for maintenance of cleanliness and fire control are equally important. Other criteria are landscaping, weigh scales and traffic odour, dust, litter and noise control. Rail haul and barging to sea also involve the use of transfer stations. They may include one or a combination of grinding, baling or compaction to enhance densities, thereby facilitating transportation effectiveness and efficiency.

9.5.13 Processing Techniques

Processing techniques are employed in solid waste management systems to bring out the following (1) improvement of the efficiency of solid waste disposal systems, (2) recovery of resources (usable materials), and (3) the preparation of materials for the recovery of conversion products and energy. Processes employed routinely to improve the efficiency of solid waste systems and to recover materials manually are considered in this section. Mechanical systems are also employed for the recovery of the materials. Important processing techniques used routinely (that is on a regular basis) in municipal solid waste systems, include compaction, thermal volume reduction (incineration) and manual separation of waste components. Factors that should be taken into account while evaluating on-site processing equipment are summarized in Table 9.7.

9.5.14 Mechanical Volume Reduction

Mechanical volume reduction is perhaps the most important factor in the development and operation of solid waste management systems. Vehicles which are equipped with compaction mechanisms are quite useful for the collection of most municipal solid wastes. To augment the useful life of landfills, wastes are compacted. Paper recycling is baled for shipping to processing centers. When compacting a broad range of municipal solid wastes, it has been observed that the final density (typically about 1100 kg/m^3) is essentially the same regardless of the starting density and applied pressure. This fact is quite important in evaluating the claims made by manufacturers of compacting equipment.

Table 9.7 Factors that Should be Considered in Evaluating On-site Processing of Solid Wastes

Factor	*Evaluation*
Capabilities	What will the device or mechanism do? Will its use be an improvement over conventional practices?
Reliability	Will the equipment in question perform its designated functions with little attention beyond preventive maintenance? Has the effectiveness of the equipment been demonstrated in use over a reasonable period of time for its reliability or merely predicted?
Service	Will capabilities for servicing beyond those of the local building maintenance staff be required occasionally? Are properly trained service personnel available through the equipment manufacturer or the local distributor?
Safety of operation	Is the proposed equipment reasonably fool proof so that it may be operated safely by tenants or building personnel with limited mechanical knowledge or abilities? Does it have adequate protection to discourage careless use?
Ease of operation	Is the equipment easy and convenient to operate by a tenant or by building personnel? Unless functions and actual operations of equipment can be performed easily, they may be ignored or "short-circuited" by paid personnel or by tenants.
Efficiency	Does the equipment perform efficiently and with a minimum of attention? Under most conditions, equipment that completes an operational cycle each time it is used, should be selected.
Environmental effects	Does the equipment pollute or contaminate the environment? Where possible equipment should reduce environmental contamination presently associated with conventional functions.
Health hazards	Does the device, mechanism or equipment create or amplify health hazards?
Aesthetics	Does the equipment and its arrangement offend the senses? Every effort should be made to reduce or eliminate offending sights, odours and noises.
Economics	What are the economic factors involved? Both first and annual costs must be assessed carefully. Future operation and maintenance cost must also been assessed carefully. All factors being equal, equipment produced by well-established companies having a proven record of satisfactory operation, should be given appropriate consideration.

Adapted from G. Tchobanoglous, H. Theisen and R. Eliassen (1977), *Solid Wastes Engineering—Principles and Management Issues*, McGraw Hill, New York.

9.5.15 Thermal Volume Reduction

The volume of municipal waste can be brought down by around 90 per cent by incineration. In the past, incineration was quite common. However, with more restrictive air pollution control requirements necessitating the use of expensive clean-up equipment, only a limited number of municipal incinerators are currently in operation. More recently, increased haul distances to available landfill sites and increased fuel costs have brought about a renewed interest in incineration and a number of new incinerator projects are now planned.

9.5.16 Manual Component Separation

The manual separation of solid waste components can be carried out at the source where solid wastes are generated, at a transfer station, at a centralized processing station or at the disposal site. Manual sorting at the source of generation is the best way to accomplish the recovery and reuse of materials. The number and types of components salvaged or sorted (e.g. cardboard and high-quality paper, metals and wood) have a bearing on the location, the opportunities for recycling and the resale market. It would be desirable if the residents separate newsprint, aluminium cans and glass on a voluntary basis. The separated components are placed at the curb for collection with a special vehicle.

9.5.17 Ultimate Disposal of Solid Waste

Disposal on or in the earth's mantle is the only viable method for the long-term handling of (1) solid wastes that are collected and are of no further use, (2) the residual matter remaining after solid wastes have been processed, and (3) the residual matter remaining after the recovery of conversion products and/or energy has been accomplished. The most commonly employed method of disposal for municipal wastes is landfilling. In the case of industrial wastes' disposal, land forming and deep-well injection, have been employed. Incineration in a strict sense is a processing method even though it is often considered a disposal method.

Landfilling with solid wastes

Landfilling involves the controlled disposal of solid wastes on or in the upper layer of the earth's mantle. Site selection and landfilling methods and operations, occurrence of gases and leachate in landfills and movement and control of landfill gases and leachate are the four important aspects to be considered in the implementation of sanitary landfills.

Criteria that must be taken into account while evaluating potential solid waste disposal sites are presented in Table 9.8.

Table 9.8 Criteria that Must be Considered in Evaluating Potential Landfill Sites

Criterion	*Remarks*
Available land area	Site is expected to have a useful life greater than 1 year (minimum period)
Haul distance	Will have appreciable impact on costs of operation
Soil conditions and topography	Cover material must be available at or near the landfill site
Surface water hydrology	Has has effect on drainage requirements
Geologic and hydrogeologic conditions	Probably the crucial factors in establishment of landfill site, especially with regard to site preparation
Climatologic conditions	There should be provisions for wet weather operation
Local environmental conditions	Noise, odour, dust, vector and aesthetic factors control are required
Ultimate use of site	Has an effect on long-term management for landfill site

Adapted from H.S. Peavey, D.R. Rowe and G. Tchobanoglous, *Environmental Engineering*, McGraw Hill, New York, 1985.

Methods and operations of landfilling

To harness the available area at a landfill site effectively, a plan of operation for the placement of solid wastes must be prepared.

A variety of operational methods have been developed, primarily on the basis of field experience. The principal methods employed for landfilling dry areas may be categorized as: (1) area, (2) trench, and (3) depression.

The area method is employed when the terrain is unsuitable for the excavation of trenches in which the solid wastes are to be placed. The filling operation is usually started by constructing an earthen levee against which wastes are placed in these layers and compacted. Each layer is compacted as the filling progresses until the thickness of the compacted solid wastes reaches a height of 2 to 3 metres. At that point of time and at the end of each day's operation, a 150–300 mm layer of cover material is placed over the completed fill. The cover material is hauled by truck or earth-moving equipment from adjacent land or from borrow-pit areas.

In some newer landfill operations, the daily cover material is omitted. A completed lift including the cover material is termed cell. Successive lifts are placed on the top of one another until the final grade called for in the ultimate development plan is reached. A final layer of cover material is used when the fill reaches the final design height.

The trench method of landfilling is ideally suited to areas where an adequate depth of cover material is available at the site and where the water table is well below surface. To start the process (for a small landfill) a portion of the trench is dug with a bulldozer and the dirt is stockpiled to from an embankment behind the first trench. Wastes are then placed in the trench, spread into thin layers and compacted. The operation continues until the desired height is reached. Cover material is got by excavating an adjacent trench or continuing the trench that is being filled. In large landfills, a dragline and one or more scrapers are employed to excavate a deep rectangular pit.

At locations where natural or artificial depressions exist, it is often possible to use them effectively for landfilling operations. Canyons, ravines, dry burrow pits and quarries have all been employed for this purpose. The techniques to place and compact solid wastes in depression landfills vary with the geometry of the site, the characteristics of the cover material, the hydrology and geology of the site and the access to the site. In a canyon site filling starts at the head end of the canyon and ends at the mouth, This practice prevents the accumulation of water behind the landfill. Wastes usually are deposited on the canyon floor and from there are pushed up against the canyon face at a slope of about 2 to 1. In this way, a high degree of compaction can be accomplished.

Owing to the problems associated with contamination of local ground waters, the development of odours, and structural stability landfills in wet areas are seldom employed. If wet areas, such as swamps and marshes, tidal areas, ponds and pits or quarries must be used as landfill sites, special provisions need to be made to contain or to eliminate the movement of leachate and gases from completed cells. Usually this is accomplished by first draining the site and then lining the bottom with a clay liner or other appropriate sealants. If a clay liner is used, it is important to continue the operation of the drainage facility until the site is filled to eschew the creation of uplift pressures that could cause the liner to rupture from heaving.

Land farming

Land farming is a waste-disposal method in which the biological, chemical and physical processes that take place in the surface of the soil are employed to treat biodegradable wastes. Wastes to be treated are either applied on top of the land which has been prepared to receive the wastes or injected below the surface of the soil.

When organic wastes are added to the soil, they are subjected simultaneously to the processes given hereinafter. (1) bacterial and chemical decomposition, (2) leaching of water-soluble components in the original wastes and from the decomposition products, and (3) volatilization of wastes and from the products of decomposition.

Factors that must be taken into consideration while evaluating the biodegradability of organic wastes in a land forming application include (1) composition of the waste, (2) compatibility of wastes and soil microflora, (3) environmental requirements including oxygen, temperature, pH and inorganic nutrients, and (4) moisture content of the solid-waste mixture.

Land forming is suitable in the case of wastes that contain organic constituents that are biodegradable and are not subject to significant leaching while the bioconversion process is taking place. For instance, petroleum wastes and oily sludges are ideally suited for disposal by land forming. A variety of other organic wastes with similar characteristics are also suitable. Properly managed land farming sites can be reused at frequent intervals without any adverse effect.

Deep-well injection

Deep-well injection for the disposal of liquid solid wastes involves injecting the waste deep in the ground into permeable rock formations (typically limestone or dolomite) or underground caverns. The installation of deep wells for the injection of wastes adopts the practices employed for the drilling and completion of oil and gas wells. To isolate and protect potential water supply aquifiers, the surface casing must be set well below such aquifiers and it should be cemented to the surface of the well as shown in Figure 9.1.

The drilling fluid should not be allowed to penetrate the formation that is to be employed for waste disposal. To prevent the occurrence of clogging of the formation, the drilling fluid is replaced with a compatible solution. Also in some situations, it may turn out to be necessary to subject the formation to acid treatment before injection of wastes commences.

Deep-well injection has been employed principally for liquid wastes that are hard to treat and dispose of by more conventional methods and for wastes which are hazardous. Chemical, petrochemical and pharmaceutical wastes are most commonly disposed of, with this method. The wastes may be liquid, gas or solid. The gases and solids are either dissolved in the liquid or are carried along with the liquid.

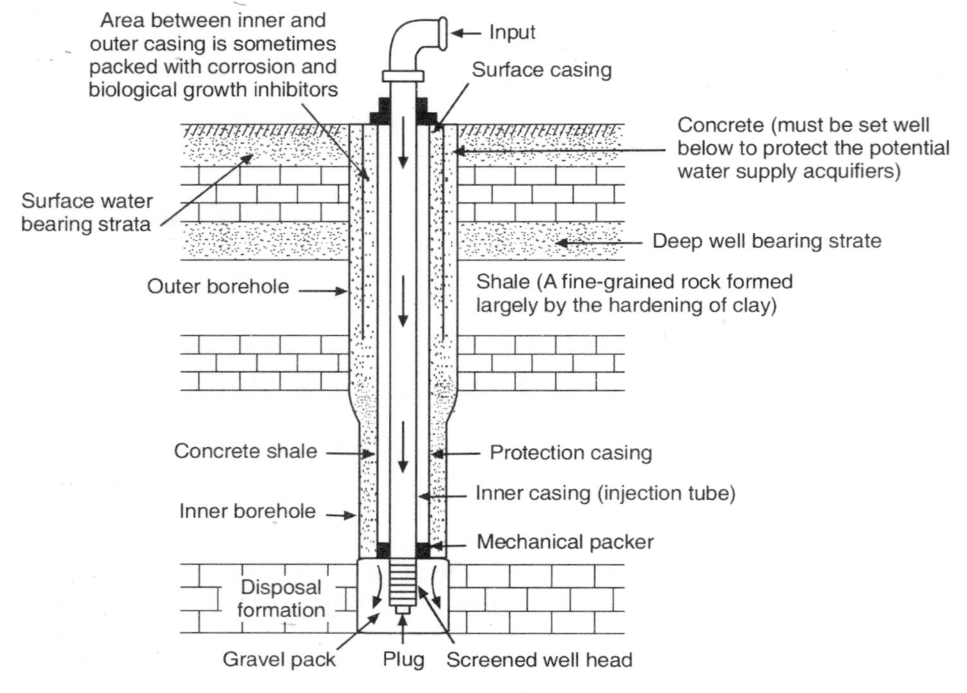

Figure 9.1 Well used for injection of liquid and solid wastes.

A comparison of the removal methods for solids from liquid effluents is given in Table 9.9.

Table 9.9 Comparison of Solids' Removal Methods from Waste Water Effluents

Method	*Application*	*Advantages*	*Disadvantages*
Settling (Sedimentation technique)	Readily settling solid particulates of 1 micron size; easily flocculated matter, design, scale-up possible from laboratory tests	Liquid effluent chemically treated and reused, low operating cost, abrasive solids handled	Large areas compared to volume of particulate required, pond or groundwater contamination, natural evaporation
Liquid cyclones	Concentration of solids	Low initial cost, low maintenance, capability to handle abrasive solids, low space requirements	High solids' filtrate an overflow, high power requirements
Continuous centrifuging	Sub-micron slurries	High collection efficiencies, low space requirements, large design variety	High capital and operating costs, susceptible to abrasion and corrosion
Continuous filtration technique	Where solids have some recovery value; are porous or are incompressible	Dewatered waste product, moderate space requirements	High initial cost, high maintenance and operational costs.

Adapted from N. Paul Cheremisinoff, and Richard A. Young, *Pollution Engineering Practice Handbook*, McGraw Hill, New York, 1990.

The biological and thermal processes for recovering the conversion products from solid wastes are presented in Tables 9.10 and 9.11 respectively.

Table 9.10 Biological Processes for the Recovery of Conversion Products from Solid Wastes

Process	*Conversion product*	*Preprocessing*
Aerobic conversion	Compost (soil conditioner)	Separation of organic fraction, particle size reduction
Alkaline hydrolysis	Organic acids	Separation of organic fraction, particle size reduction
Anaerobic digestion (in landfill)	Methane	Only placement in containment cells
Anaerobic digestion	Methane	Separation of organic fraction, particle size reduction
Fermentation (following acid or enzymatic hydrolysis)	Ethanol, single cell protein	Separation of organic fraction, particle size reduction acid or enzymatic hydrolysis to produce glucose

Adapted from H.S. Peavey, D.R. Rowe and G. Tchobanoglous, *Environmental Engineering*, McGraw Hill, New York, 1985.

Table 9.11 Thermal Processes for the Recovery of Conversion Products from Solid Wastes

Process	*Conversion product*	*Preprocessing*
Combustion (incineration technique)	Energy in the form of steam or electricity	None in mass fired incineratory; preparation of refuse derived fuels for suspension or semi-suspension firing in boilers
Gasification	Low-energy gas	Separation of the organic fraction, particle size reduction, preparation of fuel cubes or other RDF (refuse derived fuels)
Wet oxidation	Organic acids	Separation of organic fraction, particle size reduction, preparation of fuel cubes or other RDF
Steam reforming	Medium energy gas	Separation of the organic fraction, particle size reduction, preparation of fuel cubes or other RDF
Pyrolysis	Medium energy gas, liquid fuel, solid fuel	Separation of the organic fraction, preparation of fuel cubes or other RDF particle size reduction
Hydrogasification/ Hydrogeneration	Medium energy gas, liquid fuel	Separation of the organic fraction, preparation of fuel cubes or other RDF particle size reduction

Adapted from H.S. Peavey, D.R. Rowe and G. Tchobanoglous, *Environmental Engineering*, McGraw Hill, New York, 1985.

Some of the important factors that should be taken into account in the design and operation of solid waste landfills are given in Table 9.12.

Table 9.12 Important Factors that Must be Considered in the Design and Operation of Solid Waste Landfills

Factor	*Remarks*
Design access	Paved all weather access roads to landfill site: temporary roads to unloading areas
Cell design and construction	Will vary depending on terrain landfilling method and whether gas is to be recovered
Cover material	Maximize use of on-site earth materials approximately 1 m^3 of cover material will be required for every 4 to 6 m^3 of solid wastes; mix with sealants to control surface infiltration. In some designs, no intermediate cover is employed
Drainage	Drainage ditcher to divert surface water run-off; maintain, 1 to 2 per cent grade on finished fill to prevent ponding, has to be installed
Equipment requirements	Vary with size of landfills used
Fire prevention	Water on site: if non-potable, outlets must be marked clearly, proper cell separation prevents continuous burnthrough if combustion takes place
Protection of groundwater	Divert any underground springs; if need be, install sealants for leachate control, install wells for gas and groundwater monitoring
Land area	Area should be large enough to hold all wastes for a minimum of 1 year, but preferably for longer period of 5 to 10 years
Landfilling method	The terrain and available cover will have a bearing on the choice
Litter control	Use movable fences at unloading areas, crews should pick up litter at least once per month or as per the requirements
Operation plan	With or without the co-disposal of treatment plant sludges and the recovery of gas
Spread and compaction	Spread and compact wastes in 0.6 m (2 ft) layers
Unloading area	Keep small, generally under 30m (100 ft)
Operation	
Communications	Telephone for emergencies situations
Days and hours of operation	Usual practice is 5 to 6 days/week and 8–10 hours/day
Employee facilities	Rest rooms and drinking water should be provided
Equipment maintenance	A covered shed should be provided for field maintenance of equipments
Operational records	Tonnage, transactions and billing if a disposal fee is charged
Salvage	No scavenging; salvage should occur away from the unloading area no salvage storage on site
Scales	Essential for record keeping

Adapted from H.S. Peavey, D.R. Rowe and G. Tchobanoglous, *Environmental Engineering*, McGraw Hill, New York, 1985.

The Incinerator Institute of America has established a classification system for wastes based partly on content, and partly on BTU valve and moisture content.

Type 0—Trash: Mostly paper, cardboard, wood, etc. This type of waste contains 10 per cent moisture. 5 per cent incombustible solids and a heating value of 8500 Btu/lb.

Type 1: Rubbish paper, wood scrap, foliage, floor sweepings. This type of wastes contains 25 per cent moisture, 10 per cent incombustible solids and has a heating value of 6500 BTU/lb.

Type 2—Refuse: An even mixture of rubbish and garbage. This waste contains upto 50 per cent moisture, 7 per cent incombustible solids and has a heating value of 4300 Btu/lb.

Type 3: Garbage, animal and vegetable wastes. This waste contains upto 70 per cent moisture, upto 5 per cent incombustible solids and has a heating value of 2500 Btu/lb.

Type 4: Pathological waste, animal remains. This waste contains upto 85 per cent moisture, 5 per cent incombustible solids and a heating value of 1000 Btu/lb.

Type 5: Industrial gaseous liquid, or semi-liquid residues.

Type 6: Industrial solid wastes.

Wastes 5 and 6 are undefined because of the wide variety in material. The only information given is that the characteristics of each waste vary and must be determined individually in order to engineer an effective disposal system. However, more data are needed in a solid wastes classification system than the BTU values, moisture content and ash content in order to evaluate the waste.

Some of the important and frequently used techniques for processing solid wastes to recover materials and to prepare the waste for further processing are summarized in Table 9.13.

Table 9.13 Processing Techniques used to Recover Materials and to Prepare Wastes for Subsequent Processing

Process techniques employed	*Desired Function*	*Representative equipment and/or facilities and their applications*
Mechanical alteration of size and shape	Alteration of the size and shape of solid waste components	Equipment used to bring about a reductiion in the size of solid waste includes hammermills, shredders, roll crushers, grinders, chippers, jaw crushers, rasp mills and hydropulpers
Mechanical component separation	Separation of materials that may be recovered usually at a processing facility	Trommels and vibrating screens and processed wastes; and processed wastes; disc screens for processed wastes, zigzag, vibrating air, rotary air and air knife classifiers for processed wastes. Jig pneumatic sink/float inertial, inclined or shaking table floatation and optical sorting are employed to separate the light and heavy materials in solid wastes
Magentic and Electro-mechanical separation technique	Separation of ferrous and non-ferrous materials from processed solid wastes	Magnetic separation is used for ferrous materials, eddy current separation for aluminium, electrostatic separation for glass in wastes free of ferrous and aluminium scrap, magnetic fluid separation for non-ferrous materials from processed wastes.
Drying and dewatering	Moisture removal from solid wastes	Convection, conduction and radiation dryers have been employed for solid wastes and sludge, centrifugation and filtration are employed to dewater treatment plant sludge.

Adapted from H.S. Peavey, D.R. Rowe and G. Tchobanoglous, *Environmental Engineering*, McGraw Hill, New York, 1985.

EXERCISES

9.1 What are the sources of solid wastes within a community?

9.2 Describe the functional elements of a solid waste management system.

9.3 Discuss the on-site handling and storage of solid wastes.

10 Waste Minimization and Pollution Prevention

In order to make pollution control most effective and economical it should be thought of as an integral part of process design irrespective of whether the design is for a large sized new plant or for altering an existing facility. In this context, it may be added that the design engineer's contribution will be reflected not only in the selection of process and equipment, but also in minimizing pollution. If the engineer is not aware of this aspect, his efforts can create waste streams that may either be technologically impossible to treat or too expensive for such treatments. The management concerned should also extend the necessary support as this will go a long way in setting the tone and atmosphere for such waste treatment technology incorporated in initial design. The management can effect this by

1. Setting a policy for waste minimization that goes a longer way than just meeting governmental stipulations.
2. Funding only such of the projects that include waste minimization considerations.
3. By making waste treatment technology available to the process design group. This information could include knowledge of governmental requirement, environmental effects and waste treatment techniques as such.
4. By reviewing the process engineer's performance for minimizing waste streams in the same fashion that the project is evaluated and reviewed for other usual factors like costs, schedules and product quality.

10.1 PROCESS DESIGN CONSIDERATIONS

At the outset, all sources of pollution need to be evaluated during process design to ensure the minimum environment impact of the waste effluents. The effect of each source on the quantity and composition of waste streams, gaseous, aqueous or solids should be evaluated and facilities provided for satisfactory disposal. The sources listed out here may vary with the type of operations and activity in a manufacturing facility.

Process operations may give rise to continuous effluents and intermittent effluents. Continuous effluents are discharges that occur continuously in normal operation. Their minimum, average and maximum values of composition and quantity need to be evaluated.

Intermittent effluents, on the other hand, include wash up for equipment, turn around on batch systems, filter clean up, ion-exchange regeneration, etc. For such discharges a holding tank for averaging the flow should at least be provided.

The source can also a leaking equipment and piping and operating steps that discharge organics to the ground which eventually contaminate the run-off from the rainfall. Typical discharges include the ones from pump gland packing, sampling points and product transfer points.

Remedial measures would include collection and channelization of such waste streams to a central collection system for recovery or final treatment.

Pollution due to spills and upsets

Even during the design, attention must be focused on the elimination of operating actions and conditions resulting in abnormal waste loads. If these cannot be eliminated at least some provision for holding or for recycling may be required.

Startup

The startup procedures for getting a new system online must include the consideration of startup conditions on waste effluents.

Maintenance

Actions necessary to prepare equipment for the repairing work should be anticipated and provision made to minimize waste effluents. Draining and cleaning of equipment preparatory to repairs may result in large waste loads.

Disposal of residual waste

Providing satisfactory facilities for the disposal of residual wastes is a part of a process design job that is the provision of waste treatment facilities. The process design engineer should explore the possibilities of reducing or even minimizing the load on them. The potential costs for such treatment in terminal facilities in some case may even favour effecting process changes to include recovery or recycle of effluents.

10.2 IDENTIFICATION OF WASTE MINIMIZATION OPPORTUNITIES FOR ENVIRONMENTAL PROTECTION

In the following section, waste stream analysis and process analysis techniques to identify waste minimization opportunities in a new or an existing process are outlined.

10.2.1 Analysis of the Waste Stream

The first step in implementing waste minimization is to identify the options that will eliminate or minimize the waste stream's volumetric flow which has the greatest influence on investment and operating costs. It is equally important to eliminate or minimize the components of concern which necessitate treatment of the stream. To arrive at the best options, each waste streams should be analysed as follows:

1. Listing of all components in a waste stream along with the key parameters has to be done.

   For instance, for a waste water stream these could be water, organic compounds, inorganic compounds (which may include both dissolved and suspended) pH, etc.
2. (a) Identification of the components causing concern, e.g. hazardous air pollutants, carcinogenic compounds, wastes regulated under the laws enacted by the Pollution Control Boards, etc.

   (b) Determination of the sources of these components within the process.

   (c) Development of process improvement options to reduce or eliminate their generations.
3. (a) Identification of the highest volume materials—often these are diluents, such as water, air, carrier gas or a solvent. These materials frequently control the investment and operating costs associated with the end-of pipe treatment of the waste streams and have a significant impact on the cost of manufacturing process.

   (b) Determination of the sources of these high volume materials within the process.

   (c) Development of process improvement options to reduce their volume or eliminate them.
4. In case the components identified in step 2 are successfully minimized or eliminated, then one needs to identify the next set of components that have a large impact on investment and operating cost (or both) for end-of-pipe treatment. For instance, if the aqueous waste stream was originally a hazardous waste and it was incinerated, eliminating the hazardous compound[s] may permit the stream to be sent to the waste water treatment facility. However, this may overload the Biochemical Oxygen Demand (BOD) capacity of the existing waste water treatment facility, making it necessary to identify options to effect reduction in organic load in the aqueous waste stream.

10.2.2 Analysis of the Process

The only materials that are truly valuable to the commercial proposition are the raw materials for reaction, any intermediates and the final products. Apart from the feed, other input streams (e.g. catalysts, air, water, etc.) are required because of the designers' limited knowledge of how to manufacture the product without them.

The function of these input streams is to transform feed materials into products. To reduce or eliminate them, the feed materials, intermediates or products must serve the same function as these input streams or the process needs to be modified to eliminate them.

For either a new or existing process, an analysis of the process comprises the following steps:

1. Listing of all feed materials reacting to salable products, any intermediates and all salable products. This may be named List 1.

2. Listing of all other materials in the process, such as non-saleable by-products, solvents, water, air, nitrogen—acids, bases, and so on. This may be named List 2.
3. For each compound in List 2, one may explore the possibility of using a material from List 1, to perform the same function as the compound in List 2. One may also examine as to how the process may be modified to eliminate the need for the material in List 2.
4. For those materials in List 2 that are the result of producing non-saleable products (i.e., waste by-products) one may examine the possibility of modifying the chemistry or process to minimize wastes or eliminate wastes (for example, 100 per cent reaction selectivity to a desired product).

When coupled with the application of fundamental engineering and chemistry practices, examining a manufacturing process by these waste stream analysis and process analysis techniques, will often result in a technology plan for a minimum waste generation process.

Good opportunities can be found in almost every production process, and even some projects that involve basic process changes can be implemented in a relatively short period of time.

10.3 WASTE MINIMIZATION PROGRAMME

The steps suggested by Rittmeyer, the salient features of which are presented here can enable any organization to get started on establishing a successful waste minimization programme. It is desirable if the companies that have relied on traditional treatment and disposal method re-evaluate their environmental protection strategies and identify alternative means to prevent the generation of contaminants. Any effort towards pollution prevention and waste minimization should help by serving as a guide through the steps, to establish a successful programme and comply with the pollution preventions norms. It should also motivate organizations to design a process or plant with waste minimization in mind. A check list of actions can be provided to the organization, and the personnel concerned can employ them in the working of the plant.

The waste management hierarchy is as follows:

Source Reduction

(The reduction or total elimination of the waste at the source, usually within the process in question. Possible ways of achieving this, include process modifications, substitutions of feedstock, improvements in the purity of feedstock, better housekeeping and management practices. Increases in the machinary efficiency and recycling within the process would also help reduce the waste at the source)

↓

Recycling

(Herein comes the use or reuse of hazardous wastes as an effective substitute for an industrial product or as an ingredient of feedstock in a manufacturing process. It can take place on or off-site and would include the reclamation of useful constituents within a waste material, the removal of pollutants from an effluent to allow it to be used, or the use of waste as a fuel supplement or fuel substitute

↓

Less Desirable Strategies

(Like waste separation and concentration, waste exchange and energy/material recovery)

↓

Effluent (Waste) Treatment

(Any method, technique or process that modifies the physical, chemical or biological nature of any hazardous waste in such a way that it neutralizes the waste, recovers energy or material resources from the waste or tenders such waste non-hazardous, less hazardous, safer to manage, amenable for recovery and storage or reduced in volume.)

↓

Disposal of Wastes

(Measures like discharging, depositing, injecting, dumping, spilling, leaking or placing of hazardous waste into or on any land or water so that such waste or any constituents can get into the air or be discharged into any water including groundwater)

Adapted from Robert W. Rittmeyer, Chemical Engineering Progress, NY, May (1991).

It can be observed from the hierarchical presentation that source reduction is the most desirable technique and disposal is the least desirable.

Source reduction and onsite, closed-loop recycling are the recommended methods. The hierarchy shows other, less desirable strategies in order of decreasing preference. Both the Pollution Prevention Board and law require the use of source reduction and recycling whenever feasible, both also take into consideration the fact that many companies will need to continue to rely on traditional treatment and disposal method to a certain extent.

As a multimedia approach to environmental protection, pollution prevention represents two advances beyond industries' traditional waste management methods. First it shifts the focus away from control, transfer between media and waste treatment to proactive generation avoidance. Second, it concentrates on reducing the quantity and toxicity of multimedia wastes and not only those that are hazardous and/or regulated.

10.3.1 Making a Beginning

Wastes' generation can potentially be made less in virtually every aspect of a plant's operation. Therefore, to launch a pollution prevention programme, one must secure the support and participation of plant personnel. The workers' hands-on-knowledge of plant processes may enable them to develop creative pollution prevention strategies.

To initiate interest and instill confidience in the programme, a written statement affirming the company's commitment to pollution prevention and enlisting the help of all employees in launching the programme, may be issued. This initial communication should inform workers about how the reduction of waste can bring about an improvement in the company's business, protect them and the community and motivate them to get involved. Then as the next step, a plant meeting can be held to augment the written memo, to reinforce the organization's commitment and to answer workers' questions.

10.3.2 Organizing a Pollution Prevention Task Force

The broad scope of a pollution prevention programme makes a team work essential. Therefore, a task force can be constituted to help in implementing the programme after communicating the organization's commitment to pollution prevention.

The size of the task force is not of much importance. The team, must, however represent all manufacturing and administrative areas. Members do not need an environmental background to participate. The team may comprise process and environmental engineers, health and safety experts, public relation and marketing professionals, maintenance personnel, management information systems' (MIS) professionals, lawyers, accountants, purchasing personnel, administrators and others as available, as members. Large facilities may have adequate in-house resources to constitute the team; if not outside assistance can be enlisted.

Next, "*a second-in-command*" to help administer the programme can be appointed. It has to be borne in mind that this person must have the authority to enforce his or her designated responsibilities. These responsibilities can be assigned with some discretion, but may include the following.

1. Acting as a central contact person for information
2. Record keeping
3. Scheduling team meetings
4. Supervising the implementation of the pollution prevention programme
5. Tracking and measuring results or outcome of the efforts made

10.3.3 Establishing Plant Level Goals and an Implementation Schedule

The team's first task will be to establish plant level waste quantity and toxicity reduction objectives. For the base year, we may rely on the experience and knowledge of plant operations to set objectives. The objectives may be stated in consistent, quantifiable terms (e.g. percentages, kilograms, etc.). Some sample goals that might be set over a specific time-frame say a two-year period, include

1. reducing hazardous waste by 50 per cent
2. reducing toxic releases by 50 per cent
3. reducing emissions of say, chloroflurocarbons (CFCs) by 50 per cent
4. reducing the use of chlorinated solvents by 50 per cent or at a minimum substituting less toxic solvents.

It is expected of the team to develop an implementation schedule. This will allow the facility to track and report progress, measure rate of progress, demonstrate results to management and workers, provide motivation to the staff and compare results at similar facilities. The schedule may be based on the professional judgment of the employees.

10.3.4 Performing an Opportunity Evaluation

When the team has been selected, the goals stated and a schedule established, the next step is to perform an opportunity assessment. This process involves compiling an inventory of waste streams, investigating the streams and/or the processes that generate them to develop alternatives for pollution prevention, screening alternatives based on economic and technical feasibility and generating a final set of options for pollution prevention.

10.3.5 Compilation of Effluent Streams' Inventory

In the assessment of opportunity one has to determine the following:

1. The origin of hazardous and non-hazardous effluent streams
2. The cause for the existence of these streams
3. The quantity of emission of each stream
4. The frequency of the emission of each stream
5. The characteristics of streams
6. The sources of liquid effluents and air emissions
7. The method of managing the streams; on or offsite

Conducting a plant record survey is a good beginning and the sources that are required to be consulted include the following:

1. Hazardous waste manifests
2. Annual hazardous waste generator reports
3. Environmental evaluation reports
4. Permits
5. Laboratory reports and characterization data
6. Material safety data sheets
7. Production records of the plant
8. Internal effluent tracking reports
9. Vendor transport, treatment, disposal and recycling contracts of the wastes

A brief questionnaire may then be circulated to survey workers who understand major site processes. This will supplement and verify the information obtained during the record review, eliminate data gaps, create a picture of why waste is generated and how process variations can affect its volume and characteristics.

Next, the facility has to be usually inspected. Plant size and complexity will dictate how many people are required to participate and should be based on professional judgment. Before beginning, the team may be asked to formulate a list of questions for the inspection. The questionnaire may comprise questions to document each plant waste stream from its source to final disposal, including information about its collection and storage, treatment, recycling transport discharge and disposal. The team should accurately record all observations during the plant inspection.

A process flow diagram provides a convenient format for summarizing individual waste and emission source found through the records survey, facility inspection and personnel questionnaire. The final step in developing the waste inventory is to summarize and document information for each plant waste stream.

If the facility in question is large, it may be worth while to computerize these data. This will facilitate sorting by plant area, type of waste, disposal method and cost of disposal. Computerizing data also allows the information to be updated readily. When operation or process changes affect waste generation volume, rates or composition, computerization also helps with the tracking of wastes and generation of reports for company management and government agencies.

10.3.6 Evaluating the Waste (Effluent) Streams

Next, an evaluation of the information that has been gathered to identify waste streams or emissions that are the most critical or most amenable to pollution prevention techniques. Several criteria can be employed to prioritize the streams, which include

1. Regulatory considerations
2. Toxicity of the waste streams
3. Quantity of the waste streams
4. Management costs
5. Safety and health risks involved
6. Potential for success
7. Concerns about potential environmental liability

If additional assistance in prioritizing streams is required, sources such as technical literature trade associations, equipment vendors and environmental engineering consultants, can be consulted.

10.3.7 Implementing 'Easier Methods'

It is likely that some areas where it will be easy to prevent pollution without investing a large amount of capital, may be available. Such changes may include the following·

1. Segregation of effluents to prevent mixing hazardous and non-hazardous wastes.
2. Improved material handling and inventory practices to reduce the quantity of materials whose shelf life has expired.
3. Preventive maintenance (to prevent spills).
4. Production scheduling (to reduce quantities of batch-generated wastes or unused raw materials).
5. Incorporating minor operational changes.

Small adjustments often reduce not only pollutant production, but can save time and money. Any such options may be implemented at the earliest.

10.3.8 Screening of the Alternative

After the streams have been ranked, the next step is to identify pollution prevention options.

A beginning can be made with high-priority streams. Some processes may have to be re-examined again and again before a solution can be identified.

Once the team has generated a comprehensive set of options, screening and narrowing of the alternatives can be done to arrive at the most feasible ones. Technical and economic evaluations will help in eliminating impractical solutions.

The technical evaluation measures the feasibility of options against many criteria. More often than not these include

1. Quality of a product
2. Safety of product
3. Personnel health and safety
4. Space requirements in the facility
5. Installation schedule
6. Production downtime in the operating plant
7. Reliability
8. Commercial availability
9. Permitting requirements
10. Regulatory restrictions
11. Effects on other environmental media
12. Personnel skill requirement
13. Customer acceptance

If an option appears impracticable based on any of these criteria, it can be postponed. It may be revised later if operations, personnel, products or environmental legislation change.

In the economic evaluation of the programme, remaining options can be screened for economic viability. Standard economic measures include

1. Capital requirement for the pollution prevention programme
2. Payback period
3. Net present value
4. Return on investment

In addition to these measures of 'real' savings, one should be careful not to overlook intangible or potential benefits. These may be associated with reduced compliance costs or avoiding penalties for improperly handling and disposing of wastes. If any option appears to be infeasible based on cost, it may be set aside and revised later.

10.3.9 Tracking Systems for Pollution Prevention programme

Many factors help in determining the best way to track and report on the results of the pollution prevention programme. These include facility type, complexity, size, internal resources (such as personnel and computer systems) and individual objectives. Tracking tactics will also be governed by how individual waste streams in the plant are generated and the scope of the programme.

The waste survey table should enable a comparison of the quantity of waste generated before and after implementation of a pollution prevention option on a facility-wise basis. But while this basic format and approach may work in most operations, several factors may make it impractical. These include the considerations given here.

1. Changes in the production rate; when this is a matter of concern, the best option is to use ratio of the waste generated to the production rate to accurately measure the effectiveness of the programme.
2. Multi-product manufacturing lines—in this case it is more efficient and effective to track pollution reduction by product line. This too may, however, become impractical in case there are too many products.
3. Areas where waste is not directly related to the production rate—wastes that are, for instance, generated infrequently or by maintenance activities may be more accurately measured on the basis of time or by plant area.

To develop an effective tracking system it would be desirable to start by examining plant records to see if some kind of tracking system already available. If tracking systems for air permits, solid waste disposal and waste water discharges already are available, it will be only necessary to combine and supplement this information into a comprehensive multimedia plant pollution generation database.

The new tracking system may be very helpful in performing many tasks which include the following:

1. Estimating the costs involved in waste management, including estimates for future waste related liability.
2. Ranking additional waste streams based on their volume, the cost to manage them and their toxicity.
3. Tracking and reporting on pollution prevention activities.
4. Combining and relating all waste effluent streams to a specific process.
5. Providing data for pollution control board reporting requirements.

10.4 ENHANCEMENT OF THE EFFECTIVENESS OF HAZARDOUS WASTE TREATMENT USING SPECIALITY CHEMICALS

It has been recently observed that various specialty chemicals have the ability to enhance the performance of on site waste treatment methods either by making a treatment technique feasible or by improving a technique's economic viability.

The methods employed for on-site treatment and the types of effluent streams needing treatment vary widely. The performance and economics of many processing equipments, such as tanks, clarifiers, belt filter presses, plate and frame filter presses, centrifuges, oil collectors, vibrating screens, dissolved air floatation units, separators, thickners and biological reactors are dependent to a large extent on chemical conditioning of the waste.

Acid, caustic substances, and inorganic and organic substances are among the chemicals commonly employed to condition waste materials prior to their treatment. Most of these pre-treatment chemicals have been extensively covered in the literature. Some more recent, less publicized developments reported by Collett are presented here.

These treatment methods include three basic chemistries; coagulation, flocculation and demulsification. The coagulants and flocculants involve particle surface chemistries, such as charge neutralization and particle size development which are required for solid-liquid separation. The demulsification chemistries which are employed more for liquid-liquid separation, employ varying technologies of charged surfactants, particle wetting agents and flocculants.

The terms coagulations and flocculation are normally used interchangeably to describe the formation of aggregates. For the purposes of understanding the points presented in this section, coagulation is defined as charge neutralization to overcome the repulsion between particles to form microflocs ranging in size from micrometres to millimetres, whereas flocculation is defined as the formation of inter particle bridges to form macroflocs, which range in size from millimetres to centimetres.

10.4.1 Coagulation

Coagulation of suspended solids, i.e., the neutralization of charge to get rid of the repulsion between particles destabilizes the particles in suspension. The destabilized particles may or may not form large aggregates, depending on the nature of the particles and the type of coagulant employed. This section focuses on the synthetic organic coagulants that yield smaller aggregates or microflocs because they have been developed recently and may have an impact on and influence the treatability of liquid effluents. These coagulants are employed in situations where small flocs are opted for, such as in mixed granular filtration and they are used in conjunction with flocculants where a high degree of destabilization is required.

These synthetic organic coagulants frequently referred to simply as polymers, fall into the general family of polyamines. These are positively charged or cationic. Their most important properties are molecular weight, charge density, quaternization and molecule type. There are four common molecular types of polyamines—melamine formaldehyde, epichlorohydrin, dimethyl diallyl ammonium chloride and amino methylated polyacrylamide. All the four of them are very highly charged liquids the molecular weights of which range from very low to high; 5000 to 5,000,000.

Melamine formaldehyde polymer (melamine)

This is commonly referred to as melamine and is a partially water-soluble colloidal suspension. Its molecular weight is very low (10,000) and it is very highly charged.

The molecule is not quaternized (that is, the nitrogen atom is not surrounded by other components on all four sides); so it is sensitive to pH and to chlorine and dissolved solids in the effluent stream. Most melamine polymers are available in concentrations of 5 to 10 per cent actives (that is 5–10 per cent of the solution is active melamine polymer).

Its major applications are in surface water clarification and oily waste water separation. In the recent past, the quantity of free formaldehyde in some melamine polymers has been lessened, so that these coagulants have been approved by the US Environmental Protection Agency (EPA) for use with potable water.

Epichlorohydrin (EPI) polymer

It is water-soluble and is available in varying molecular weights ranging from 5000 MW to 500,000 MW and is fully quaternized. Epichlorohydrin polymers are highly cationically charged and are commonly available in concentrations of 25–50 per cent actives.

The EPI polymers have a wide range of industrial applicability, including biological solids' separation, primary waste water clarification, soluble oil waste coagulation, latex waste separation and oil separation. Recent reductions in the quantity of contamination from by-products and side reactions have enabled some EPI polymers to receive EPAs potable water approval. The higher molecular weight EPIs are sometimes used as flocculants in situations where larger flocs are desired. EPIs are also used as demulsifiers.

Dimethyl Diallyl Ammonium Chloride (DMDAAC) polymer

It is water soluble and is available in varying molecular weights ranging from 100,000 MW to 500,000 MW. It is highly charged, fully quaternized and commonly available in concentrations ranging from 10 to 40 per cent actives.

These polymers have a variety of industrial applications including surface water clarification, coal refuse clarification, primary waste water clarification and biological solids coagulation. Some DMDAAC polymers have EPAs potable water approval.

Amino methylated polyacrylamide polymers

Amino methylated polyacrylamide polymers are commonly known as Mannich solutions. They are available in varying charges and with molecular weights that are generally considered to be in the very high range viz. 2,000,000 to 6,000,000 MW. These polymers are commonly available in concentrations of 1–6 per cent actives, and they tend to be very viscous (with a viscosity of 20,000 to 50,000 centipoises). They are not quaternized, so they are sensitive to pH, chlorine and dissolved solids.

Mannich solutions are employed in many industrial applications which include biological solids clarification, oily water separation, paper waste clarification and dewatering of biological solids and oily sludges. Owing to their high molecular weights, they are also employed as flocculants.

Evaluation of the different types of coagulants and processes calls for comprehensive laboratory evaluation. The first step is to obtain representative samples of the effluent, which often dictates the processes to be considered. Once a test protocol has been established, testing all the types of coagulants is often the best. The laboratory testing of the samples with the technical assistance provided by the supplier, will form the basis for evaluating costs and feasibility of treating a waste.

10.4.2 Flocculation

Flocculation of solids generally implies the formation of large open network aggregates or macroflocs. This is believed to take place by five mechanisms (sweep flocculation, specific ion absorption, double layer compression, polymer bridging and polymer charge patching), more than one of which may often take place at the same time. However, knowing the type of mechanism that prevails for a specific waste is not necessary; so tests are generally not performed to determine the controlling mechanism.

Mechanical particle transport gives rise large flocs, which are a prerequisite for efficient solid-liquid separation to be carried out in most dewatering equipment. The size of the flocs formed and the ability of these flocs to withstand the shear imparted by the processing equipment are highly governed by the nature of the solids and the type of flocculant employed.

The flocculants discussed in this section are water-soluble synthetic organic polyacrylamide copolymers usually referred to as polyacrylamide polymer. Significant developments made in the past decade are in respect of both applications and molecular properties.

Polyacrylamide polymers are available with various properties. They come in many forms including solids, gels, liquids, suspensions and emulsions with the choice of form dependent on the user's preference.

The polymer properties that are of much importance are molecular weight, charge density, charge type and molecular structure. The molecular weights of these polymers are considered high to very high (5,000,000 to 15,000,000 MW). The charge can vary from zero to 100 per cent, with the type of charge ranging from non-ionic to anionic to cationic. Also the molecular structure varies owing to the fact that several molecules are routinely copolymerized with polyacrylamide for application as waste treating polymers.

The manufacturers of water-soluble polyacrylamide polymers have concentrated their development efforts on eight monomers that are polymerized with acrylamide. These types have been classified into two basic categories, non-ionic anionic and cationic.

Polyacrylamide polymers have been employed in nearly all dewatering processes commonly found in effluent treatment, including belt, filter presses, plate and frame filter presses, centrifuges, vibrating screens and thickeners. The performance of these processes has a bearing on the type of solids being dewatered and the type and quantity of flocculants employed. Despite the fact that the selection of the optimum polymer is usually based simply on cost versus performance, testing and evaluation are complex. These procedures must represent specific process conditions in order to accurately gauge performance. For instance, the procedures for evaluating a polymer for belt press dewatering are different than those for centrifuges or thickeners. Besides, the desired results of the processes must be considered as well.

The most significant step in commencing the evaluation of processes and flocculants is obtaining a representative sample of the waste. All the results from comprehensive laboratory evaluations are dependent on this sample. A testing protocol must be established for each process being taken up for consideration. Then all the flocculant's uses must be evaluated, followed by process optimization. A reputable supplier can be of significant help in providing polymer samples and technical assistance.

10.4.3 Demulsification

Demulsification of a stable emulsion will call for some form of treatment. An emulsion is a mixture of two immiscible liquid phases where one phase is dispersed into the other phase. Fine solids, such as metal residues, coke fines and silica, may also be present to stabilize the mixture. The emulsion stability often has a bearing on the type and quantity of emulsifying agents, viscosity, quantity of water, pH, shear and agitation.

The first step in choosing a demulsifier is to determine the type of emulsion that must be destabilized. The two common types are: oily waste water emulsion (where water is the continuous phase) and waste water oil emulsion (where oil is the continuous phase). The distinction is normally easily determined by adding a drop of waste into a cup of water; if the waste and water mix readily, then the waste's continuous phase is water, that is, the waste is an oily waste water emulsion.

Oily waste water emulsions are colloidal systems of electrically charged oil droplets surrounded by an ionic medium. The emulsions are most commonly broken by the combination of the processes viz; chemical treatment and mechanical processes, such as dissolved air floatation, induced air floatation and separation.

The mechanisms by which the chemicals cause destabilization of these emulsions generally include neutralization of the repulsive charges between particles, precipitation of the emulsifying agents and changing the interfacial film. This film may be broken by providing enough mixing to ensure that the treatment chemicals collide with the emulsified oil droplets. The chemicals often used to break oily water emulsions include the ones discussed in the coagulation and flocculating sections besides many surfactants.

The testing protocol for choosing the best demulsifying chemical treatment for a process must accurately simulate that process in order to attain the most reliable results. In general, jar tests are employed to determine, which chemicals are best, and to approximate dosage requirements, the degree to which suspended solids are reduced, the achievable water quality and the amount of oil left over. For processes like induced air floatation, dissolved air floatation and granular filtration other tests may be involved.

Waste-oil emulsions or water-in-oil emulsions are normally viscous; concentrated mixtures are formed when oil comes into contact with solids and water. The water is dispersed in the oil as a finely divided emulsion in which the oil-coated solids act in order to maintain the emulsion stability. These emulsions are often broken by physical methods or chemical treatments or a combination of the two. The three methods most often employed to break these emulsions are:

1. Removing the emulsifying agents from the film interface by the techniques of precipitation, decomposition or changes in solubility.
2. Forming an emulsion opposite the existing one.
3. Disrupting the interfacial film.

The chemical treatment's efficiency is frequently enhanced by heat and mixing. The chemicals to demulsify waste-oil emulsions are usually non-ionic or anionic, oil-soluble and highly surface active. Also they tend to be multi-component chemical systems to facilitate a wide range of variation in the emulsion composition.

After an emulsion is broken, the specific gravity difference between components enables a distinct phase separation. The type of emulsion and chemistries employed often dictate the type of process that is suitable for efficient separation. Processes commonly employed include tankage, centrifuges and filter presses.

Selection of the optimum chemical treatment is complex and calls for vigorous laboratory scale evaluation. It is of much significance to remember that each oily waste is unique and may contain various emulsifying agents, biocides, metallic and non-metallic solids, antioxidants and other chemical additives. Thorough testing of all emulsion-breaking chemicals is recommended.

The testing is the basis for determining whether a specific waste can be treated in a particular manner and the cost involved.

Many of the reputable suppliers of demulsifiers will provide technical support and samples of demulsifiers for laboratory evaluation. Once a testing protocol is established and representative samples of waste are obtained, then comprehensive testing is needed for establishing the most cost-effective treatment of an effluent.

10.5 DISPOSAL METHODS FOR WASTE WATER CONCENTRATES

Municipal waste water flow and most of the liquid effluents carry organic and inorganic solid materials in the suspended form. Dissolved organic matter and inorganic ions are also present in liquid effluents. One of the major objectives of liquid effluent treatment is to bring about a change in phase of dissolved materials to gases or solids and then to remove both the original and the converted suspended solids by processes like gravitational settling. The conversion of dissolved organics may be brought about by the synthesis of a mixed culture biomass by techniques like biological or chemical flocculation, precipitation and agglomeration, by solid adsorption, ion-exchange, reverse osmosis and others.

Secondary sewage treatment as it is now practised employs biomass synthesis and biological flocculation as the prominent conversion mechanisms. The removal of inorganic ions and refractory organics is presently achieved in tertiary treatment by adsorption and precipitation. The suspended solids are separated from the water flow in tertiary treatment by the process of gravity settling and sand filtration. The concentrated solids' slurries are termed concentrates or sludges.

Getting rid of the contaminants from the liquid effluent flow is one of the purposes of treatment, but the disposal of the pollutant concentrates after their removal is effected, and is of equal significance. Although the solids present may represent only a few hundredths of one per cent of the effluent flow on a mass basis, the solids processing costs in a treatment plant are approximately one fourth to one-half of the total capital and operating costs. The majority of malfunctions and operational problems of a treatment plant are also associated with the sludge disposal system (Edwin, R. Bennet).

The concentrates are also of importance in the protection and conservation of the local environment. The water phase is released to the rivers to be reused, evaporated or returned to the ocean and also to dynamically re-enter the hydrologic cycle. The solid contaminants that are removed are a static material that must be introduced into the immediate environment with a minimum of contamination of the air or land resources, without objectionable odour and with some benefit, if possible.

It should also borne in mind that effluent treatment equipment and process must be planned on a long-term basis, to obtain reasonable cost and that the large number of plants of various sizes, implies that in some plants, operator skills may be quite low. In general, long-term reliability is more important than scientific sophistication.

A flow diagram of an effluent treatment plant in shown in Figure 10.1. The primary and secondary process layouts are quite typical of current practice. The tertiary treatment section if required, is tailored to meet environmental requirements and the processes indicated in Figure 10.1 are one possible layout. Secondary treatment is now employed by most cities and industrial units. Tertiary treatment is now beginning to be employed in areas of specialized environmental requirements. The primary and secondary treatment processes indicated, are quite common for municipal treatment plants. Industrial effluent pollutants are quite different from one industrial plant to another. In general, industrial effluent treatment plants include some of the unit operations shown, although the quantity of materials to be removed may vary widely. The considerations given here apply most directly to municipal processes.

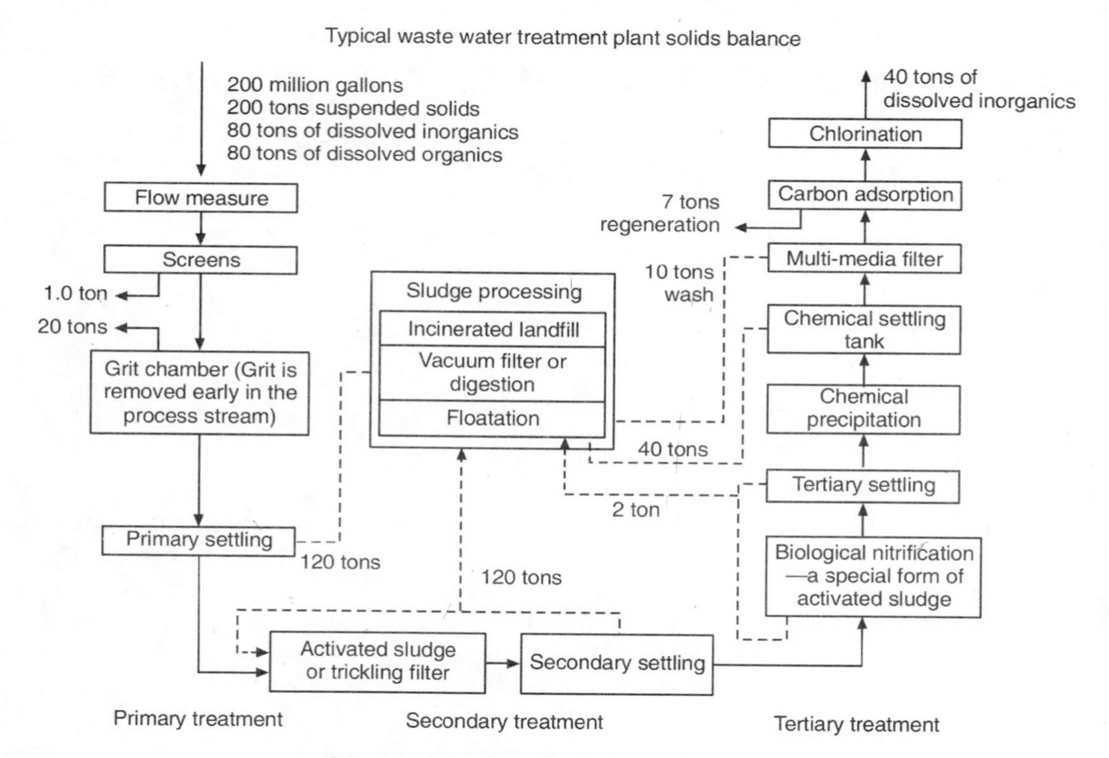

Figure 10.1 An effluent treatment plant.

Adapted from Edwin, R. Bennett, 'Air and Water Pollution', *Proceedings of the Summer Workshop*, University of Colorado, August 1970.

10.5.1 Sources of Concentrates

Screenings are of small volume and are usually handled by the process of grinding or by employing small on-site incinerators.

Grit is made up of the large, heavy particles of sand, cinders, coffee grounds and cellulose organics from home garbage grinders.

Washed grit can be landfilled without giving rise to odour problems. Grit is removed at an early stage in the process stream to protect the machinery of the treatment system.

More often than not the primary sludge from the underflow of the primary settling tank is the solids resulting from plain sedimentation of the suspended solids in the inflow. The concentrate is five to eight per cent from solids of which about eightly per cent is organic and about one-half are biodegradable. The remaining organic fraction essentially comprises paper materials.

Activated sludge or trickling filter humus comes out of secondary settling tank as a two per cent solid concentrate. The solids are around seventy five per cent organic and thirty five per cent of them are biodegradable.

Biological nitrification is a special form of activated sludge and the concentrates are very similar to those of activated sludge of the secondary process.

The chemical precipitation sludges of the chemical settling tank and filter backwash are in the two per cent solid range and are nearly totally inorganic precipitates.

The adsorbed organics on the activated carbon are usually volatilized during regeneration and do not require concentrate processing.

It should be mentioned that all of the sludges generally contain pathogenic microorganisms.

10.5.2 Disposal Techniques for the Concentrates

The disposal of the solid contaminants normally involves some form of concentration or dewatering to reduce the volume of fluid handled and destruction of organics, either total or putrescible fraction. This is followed by transportation and finally safe, nuisance-free land disposal of the remaining material is carried out.

The concentration methods are employed to get rid of some of the water from the sludges. For the thin sludges of the secondary and tertiary processes, this is carried out in two steps. In the first step, the solids are concentrated to approximately the same five to seven percentage as the primary sludge. This is carried out with a slowly-stirred gravity thickener or with an air floatation device.

A sludge floatation unit functions on the principle that the sludge is aerated under a pressure of forty to sixty pounds per square inch and dissolves on equlibrium concentration of the gases of the air. Poly electrolyte is added to cause agglomeration of the solid particles and the solution is released to atmospheric pressure in a floatation tank. The dissolved gases coming out of solutions get attached to the agglomerated particles and float them to the surface. The floated sludge at a solids concentration of above five per cent is removed by a surface skimmer.

Techniques such as vacuum filtration, pressure filtration or centrifugation can be employed to raise the solids concentration to approximately twenty per cent. Each of these processes requires some form of favourable conditioning to agglomerate the small particles. These processes remove about seventy five per cent of the water from the sludge to facilitate transportation and burial or to make the sludge nearly self-sustaining from a heat balance standpoint in an incineration process.

In vacuum filtration, a drum covered with a plastic or wire mesh filter medium is slowly rotated while partially submerged in vat of conditioned solids slurry. A vaccum is applied to segments of the inside of the drum and the water is drawn through the medium and the solids are retained to form a filter cake. The drum is rotated in such a manner that the formed cake is out of the vat for a portion of the cycle and air is pulled through the pores of the dewatered filter cake. The theory of this operation follows the principle of flow through porous media. A filter press operates in a similar way except that the conditioned sludge is put into the press and a piston forces the water through the filter medium, leaving dewatered cake in the press. A centrifuge concentrates a conditioned slurry by spinning a cylindrical-conical container. The concentrated solids suspension is removed from the outer edge of the device. Each of the aforementioned devices produces a cake of 80 to 85 per cent moisture content with 80 to 90 per cent capture of solids. For cost and operational reasons, the vacuum filter is the most widely employed equipment.

Chemical conditioning can be carried out with ferric chloride and lime or with organic polyelectrolytes. Lime and iron cause disinfections, but large quantity as high as thiry to thirty-five per cent of the solids weight are needed. This also enhances the amount of sludge requiring ultimate disposal. Polyelectrolytes are the more desirable coagulant chemicals for many reasons, but they are not effective for use in all types of effluent sludge.

Heat treatment of the sludge is of recent origin. The sludge is heated to about 200°C at a pressure a 150 psi for ninety minutes. Some lipids and carbohydrates are solubilized and denaturation of proteins takes place. The cooked sludge can then be subjected to decantation and filtration without the use of chemicals. Resolubilization of organics and potential odours are problems that may need further consideration. Completely sterilized sludge comes out of this process.

New processes, improved conditioning chemicals and catalysts for the heat treatment process could be some of the areas of possible research and development in sludge dewatering.

Destruction of organic matter or at least the putrescible fraction is the objective of at least one operation in any sludge disposal process. This objective may be accomplished with the processes viz., incineration, anaerobic digestion, aerobic digestion, composting or wet combustion. These processes are briefly presented here.

Incineration: It is the drying and burning of the organic matter of the solid materials in a concentrate. It is only practised with dewatered sludge where only small quantities of auxiliary fuel are needed. Air pollution control is one of the problems associated with this method. The ash can be disposed of by open landfill technique. Complete sterilization takes place in the process.

Anaerobic digestion: It is an oxygen-free biological process with a mixed culture of bacteria at a controlled temperature of 35° C. The bacteria operate in two sequential steps converting organic matter to a short-chain organic acids. Then a different group of bacteria effect the conversion of the acids to a burnable gas with 70 per cent methane and 30 per cent carbon dioxide. The gas is used in a heat exchanger to provide the temperature control. The process is able to support itself. The digestion period ranges from one month to three months and approximately half of the solids are converted to gas. The digested sludge can be air dried on open sand beds and landfilled usually without giving rise to the odour problem.

Aerobic digestion: Aerobic digestion of biological sludges requires a digestion time of two to three weeks. This converts the putrescible organics, approximately thirty per cent of the solids to carbon dioxide and permits land disposal of the digested sludge.

Composting: It is a biological process where fungi and bacteria convert the organic matter to carbon dioxide and water under aerobic conditions at 60° C. The process is exothermic and will maintain the optimum temperature by itself. The moisture content should be adjusted from forty to sixty per cent daily. The material must be provided aeration by turning at least every day. The process needs two to six weeks for stabilization. It should be employed in conjunction with municipal refuse disposal. The solids reduction is about thirty to fifty per cent and the material may be used is a soil conditioner.

Wet combustion: It is a high temperature (300° C) and pressure (1200 psi) combustion of the organic matter in liquid solution. Air is used for the supply of oxygen. Nearly complete combustion can take place within two hours.

Research areas for solids destruction include development of new processes, improved biological conversion rates and the effect of catalysts in wet combustion.

All solid residues must eventually be placed on the land. This may be carried out by burial or landfill of dewatered raw sludges or spreading of liquid digested or composted sludges. From a conservation perspective, it would appear that returning digested organic matter to the earth would be advantageous. There appears to be three reasons why the practice is not frequently followed: (1) The fertilizer value of sludge is low. Inexpensive high strength chemical fertilizers can be transported and applied at a lower cost. (2) An excess of competing soil conditioners such as animal manure exists closer to the farms. The great excess of manure is a nation-wide problem by itself. (3) Solids that are so poor and sandy as to need soil conditions, are usually in arid areas where irrigation water would not be available, even if the soil were conditioned for agricultural use.

One of the most beneficial and productive areas of research would be a thorough agronomy study of the proper method of application of digested liquid sewage sludge to agricultural land.

10.6 TECHNOLOGIES FOR PREVENTION OF POLLUTION IN BATCH PROCESSES

Batch processes are normally a preferred operating method for manufacturing small volumes of high value products, such as many pharmaceuticals, agricultural materials and speciality chemicals. Unfortunately batch operations are also associated with the generation of unacceptably large amounts of waste per unit of product. In the past, many industrial organizations could afford this wastefulness because the high value of the final product was often more important than the cost of treating the waste and because some of the compounds in the waste streams were previously unregulated.

However, in the recent past two things have changed. First, industrial organizations are becoming aware of the true cost of the waste production, which is always greater than the cost for disposal or treatment. The true cost of waste production includes the monetary value of lost products, the cost involved in solvent purchases, the fees for permitting and monitoring emissions and the increased exposure to safety and environmental pollution risks.

Second, with increased regulations of emissions to air, water and land, many manufacturing sites can no longer handle the often concentrated toxic waste streams (especially waste water stream) coming forth from batch processes with the existing control equipments. As a consequence, facilities relying on batch processes are now facing significant new capital investment to pre-treat concentrated waste water before it can be sent to a biological treatment unit (with an idea of enlarging biological effluent treatment capacity), to install state-of-the art hazardous waste incinerators for liquid wastes and to provide point source treatment of air emissions.

It hence turns out that the true cost of waste generation provides an incentive for batch operations to identify and implement pollution prevention solutions.

Here some of the techniques and technologies suggested by James Dyer, et al. that can be employed by engineers to bring about a reduction in the waste generation at the source in batch operations, have been outlined. This section covers strategies for reactor charging, operation, discharging and cleaning; modelling batch processes; and batch versus continuous operation. These strategies can help to reduce waste generation by effecting an improvement in reactor charging, operation, discharging and cleaning practices. Three examples have been discussed: (1) Replacing an organic solvent with aqueous solvent for cleaning. (2) Reducing methylene chloride emissions by sealing atmospheric mix tanks. (3) Converting a process from batch to continuous operation.

10.6.1 The Sources and Nature of Waste Emissions

A batch operation is by definition a discontinuous process; process parameters such as temperature, pressure and concentration vary throughout the batch cycle.

In the same manner, waste emissions associated with these cycles are likely to vary. For example, a batch reaction step may generate a nitrogen purge stream whose Volatile Organic Compound (VOC) concentration varies with time. The emission rate attains a peak value in the first hour then decreases with time as the reaction gets near to completion.

Such time-dependent behaviour makes it hard to effectively specify a continuous gas-abatement system. To meet permit limit requirements, the pollution abatement equipment must be sized to treat the maximum achievable waste flow and concentration. In the case of some emissions, this maximum may last only for a few minutes. It is easy to observe how uneconomical the 'end of pipe' approach can turn out to be a commercial venture. For this reason, source reduction of these emissions by pollution prevention strategies can effect a reduction in the dependence on such expensive control technologies.

Every stage of a batch cycle has some quantity of waste emissions associated with it. Non-condensibles (such as, nitrogen and air) escape the process during the process of reactor charging and discharging carrying with them high concentrations of volatile organic compounds. Reactor operation may generate unwanted by-products that contribute to product losses. Most significant are large quantities of cleaning wastes generated during vessel cleaning between batch cycles and product campaigns. The pollution prevention strategies which can to a large extent reduce emissions for all steps of batch operations are presented here.

10.6.2 Strategies for Pollution Prevention

Most waste generation in batch operations takes place in the reactors. For this reason, the pollution prevention technologies and practices described here focus on the batch reactor. Four key reactor process steps are presented—charging, operation, discharging and cleaning.

Charging of reactors

Solvent vapour losses during raw material addition take place for two primary reasons—leaks from the process equipment, and vapour displacement as the reactor is filled with liquids or solids. To prevent these two problems from occurring, the following pollution prevention strategies can be considered.

- Feeding solvents by gravity into the reactor vessel instead of using a centrifugal, positive displacement or diaphragm pump to reduce the pressure build-up associated with pumping. This can often be accomplished in new designs by elevating raw material feed tanks above the reactor instead of besides it to take advantage of gravity.
- Installation of the closed-loop vapour recycling systems for pumping operations. As shown in Figure 10.2, this involves piping the displaced vapour from the reactor (as it fills) back to the vessel containing the feed material, e.g. a solvent storage tank. Care has to be exercised regarding the compatibility of materials.

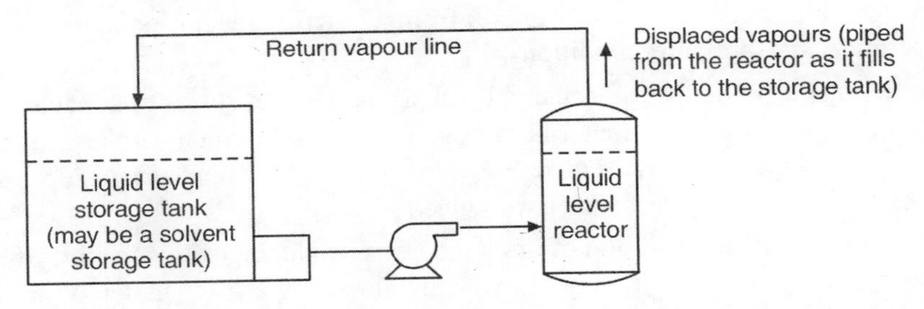

Figure 10.2 Closed-loop vapour recycling system for pumping operations.

Adapted from James, A. Dyer, et al., *Chemical Engineering Progress*, May 1999.

- Charging solids before liquid solvents to minimize solvent displacement from reactors. A more general strategy is to reconsider the sequence of raw material addition to minimize unwanted off-gassing due to vapour displacement and/or chemical reactions.
- Charging solids using lock hoppers in lieu of rotary valves or manual dumping through open lids. Lock hoppers isolate the reactor from the open atmosphere so that organic vapour emissions are kept to the minimum (Rotary valves, on the other hand, often allow vapours to escape as they rotate from the reactor vapour space back to the solids feed bin). Lock hoppers are typically employed when solids must be added to a reactor containing a volatile solvent that is at an elevated temperature and/or pressure condition. Besides, they can be designed to eliminate air infiltration into a reactor for cases in which

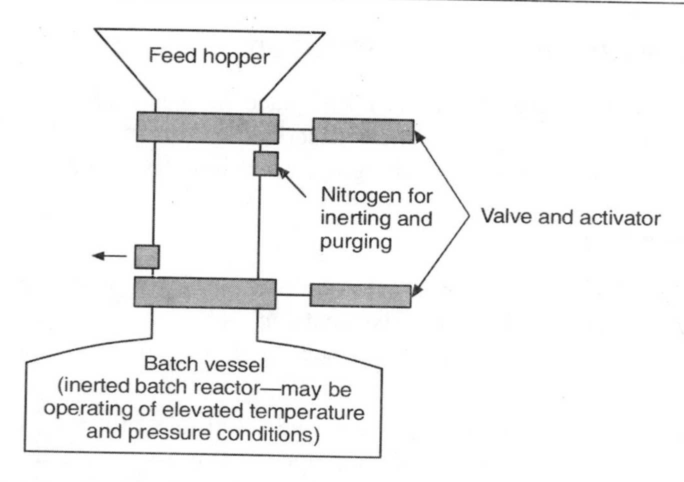

Figure 10.3 Lock hopper arrangement with two sliding gate valves for charging solids to an inerted batch reactor.

Adapted from James, A. Dyer, et al., *Chemical Engineering Progress*, May 1999.

oxygen and moisture would create an unsafe or undesirable condition. Figure 10.3 illustrates a typical lock hopper arrangement in which two sliding gate valves isolate the solids feed bin (open to the atmosphere) from an inerted reactor operating at the elevated temperature and pressure conditions.

- Considering cut-in hoppers for manually charging bags of solids to a reactor. Rather than dumping bags of solid raw material through an open lid (which can lead to much dusting and solvent vapour losses), cut in hoppers, 'cut open' the bags inside a closed feed hopper to contain any dust that is produced.
- Elimination of manual addition of dry solids by introducing solids in slurry form or using dense phase conveying. It may be worthwhile using a raw material or intermediate already in the process to serve as the carrier fluid for the slurry, rather than introducing a new chemical to the process. Dense-phase conveying uses less conveying gas to move the solids, thereby reducing the quantity of solvent evaporated from the reactor as the conveying gas is vented.
- It may be desirable to vent displaced vapours through a refrigerated vent condenser to recover and recycle condensable solvents.
- Consideration of solvents with a lower vapour pressure to minimize evaporation and facilitate recovery by condensation. Care has to be exercised while dealing with liquid solvent mixtures in which small weight percentage of a volatile solvent can account for the majority of the vapour emissions. For instance, the composition of a flammable solvent mixture employed for cleaning solvent is shown in Table 10.1.

Table 10.1 Composition of a Flammable Solvent Mixture for Cleaning Parts

Component of the solvent mixture	*Liquid composition, weight %*	*Vapour vomposition vol % or mole %*
Acetone	10%	44%
Methyl ethyl ketone	30%	43%
Toluene	30%	10%
Xylene	30%	3%

Even though, the liquid mixture contains only 10 per cent acetone by weight, the saturated vapour phase is more than 40 per cent acetone, at 26° C and 1 atm total pressure. Removing or reducing the weight fraction of acetone in the solvent mixture would in turn help in reducing vapour emissions significantly.

Reactor operation

The strategies in respect of the inert blanketing, sampling, control, sequencing and heating of batch reactors are presented here. A more exhaustive discussion of the reactor design and operation for both continuous and batch chemical reactors can be found in literature. These strategies are:

- Reducing vapour loss by enclosing open air tanks. Usage of lids with gaskets that seal tightly in case manual access is a necessity. Otherwise, a closed reactor with a permanent top and closed raw material addition system may be considered.
- Reducing vapour losses by improving seals on agitators and lids. In some situations, the agitators can be replaced with a jet mixing system which comprises a jet nozzle submerged in the reactor fluid, a pump and a liquid re-circulation loop.
- Sequencing the addition of reactants and reagents to optimize yields and bring down emissions. The following process is considered—the process in which a base, such as sodium hydroxide is added at the end of a batch reaction cycle to neutralize the reaction mass before feeding the next process vessel; the pH in the reactor is allowed to vary with the progress of reaction. Adding sodium hydroxide throughout the reaction sequence to maintain a constant pH, may result in improvement in yield and minimize the formation of by-product.
- Controlling nitrogen purge rates with automatic flow control devices in lieu of manual throttling valves. It is common to see the manual valves wide open which results in wastage of nitrogen and increase in volatile organic chemical losses.
- Using a nitrogen pad system, that is, pressure control of nitrogen in the reactor vessel head space instead of a purge based on flow control. Figure 10.4 shows a typical piping diagram for a nitrogen pad system in which the nitrogen supply pressure is greater than 60 psig. Dual pressure regulators are employed to improve low-pressure control in the reactor head space.
- Optimizing existing reactor design based on reaction kinetics, mixing characteristics and other variables to reduce by-product formation.

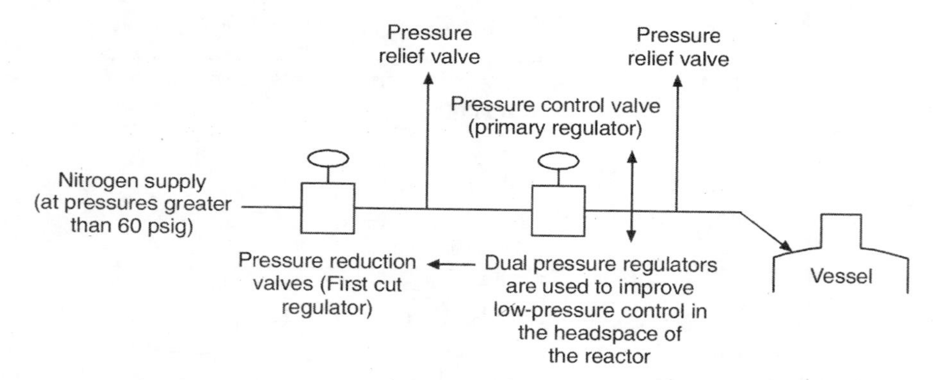

Figure 10.4 Typical piping diagram for a nitrogen padded reactor.

Adapted from James, A. Dyer, et. al., *Chemcial Engineering Progress*, May 1999.

- Collecting and recycling excess reactants and solvents used in the reaction. It may be worthwhile considering addition or upgradation of solvent purification equipment to allow recycle.
- Using a minimum number of intermediate stage process sample to determine the reaction end point. Also, consideration of automatic, in line sampling, rather than manual sampling from vessel openings, to minimize vapour emissions from openings and the quantity of sampling waste will be worthwhile. The impact of the time lag associated with manual samples analyzed in the site laboratory on reactor yield and waste generation, may be evaluated. On many occasions, the additional turn around time required for laboratory analysis leads to excess waste production. For instance, in a process for producing an agricultural intermediate, manual sampling of a batch caustic scrubbing system for *percentage NaOH* led to unnecessary safety margins in the operating procedures for purging the spent caustic scrubbing solution. Owing to the long turnaround time from the laboratory, the spent scrubbing solution contained 3–5 per cent unused NaOH. A feasible alternative to this approach is to employ an in-line pH meter to regulate caustic addition to minimize the quantity of NaOH being wasted.
- Using statistical process control techniques to regulate the reaction in lieu of relying on intermediate sampling.
- Elimination of excessive reactor boil-up by replaying direct steam jacket heating with an external, recirculating, heating loop comprising a pump and non-contact heat exchanger. This may also help in preventing undesirable and unwanted tar formation in reactions, in which a hot vessel wall leads to product degradation.
- Elimination of fugitive emissions form pressure relief values by employing an upstream rupture discs in series with the relief valve.

Discharging of reactors

- Replacement of nitrogen blowcasing with a pump. The pressurized nitrogen employed to push the reaction mass from the reactor, must be vented downstream. This nitrogen will leave saturated with solvent and possibly, valuable product and reactants as well.

- Designing and installing discharge lines on an incline to take advantage of gravity flow to the downstream equipment. Avoiding a pump brings down both investment and operating costs and also fugitive emissions from leaks.
- Reducing the reactor batch temperature before discharge to lower the vapour pressure of volatile organic compounds, which may include solvents and in some cases reactants and products.

Cleaning of reactors

- Optimizing the product manufacturing sequence and product campaigns to minimize washing operations. This may imply structuring product campaigns to manufacture clear or light coloured product before heavily pigmented products or products with relaxed quality specifications before those with tight specifications. Within a single product campaign, it may imply better understanding of the chemistry that leads to equipment fouling and then effecting a change in the sequence of reactions to minimize the formation of tar, equipment scaling, polymer build-up, and so on.
- Maximizing production runs to decrease the frequency of washes. One may talk to customers to see if product specifications can be loosened. In case some customers can accept higher levels of impurities their order may later be filled in the product campaign, if possible.
- Minimizing the number of solvents rinses employed to clean reactors by understanding the level of acceptable contamination.
- Increasing the smoothness of the internals of the vessel to effect a reduction in the quantity of cleaning solvent employed. It may be worthwhile considering the usage of non-stick; non-porous linings, eliminating baffles and minimizing other difficult-to-access points where there is a possibility of material accumulating and hardening.
- Employing high-pressure rotary nozzles to effect reduction in the quantity of cleaning solvent used.
- Positioning drain values at the lowest point on the reactor vessel to improve removal of residual solvent and product before drying or the next product campaign.
- Replacing volatile chlorinated solvents with lower volatility, non-chlorinated solvents.
- Replacing solvent-based cleaning with aqueous-based cleaning that employs detergents or other surfactants, if necessary, to effect an increase in the cleaning effectiveness. It has to be ascertained as to how pH affects cleaning.
- Collecting the final wash rinse for reuse as first-pass rinse during the next cleaning cycle. Lower quality solvent or water from elsewhere in the plant may be suitable for use as the first rinse.

10.6.3 Modelling Batch Process

Changes in standard operating conditions can have impact on the quantity of waste generated by a batch process. However, few plants can accept the risks that go along with making experimental batch runs. Recent development in computer modelling software enable us to analyze changes in batch operations to greater extent than what was possible earlier.

Several major organizations that sell process flow sheet simulation software, have already developed and/or are greatly expanding their capability to simulate batch unit operations and overall batch processes.

10.6.4 Comparison of Batch and Continuous Operations

For some processes, the conversion from batch to continuous operation is an effective way to bring about a reduction in waste generation as well as to increase total throughput.

Cleaning waste is most significantly brought down by switching to a continuous process. Continuous process operations are seldom interrupted for cleaning, because the equipment is dedicated to the production of one or only a few products. Hence, switching to a continuous process may imply practically eliminating cleaning waste.

Besides, solvent recovery in a continuous process is feasible, owing to the fact that it turns out to be more economical to install dedicated distillation equipment to handle the necessary chemical separations. Recovered solvent can then be reused in the process, instead of being incinerated or disposed of as a hazardous waste.

Examples The three case studies given below illustrate successful pollution prevention in batch operations.

Replacing an organic solvent with an aqueous solvent for cleaning: The Small Lots Manufacturing (SLM) area at an industrial facility was converted into aqueous cleaning in order to reduce solvent waste generation. SLM is an agricultural product manufacturing area employing batch organic synthesis involving reaction vessels in series. Past operations involved several rinses with a variety of organic solvents to clean the vessels between product campaigns. The spent organic wastes were not recycled, but were discarded off-site as hazardous waste.

To effect reduction in the generation of organic solvent wastes, SLM replaced a number of organic solvent flushes with detergent flushes and water rinses. Solvent usage per cleaning cycle was reduced by 16000 litres—a 60 per cent reduction over organic solvent cleaning. Moreover the resulting aqueous waste can be processed in the site's waste water treatment facility. To allow this, it is important to select a biodegradable detergent and to ensure that the additional organic and hydraulic load on the biotreatment basin could be handled.

Significant benefits other than the environmental benefits could be realized as well. There can be a substantial saving on solvent waste disposal cost and in solvent raw materials cost. Cleaning effectiveness, measured by the quantity of material carrying over into the next product was increased tenfolds. The increase in production capability resulting form the reduction in cleaning time can effect a substantial saving for the manufacturing facility concerned.

Reducing methylene chloride emissions by sealing atmospheric tanks: Methylene chloride was used as a coating and cleaning solvent in the manufacture of several graphic arts and electronic photopolymer films. In the late 1980s, the manufacturing site released more than 6,000,000 kgs/year of chlorinated solvents in the air.

The coating solution preparation areas accounted for the majority of the air borne contaminants. Coating solutions were prepared batch wise in agitated vessels employing a mixture of polymers, monomers, photoinitiators, pigments and solvents. These atmospheric mix tanks were not well sealed which led to large fugitive and point-source emissions of methylene chloride.

To enclose the batch vessels to the extent possible, the mix tanks were fitted with bolted gasketed lids. The vessels were also designed with pressure/vacuum conservation vents to allow the vapour pressure to rise to 3 psig before the tanks breathed.

By sealing up the process, the manufacturing unit reduced air emissions by around 40 per cent and there was a substantial saving in methylene chloride costs.

Converting from batch to continuous operation: The manufacturer of herbicide intermediate was unable to meet demand when operating its batch process at full capacity. To overcome the production shortfall, the management bought the intermediate from a competitor at a price higher than its cost of manufacturing.

Duplication of the existing batch process appeared to be the conservative approach to increasing production. However, the business team of the company saw an opportunity to meet the expansion objectives with minimal investment by changing from batch to continuous reaction technology. The team made a lot of effort through many intensive technology and project reviews and concluded that the continuous process appeared inherently safer and technically viable. Within one year from concept to commissioning, the business team achieved a 250 per cent increase in production capacity over the original batch process. Besides this, methanol emissions per kilogram of the product from the continuous process were 30 per cent less than those of the batch technology.

10.7 WASTE MINIMIZATION DURING DESIGN OF NEW PROCESSES AND OPERATIONS

When technology is being defined, that is, while planning new processes and operations there are plenty of opportunities to reduce waste. Long-term benefits in waste minimization can be obtained, in case old operations are phased out and new processes and operations emerge with a greater thrust or emphasis on waste minimization. To accomplish this objective any new project must go through a detailed Waste Minimization Opportunities Review (WMOR) as a part of the Definition of Technology (DT) programme. During this review, an extensive examination of waste reduction opportunities and waste management issues has to be done. Many organizations should have to take this as policy and see to it that every major new process goes though the WMOR as a part of the project approval process. In this section a summary of the waste minimization programmes adopted by a leading chemical manufacturing company is presented. Also some of the areas that can give rise to further benefits in waste minimization efforts (highlighted by Bergland, et al.), are discussed briefly.

The main objectives of WMOR process are:

1. To assist process R and D engineering personnel in understanding the reason for minimizing wastes and the concepts and technology available for effecting such waste minimization.
2. To conduct review new processes and production schemes, to ascertain if additional opportunities are available, for waste reduction.

3. To assure that a waste minimization programme does not violate the regulations suggested by the pollution control agency or result in any additional safety hazard to the process unit.
4. To document waste minimization activities and constraints for possible use in existing process and also projects to be taken up subsequently.

10.7.1 Development of New Technology

Optimum waste management is a fundamental research objective from the conceptual stage of the process through process development, engineering and design to construction and start-up, and operation. It is also a continuing objective of the plant personnel (engineers and operators), once the unit commences production. In the new technology development programme, from the time a new process is conceived until the facility commences production there are many opportunities in which waste minimization can be considered (Figure 10.5).

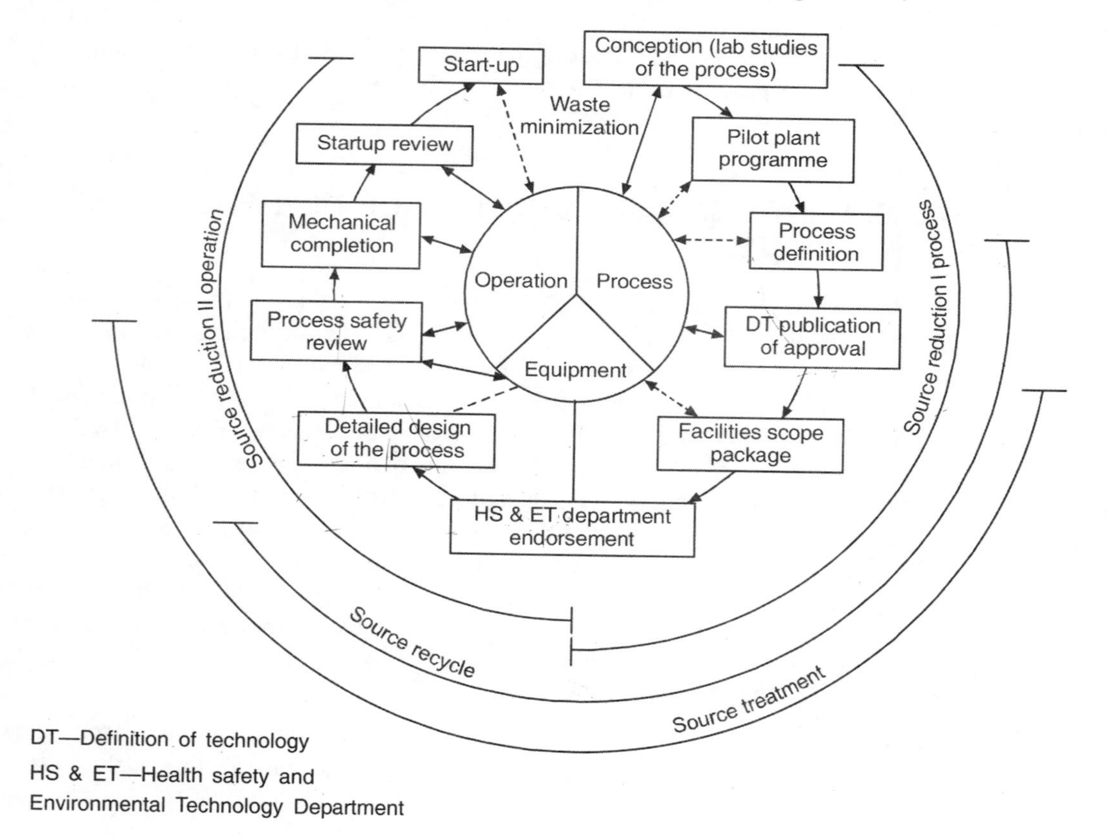

Figure 10.5 Waste minimization and technology development.
Adapted from Berglund, R.L., and Snyder, G.E., *Hydrocarbon Processing*, April 1990.

These opportunities include

1. When the original laboratory and pilot-scale studies are carried out and the process is conceptually defined.
2. When the process is being developed.
3. When the final process technology is defined.
4. When the process unit is designed and engineering specifications prepared.
5. When the unit is constructed and production commenced.

Yet, the emphasis at each stage of new technology development programme varies. After the process is conceived and laboratory and pilot plan-scale studies are initiated to define the process, the emphasis is on the process itself. Once the process is defined, and the facility is being planned and designed, the emphasis shifts to the equipment needed for the production process. With the completion of the engineering design and purchase of equipment, emphasis moves to successful installation of the equipment and operation of the process. Each of these stages of new technology development concentrates on a different aspect of the process. Each provides an optimum period for a different approach to waste minimization, to be emphasized.

10.7.2 Conception of Technology

After exploratory research personnel define a desired product and during the early stages of process development (conception, laboratory investigations, pilot-scale studies), opportunities for source reduction of waste can be most effectively considered. At this point of time, key aspect of the waste minizmation programme is to develop an appropriate mass balance around the process. It is desirable if this mass balance is as comprehensive as possible including factors such as the fate of trace contaminants in raw materials, degradation products, catalysts, sorbents and cleaning solvents. An assessment of the wastes that would be coming from the process is made. Then, for each effluent stream the following check list is completed:

1. The method or the way the waste gets generated.
2. The cause for its generation.
3. The reason for its being hazardous.
4. The possible ways of reducing its volume or quantity.
5. The cost involved to bring about a reduction in its volume and toxicity.
6. The economic feasibility of reducing its volume and toxicity.

An objective of the research personnel at that point of time should be minimal non-usable by-product formation. Thus, source reduction of the wastes would be emphasized at this stage of the operation (even though, source recycle might also be considered). The process modifications that can be incorporated at this stage of technology development are improved controls, minimizing water or solvent use, waste stream segregation or concentration, optimization of reaction conditions, improved catalyst, change of reactants, internal recycle and eschewal of contact between water and organics.

10.7.3 Definition of Technology (DT)

It is well understood that continued consideration of waste reduction opportunities during process development and definition is of crucial importance to an effective and successful waste minimization programme. As a matter of fact, a significant milestone in the development process takes palce when the Definition of Technology (DT) document is prepared for the new technology under consideration. It is prepared by the research and engineering team in charge of developing the process for review by project management for commercial viability and by Health, Safety and Environmental Technology Department, for potential concerns in these aspects. During development and DT review, a more extensive examination of the additional waste reduction and waste management issues can occur. In particular, the development and review of the DT's Health, Safety and Environmental Affairs Section will carry out the following:

1. Identification of critical data and technology necessary for the design and operation of a safe health-conscious and environmentally sound manufacturing facility.
2. Collection of the available critical data and documentation of the resolution of any HS and EA critical isues that can be resolved at the Definition of Technology stage.
3. Development of action plans for obtaining any unavailable critical data and/or for determining and solving any outstanding critical issues.

This stage in the development process comes about, however, when most of the options for reducing waste through process or raw material changes should have already been considered. This stage of the programme serves as a formal review of these activities and provides additional guidance for effecting a reduction in waste through modifications in

1. Configuration of the process
2. Selection and specification of equipment employed
3. Waste recycle
4. Waste treatment at the source
5. Waste management methods.

The two questions that are asked to the researcher in the preparation of the DT documentation are:

1. As a consequence of a waste minimization opportunities review process, have process wastes streams or gaseous emissions been eliminated, recycled or treated to mitigate toxicity? If so, the researcher is required to summarize these streams, and the waste reduction practice initiated.
2. For all remaining process wastes, effluent streams and gaseous emissions, the researcher is asked to summarize the alternatives for reduction, recycling or treatment considered.

A summary of possible waste reduction options is provided to the research personnel which will serve as a starting point for evaluating the waste reduction opportunities. These opportunites are listed under the general categories of (1) Process changes, (2) Recycle, (3) Treatment at source, and (4) Administrative controls.

In reviewing the applicability of these waste reduction options, the research personnel and review teams are asked to use a detailed check list which is hereinafter.

Check list for identifying critical issues during Definition of Technology (DT) document preparation to analyze waste management options

1. *Technical suitability:*
   - Is technology available and usable without modification?
   - What major equipment modifications are required?
   - Are there major waste modifications or pre-treatment requirements?
2. *Environmental suitability:*
   - How will the effluent be reduced in volume or hazard?
   - Will secondary releases, now or later, give rise to new air water or solid waste pollution problems?
   - Could the technology result in any new worker safety problems?
3. *Regulatory suitability:*
   - Will the technology result in wastes of less regulatory concern?
   - Can permits realistically be obtained in a reasonable timeframe for the technology under consideration?
   - Will additional regulations be imposed which could lead to additional air, water solid waste controls?
4. *Public acceptance suitability:*
   - Will the use of the technology to reduce the waste at the proposed location be acceptable to the residents, affected by the operation?
5. *Economically suitability:*
   - What is the cost involved in this compared to other technologies?

It is obvious that, the proper completion of this check list requires the interaction of individuals knowledgeable of the process, waste reduction technology and environmental regulations; hence the development of Waste Minimization Opportunities Review (WMOR) team.

To document any waste reduction considerations that have been already incorporated into the proposed process configuration and to serve as a focus for the DT review team to consider additional waste minimization opportunities (the researcher is required to complete) two special forms related to waste minimization prior to the DT review. These forms are given here in the form of Tables 10.2 and 10.3.

10.8 DEVELOPMENT OF AN EFFECTIVE START-UP, SHUTDOWN AND MALFUNCTION PLAN FOR A MANUFACTURING FACILITY

Industrial organizations that emit Hazardous Air Pollutants (HAPs) are required to apply the Maximum Achievable Control Technology (MACT) for minimizing HAP emissions. These MACT standards, also termed National Emissions Standards for Hazardous Air Pollutants (NESHAPs) in the U.S. include general provisions that require affected sources to develop and implement a written Start-up, Shutdown and Malfunction (SSM) plan. The two main objectives of this plan are to guarantee the use of good air pollution control practices and the rectification of malfunctions as soon as practicable.

Table 10.2 Condensed Sample of Form for Environmental Data Waste Reduction Description

Form G1

Process -- Submitted by --

Division -- Date --

Location --

Environmental data-description of waste reduction

Waste Stream Minimized	Total Discharge Rate	Composition of the Waste	Waste Minimization Approach Taken into Account	Reason for Rejection	Waste Minimization Approach Used	Anticipated Change
Gaseous effluents						
Aqueous effluents						
Residues solids sludge						

Table 10.3 Condensed Sample of Form for Environmental Data Waste Reduction Description

Form G2

Process -- Submitted by --

Division -- Date --

Location --

Environmental data-description of waste stream

Waste Name	Normal or Abnormal	Continuous or Intermittent	Total Discharge Rate	Composition	Problem Chemical List	RCRA Hazardous Waste	Planned Disposal Technique	LBS Waste/lb per lb Product

Development of the SSM plan should be a combined effort between environmental and operations personnel; complex sources may call for external assistance. Some important aspects of the suggestions made by Chad Scott. P.E., et al. on how to prepare and how to incorporate computer based Compliance Management Tools (CMTs) into its monitoring and record keeping elements are presented here.

Suitable equipment

The first step in developing an SSM plan is to prepare a list of the equipments subject to the MACT standards. This list is expected to include equipments, such as process equipments (e.g. centrifuges, reactors, distillation columns), storage tanks, recovery devices (e.g. oil-water separators, strippers), control devices (e.g. thermal oxidizers scrubbers), and continuous monitoring systems, i.e. monitoring systems employed to demonstrate compliance with the MACT standards during normal operations.

Possible malfunctions

The next step is to prepare a list of all possible malfunction scenarios for the applicable equipment. Discussions may be held with operating personnel regarding each scenario in order to determine whether a particular malfunction could lead to excess HAP emissions. Some good resources for identifying and evaluating malfunctions are standard operating procedures and existing alarm in the process automation systems.

Examples of some potential malfunctions that may give rise to excess HAP emissions include the following: high pressure in a reaction vessel, low flow to a scrubber, low temperature in a thermal oxidizer, and high level in a storage vessel.

Identification of corrective actions

Corrective actions have to be identified for all malfunctions that have the potential for excess HAP emissions. The standards do not necessarily require facilities to control HAP emissions resulting from malfunctions to the level established in the standard, but to do their best to keep the emissions at the minimum. The corrective actions must be documented in the start-up, shutdown and malfunction plan. Operations personnel should conduct a review of the proposed corrective actions to validate that each will effectively mitigate the malfunction and the resulting excess HAP emissions, while also ensuring the provision of adequate operational flexibility.

For instance, in case a process vessel malfunctions due to high pressure, leading to excess HAP emissions, this malfunction scenario has been identified in the SSM plan and a corrective action has been specified to lower the pressure in the process vessel below the alarm point. The wording of this corrective action allows the plant operators to respond to the malfunction to minimize excess HAP emissions, achieve compliance with the recommended standard and maintain operational flexibility.

Examples of corrective action for the malfunctions include the following:

- High pressure in a reaction vessel; the pressure may be brought down in the reactor below the high level set point.

- Low flow to a scrubber; the scrubbing water flow rate to the scrubber may be increased or control of the vents may be transferred to the back-up control device.
- Low temperature in a thermal oxidizer; transfer of control of the vent stream to the back-up control device.
- High level in a storage vessel—the level in the storage tank may be reduced.

When two corrective actions are available both should be in the Start-up, Shutdown and Malfunction (SSM) plan. This prevents the facility from deviating from the plan (and having to report the deviation to the pollution control agency) if one of the alternatives is not available or is not feasible when a malfunction takes place in the equipments of the operating plant.

Recording malfunctions

Part of an effective SSM plan implementation is to record the time and duration of each malfunction event identified. Development of suitable compliance management tools (CMTs), such as monitoring and record keeping system turns out to be almost essential in order to demonstrate continued compliance with the SSM requirements.

Some of the most common CMTs are databases that are custom-designed and implemented jointly by the environmental, process and information technology personnel. Some commonly employed platforms that can integrate with existing system to meet the SSM plan requirements include Microsoft Access, Microsoft SQL server and Oracle. One needs to include in the SSM CMT, the monitoring instruments (e.g pressure gauge, thermocouple, flow transmitter, valve position), that will be employed to record SSM events for each piece of equipment subject to the standard. If instrumentation is not available, visual inspections of certain equipment such as bypass lines, should perforce be performed and documented at regular intervals to demonstrate, that the start-up, shutdown and malfunction events are not taking place. If one has to consider recording time and duration information for malfunction events in a process historian system, in this case tags can be set up to record the necessary information from each malfunction event to simplify record keeping. In this way, the monitoring instruments record the process information, the data are stored in the facility's process historian and the SSM CMT queries the process historian to get back only the information that is related to SSM events.

Excess emissions from the facility

A list of start-up and shutdown procedures for the equipment subject to the MACT standards can be developed. This information can often be found in the standard operation procedures. These procedures may be discussed with the operations personnel to determine whether a particular routine start-up or shutdown activity gives rise to excess hazardous air pollutant's emissions. Any activity that results in an excess emission of hazardous air pollutants, must be documented in the start-up, shutdown and malfunction plan.

The start-up and shutdown procedures can be entered into the SSM computer-based Compliance Management Tools (CMT). If the procedures are available online, the SSM CMT can link to the electronic procedures over the facility's network, thus eliminating problems that are encountered in maintaining the most current versions of the procedures in the SSM plan.

Development of maintenance procedures

Specific maintenance procedures for the air pollution control device and the continuous monitoring systems (CMS) must be developed and documented in the SSM plan, including the frequency of implementation of them. Routine calibration of the CMS is needed. An onsite inventory of critical spare parts must be maintained, for instance, a spare thermocouple as a replacement for the one monitoring the temperature in a thermal oxidizer.

Routine maintenance of all monitoring equipment must be documented. As with the start-up an shutdown procedures, the maintenance procedures are entered into the SSM CMT or linked electronically from the CMT. When a maintenance database is already available at the facility, the maintenance system can be leveraged to extract the information needed to comply with the SSM provisions.

Reporting requirements

Two kinds of reports are required; the immediate SSM deviation report and the semi-annual SSM report. A deviation report is sent to the regulatory agency each time when SSM event takes places and the facility deviates from its SSM plan. This notification must be made within two days by phone or facsimile, followed by a written letter. The semi-annual report summarizes all of the deviations in the biannual reporting period.

Customized reports can be designed and incorporated into the SSM CMT to provide both immediate and periodic reports. The immediate deviation report, as well as the information needed for the semi-annual report, can be generated by the computer-based compliance-management tools.

10.9 BIOFILTRATION PROCESSES FOR THE CONTROL OF VOLATILE ORGANIC COMPOUNDS

Biofiltration is gaining a lot of importance as an emerging energy-efficient technology for the control of the volatile organic compounds (VOCs). It has been employed extensively in the US and Europe for the control of the odours from waste water treatment facilities, composting facilities and the other odour producing operations. During the past few years, it has been employed increasingly in the US for treating high volume low concentration air emission streams. A lot of research studies are being conducted to characterize its suitability for a wide variety of air emission control applications.

In the biofiltration process, off-gases containing biodegradable VOCs and other toxic or odourous compounds are made to pass through a biologically active bed of peat, soil and other media. Contaminant compounds diffuse from the gas phase to the liquid or solid phase in the media bed, get transferred to the biofilm layer where microbial growth takes place and subsequently are biodegraded.

Biofiltration has become an attractive alternative to conventional air pollution control technologies (e.g. scrubbers, thermal oxidizers) for the following reasons (Stephen Adler, F.):

1. Removal efficiencies of greater than 90 per cent have been observed for many of the more common air pollutants, including some of those listed by the Environmental Protection Agency U.S.A. as Hazardous Air Pollutants (HAPs).

2. Owing to lower capital and operating costs biofiltration may offer economic advantages in applications in which the air stream contains pollutants at relatively low concentrations (upto 1000 ppm although this is very contaminant-specific and varies widely) and moderate to high flow-rates which may range from 20,000 to 100,000 cft/minute, depending on the nature of the contaminant.
3. Another advantage is biofiltration does not require large quantities of energy during operation and generates a relatively low volume, low-toxicity waste stream.

However, it does not typically achieve the very high (more than 99 per cent) destruction and removal efficiencies or maintain the relative consistency of treatment demonstrated by technologies that are not dependent on microorganisms.

10.9.1 The Basics of Biofiltration Technology

Biofiltration is a general term applied to the conversion of gas phase—chemical compounds to the common biological degradation products of carbon dioxide, water and inorganic salts. It works on the basis of two primary fundamental mechanisms which are sorption and biodegradation.

Technologies that can be considered to be some forms of biofiltration include soil beds, biofilters, bioscrubbers, biotrickling filters and engineered biofilters. All these operate based on the same fundamental mechanisms of contaminant sorption and biodegradation and they of course have different designs and control parameters, operational flexibility and performance characteristics. It may be noted in this context that conventional trickling filter employed for waste water treatment is sometimes referred to as a biofilter, but it is a totally different technology. A typical biofilter configuration is schematically represented in Figure 10.6.

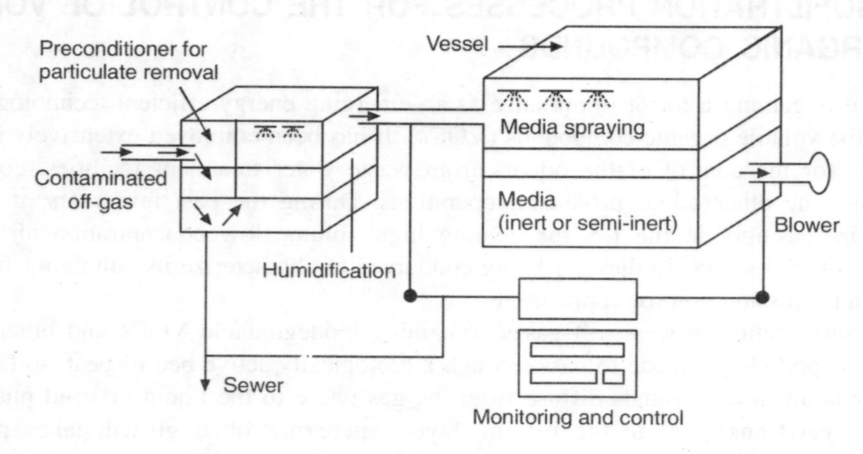

Figure 10.6 Arrangement for a typical biofilter.

Adapted from Stephen, F. Adler, *Chemical Engineering Progress*, April 2001.

The contaminated off-gas is passed through a pre-conditioner for particulate removal and humidification (if necessary). The gas stream that is conditioned is then sent into the bottom of a filter bed of soil, peat, composted organic materials (such as wood or lawn waste), activated carbon, ceramic or plastic packing or other inert or semi-inert media. The medium provides a surface for microorganisms attachment and growth. The off-gas stream is typically either forced or induced through the system with the help of a blower. A vent stack is employed when the situation warrants to meet monitoring or dischage requirements.

Mixtures of media types are sometimes used to provide operational advantages. In a soil peat or compost bed, the medium itself may provide some or all of the essential nutrients to facilitate microbial growth. Bulking agents and/or minerals can be incorporated into the media, depending on pH control requirements.

As the contaminated gas stream passes through the bed, contaminants get transferred from the gaseous phase to the media. Three primary mechanisms that are responsible for the occurrence of this transfer and the subsequent biodegradation in organic media biofilters are given here.

1. Gas stream—adsorption on organic media—desorption/dissolution in aqueous phase—biodegradation.
2. Gas stream—direct adsorption in biofilm—biodegradation.
3. Gas stream—dissolution in aqueous phase—biogradation.

Once adsorbed in the biofilm layer or dissolved in the water layer surrounding the biofilm, the contaminants are made available to the microorganisms as a food source to support microbial life and growth. Air that is free, or nearly free, of contaminants is then drawn off from the biofilter.

There are many variations to this basic approach. Biological activity in a filter will ultimately lead to degradation of a soil or compost media as organic matter is mineralized and the media particles are compacted. Degradable filter materials typically require replacement every five years if not earlier.

Proper media selection influences biofilter performance with respect to its compaction and useful life. In addition, the media largely determines environmental conditions for the resident microorganisms. These microorganisms are the most important component of the biofilter, since they produce the actual transformation or destruction of contaminants. Microorganisms can vary significantly in metabolic capabilities and preferences. Naturally occurring microbes are normally suitable and most desirable for treating most of the gas phase contaminants. However, some of the more unusual anthropogenic chemicals may require specialized microorganisms. Sometimes these organisms can simply be taken from sewage sludge and acclimated to the specific contaminants that are present. In a few cases, specially grown pure mixed or genetically engineered cultures may be preferred. Microbial cultures require a carefully controlled environment for optimal contaminant degradation. The most significant environmental factor for microbial function is the moisture in the contaminated air stream entering the biofilter. Many of the industrial or remediation off-gases have less than 100% relative humidity; hence supplemental humidification is frequently needed to minimize bed drying. This can be accomplished with an upstream humidifier (commonly a spray tower), spray nozzle humidifiers mounted within the biofilter or steam injection built into the biofilter. Bioscrubbers and biotrickling filters, which depend on a recycled aqueous phase solution, do not require pre-humidification. Humidification is also generally the single most influential parameter influencing the sorptive capacity of a biofilter, especially at lower inlet concentrations where Henry's law (for mass transfer) controls the mass-transfer rates within the biofilter.

In the past, it was a common practice to construct the biofilters as single bed system. Recently, fully enclosed biofilters have turned out to be more popular. These are frequently required to comply with emission monitoring requirements. Enclosed systems normally contain separate stacked beds in parallel or in series. This allows for a greater contaminant loading over a given footprint area. Fully enclosed systems also provide more precise control of biofilter moisture, thereby reducing the possibility of failure due to moisture level fluctuations.

10.9.2 Compounds Amenable to Biofiltration Technique

Biofiltration has been demonstrated to be efficient for the removal or destruction of many off-gas pollutants, particularly organic compounds, but also some inorganic compounds, such as hydrogen sulphide and ammonia. Many factors contribute to the overall removal efficiency. Owing to the fact that biofiltration functions through contaminant sorption, dissolution and biodegradation, contaminants that are amenable by biofiltration must possess two characteristics.

High water solubility: This in combination with low vapour pressure, gives rise to a low Henry's law constant, and thus increases the rate at which compounds diffuse into the microbial film that develops on the media surface. The classes of compounds that show a tendency to exhibit moderate to high water solubility include inorganics, alcohols, aldehydes, ketones and some simple aromatics (BTEX compounds). Compounds that are more highly oxygenated, are generally removed more efficiently than simpler hydrocarbons. However, some biofilter designs have been developed to handle less water-soluble compounds, such as petroleum hydrocarbons or chlorinated hydrocarbons.

Ready biodegradability: Once a molecule is adsorbed on the organic material in filter media or in the biofilm layer, the contaminant must then be degraded. Otherwise, the filter bed concentration may increase to levels that are toxic to the microorganisms or unfavourable to further mass transfer (sorption and dissolution). Either of these conditions will result in reduction in the biofilter efficiency or even complete failure. More readily degradable organic compounds include those having lower molecular weights and those that are more water soluble and polar. Some inorganic compounds such as hydrogen sulphide and ammonia can also be oxidized biologically.

Research now in progress aims to identify methods of treating contaminants that were previously considered to be untreatable by biofiltration process, such as chlorinated hydrocarbons. Use of innovative reactor designs, specialized or anaerobic microbes or supplemental substrates can be helpful in achieving this result.

10.10 ADOPTION OF ENVIRONMENTAL HEALTH SAFETY PERFORMANCE IN PRODUCT DEVELOPMENT

Chemical and processing industries are paying ever increasing amounts of attention to the environment, health and safety. Customers would definitely prefer product and processes with superior EH and S performance. Besides, society holds industry responsible to ensure that chemical products and processes perform safely and responsibly in all stages of their life cycle, from raw materials sourcing to disposal.

As a consequence, there is a drive and concern to incorporate EH and S considerations, into each product from its production. This is the cost-effective way to manage a product's environmental and safety performance and to avoid problems that can lead to recall, remediation or expensive redesign.

10.10.1 The Process of Product Development

An organization involved in the manufacturing activity can design EH and S considerations into a new product or process by following the methodology reported by Gregory Bond some of the salient features of which are presented in this section. This methodology can be termed as Development and Commercialization Technology (D and CT) process.

One of the advantages or merits of D and CT is its ability to evaluate the proposed product's EH and S viability before making investment of time and capital. To do this, EH and S concerns and regulatory compliance considerations are integrated into the product development process. A cross functional team guards the project through the five stages and at each stage a decision is made whether or not it is worth going on to the next stage.

As development activities are in progress, more and more data are gathered to determine both the costs of production and distribution and the cost involved in the compliance of EH and S needs. Many a time, the latter costs which may include extensive testing will be a major part of development costs. If these are prohibitive the product may be scrapped. Thus, D and CT is particularly suitable to products with high development costs and requiring extensive capital investment.

10.10.2 Identification of Showstoppers

The five stages of the D and CT process progressively review these data in detail. The early stages—concept shaping and concept analysis, provide the case for a fit with the business and a *macro-examination* to identify major problem or showstoppers. They are generally run through rapidly a few weeks or months in most situations and are structured to screen out the winners and identify showstoppers before the investment of large amounts of time or money is made.

The later stages of D and CT process viz., validation, development and implementation, need answering more sophisticated questions and require more resources to answer them. Members of the EH and S staff are involved in all stages, but they are much more active in the third or validation stage.

All evaluations and decisions at each stage of the process are the responsibility of a cross functional team, comprising representatives from research and development, marketing, manufacturing, engineering and construction, legal and patent as well as EH and S. Precisely what types of data the EH and S team members contribute as the development proceeds will differ from case to case depending on the nature of the product or process. In the following sections the types of EH and S input each stage of the D and CT process are discussed.

Stage I Concept shaping

This stage does not include creation of a concept that has already been done and may have appeared from any number of functions including Research and Development (R and D). At this stage the multifunctional team screens the 'raw' idea evaluating its suitability for the company and shapes it into a business concept.

Stage II Concept analysis

At this level of the reviewing process, the heart of *assessment before investment* the basic hypotheses of the concept are reviewed and modified depending on the need. Critical issues (showstoppers) are identified and a preliminary economic model is generated to assess value to the company and the customer.

In the second stage, the principal EH and S Question to be resolved is whether any obvious EH and S issues associated with the concepts present, an unacceptable liability; EH and S members of the team also examine questions given here.

Whether the manufacturing, distribution, use and disposal of the end product will be socially acceptable and what would be the opinion of a responsible environmental group about this product?

In case, the intermediates, by-products or end-products are toxic and/or bioaccumulative, whether there is an emissive application (i.e., a process that generates emissions) envisioned with a potential for environmental and/or human exposure?

Conversely, whether there are environmental benefits to the concept that could in turn provide a competitive advantage?

Stage III Validation

In this key stage, which is likely to take a few months, the showstoppers identified in the previous stage are resolved to an *appropriate* level. Further modelling is conducted to examine a variety of questions covering issues, such as pricing, volume, and competitive response.

In the validation stage, the principal EH and S questions include the following:

1. Whether any of the raw materials, intermediates or by-products are being targeted for phase out by industry, regulating groups or responsible environmental organizations so that they may not be available in the future? Additional questions look into such issues as stratospheric ozone depletion, tropospheric ozone formation and carcinogenecity classification, to determine if there is a possibility of this product being restricted in the future.

2. Whether we or our end-use customers can anticipate safe and cost-effective transportation, storage, use and disposal of intermediates and products? This considers such issues as whether regulations pose current or pending transportation or siting constraints; whether the proposed manufacturing location is appropriate; and whether materials involved should be used only on-site and not transported.
3. What are the critical safety concerns of this project including factors such as chemical reactivity, flammability or explosivity of the raw materials, intermediates, by-products or end-products?
4. What is the expected life cycle of the product under consideration? Is recycling an issue with the project or its packaging? What will the ultimate disposal process of the product and by-products be? Will unacceptable quantities of waste be generated from the production process?
5. What are the appropriate EH and S timelines involving regulatory compliance, toxicology studies and cost/resources? What studies must be designed to acquire the data required for these studies?

Stage IV Development stage of the D and CT process (development and commercialization technology)

This is the stage where the development work is carried out, driven and managed by a multi-functional critical path plan. Here cycle time, reduction, based on best-in class industry standards is a top priority.

This is the point at which the EH and S members of the team address the questions. Has the business resolved all the previous issues and questions or does it have a plan or timeline available to resolve remaining issues and questions prior to product scale up and market launch?

At this stage too, a pre-launch EH and S risk review is prepared and a product in-charge is assigned to the product. The duties of the product in-charge include:

- maintenance of an effective product stewardship programme
- evaluation of customer practices
- participation in business EH and S risk reviews
- development of literature and information
- investigation of incidents and implementation of corrective action

Stage V Implementation

Finally, for such of those projects that pass successfully through the first four stages, a commercial scale manufacturing facility is built (or contract manufacturing is secured), operations are fine-tuned, the supply chain is developed, and marketing and sales plans are finalized. The work of the EH and S members of the team essentially gets to completion.

It is important, however, to document remaining EH and S issues and risks, to determine whether a formal risk review is needed, to schedule ongoing EH and S risk reviews if warranted by situation and to ensure that the Material Safety Data Sheet (MSDS) labels and product stewardship literature are in place.

10.10.3 Values of EH and S Contributions

The EH and S representatives on a D and CT project team contribute specific value to the design of a new product or process. First, as the company's leading experts on matters relating to the environment, health and safety, they are able to inject their concerns early in the product development, and by their asking the right questions they are in a position to increase the company's chances of making the right decisions in the face of possible EH and S concerns. Such decisions enable the following:

- A quicker and more cost-effective new product launch, because regulatory approvals are anticipated.
- More effective utilization of EH and S resources.
- Increased product acceptability in the environmental field, since environmental questions are settled.
- Greater product sustainability in the marketplace, since potential health and safety problems are addressed ahead of time.

This value is a significant contribution to the ongoing viability of the chemical and processing industry, particularly in this period of intense scrutiny of chemical products and processes and of increasing and stringent regulations.

10.11 TECHNIQUES FOR POLLUTION PREVENTION IN EQUIPMENT AND PARTS CLEANING OPERATIONS

Needless to mention, cleaning equipment and parts is an important preparatory step in many industrial processes. Cleaning removes dirt, oil, wax, oxides, residual products, monomers and other contaminants that may adversely affect the final product quality or the operating efficiency of the equipment or part concerned. Two general types of materials are removed, soft films such as dirt, oil and grease, and hard films such as hardened polymeric films, scale and paints. Until recently, solvent strippers, such as methylene chloride were frequently employed to remove both types of films from the surfaces being cleaned.

The main concern for facilities with batch manufacturing operations that employ multipurpose equipment is cross-contamination of product in addition to the environmental concerns such as air emissions and hazardous waste disposal. The pollution-prevention engineering practices and techniques outlined by Kenneth L. Mulholland, et al. will be useful in avoiding the necessity to clean the equipments employed in chemical and processing industry. Some of the important aspects of these approaches have been presented here.

10.11.1 The Nature of the Sources of Emission

Cleaning, degreasing and coating-removal technologies generally involve the application of an organic solvent (or solvents) to the material being removed. During the process, air emissions are invariably generated from the use of the solvent. And, after the cleaning process, other effluent streams solvent and metal-containing waste water and sludges remain that require disposal. The waste water streams and sludges often contain materials that are classified as a RCRA (Resource Conservation and Recovery Act) hazardous waste.

10.11.2 The Pollution Prevention Continuum

The pollution prevention continuum in Table 10.4, indicates the relative merits of options that are available to eliminate or reduce the frequency of equipment cleaning or to minimize the waste generated during the cleaning process.

Table 10.4 The Pollution Prevention Continuum for the Cleaning of Equipment

Extent of pollution prevention	*Technology or practice to be adopted*
>95%	Minimizing the need for cleaning, draining completely dedicating a single product, flushing with process material, continuous versus batch process
>90%	Loosening product specifications, modifying surfaces to eliminate cleaning
>75%	Carbon dioxide or cryogenic blasting, burning or heat volatilization, mechanical removal
>50%	High and medium pressure water washing, blasting with plastic beads, sodium bicarbonate, wheat starch, watering with detergent or ultrasonics
>10%	Flushing with solvent from another process, decreasing number of flushes, reusing last flush from previous campaign
<10%	Solvent substitution

The decision of how far to make progress towards a zero waste and emission design will be governed by a number of factors which include corporate and business environmental goals, economics, technical feasibility and regulations, that are applicable from time to time.

Achieving more than 95% pollution prevention

In the design of a new manufacturing facility, equipment should be designed without low spot, projecting ridges, or other crevices that will collect material that later becomes difficult to remove. Properly designed vessels will contain sloping interior bottoms and piping arrangements with valved low points or valves that drain back into the main vessels.

Existing process equipment can be retrofitted with drain valves installed at strategic low points. If the low points are too low for standard tanks and pumps, then a special portable collection vessel may be designed to collect and transport the material to a tank.

Besides this, equipment can be dedicated to a single product line. This eliminates the necessity to wash the inside of the equipment between production campaigns. Alternatively, flushing contaminated equipment with the saleable product or a process intermediate, then recycling the flush back to the process where it can be purified in existing separation equipments, can be done.

A continuous process inherently generates less waste, owing to the fact that there is no need to shutdown, drain and clean the equipment.

Achieving more than 90% pollution prevention

Relaxation of specifications for finished product allows for higher levels of impurities in the final product or cross contamination of products. This brings about the reduction or elimination of the need for solvent washes between product campaigns. Many a time, specifications for products manufactured in the same equipment are different and one set of specification may be more stringent than another. Through careful planning and inventory control, product change-over can be made from products with tighter specifications to products with relaxed specifications.

Application of an antistick coating such as polytetrafluoroethylene (PTFE), an example of this is: Teflon or silicon, to the interior walls of processing equipment could make easy drainage and removal of left-over residue possible.

Achieving more than 75% pollution prevention

Carbon dioxide blasting systems (Figure 10.7) comprise a refrigerated liquid CO_2 supply, an equipment for converting the liquid CO_2 to solid pellets.

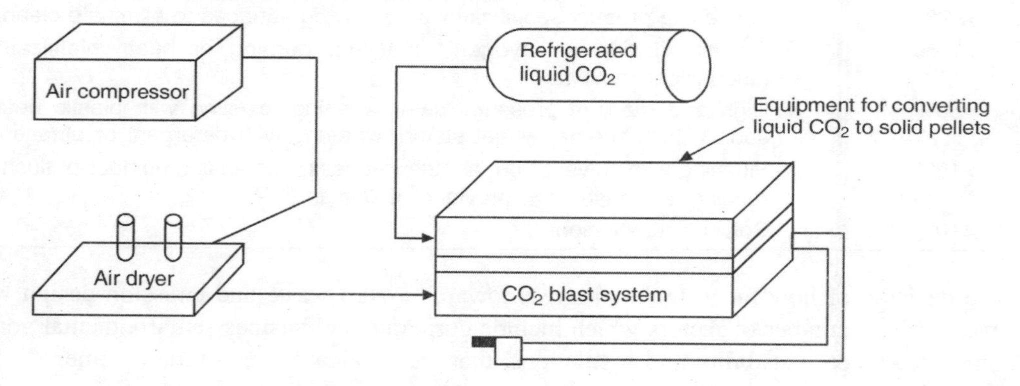

Figure 10.7 Carbon dioxide blasting system.

Adapted from Kenneth, L. Mulholland and James, A. Dyer, *Chemical Engineering Progress*, May 1999.

The solid pellets remove coatings by a combination of impact, embrittlement, thermal contraction and gas expansion. Owing to the fact that CO_2 pellets sublime, a waste water or liquid waste is not generated and a dry coating residue is collected.

Liquid nitrogen cryogenic stripping employs liquid nitrogen to cool the surface and to help to propel the plastic bead blasting media. The liquid nitrogen embrittles and shrinks the coating and the high-velocity, non-abrasive pellets crack, debond and break away the coating.

The removal of coatings using heat that is generated by flames, lasers or flash lamps gives rise to the vaporization of the coated materials. The non-volatile portions of the coating, such as metals form particulate ash that is removed and collected by a vacuum air removal system.

Where feasible, manual cleaning employing scrapers or spatulas might help in eliminating the necessity for any subsequent solvent wash.

Achieving more than 50 per cent pollution prevention

Employing high-pressure water washing. High-pressure water washing also called **water jet stripping**, employs the impact energy of the water to remove coatings. The water is collected, filtered and then recycled. The coating thickness and hardness dictate whether a high-pressure (15,000) to 30,000 psig) or medium-pressure (3,000 to 15,000 psig) water jet is needed.

A typical high-pressure water wash system for a process vessel is illustrated in Figure 10.8. The high-pressure water lance is attached to a carriage which is in turn attached to the bottom of the vessel. A chain drive moves the lance up and down the carriage as needed. A swivel joint at the base of the lance enables free rotation. The nozzle at the tip of the spinning lance has two apertures that emit cone-shaped sprays of water at 10,000 psi with a combined flow rate of 16 gal/min. Operation of the lance is controlled from a panel well removed from the vessel. The process is designed in such a way that, no high-pressure spray leaves the interior of the vessel. These precautions guarantee operator's safety during washout.

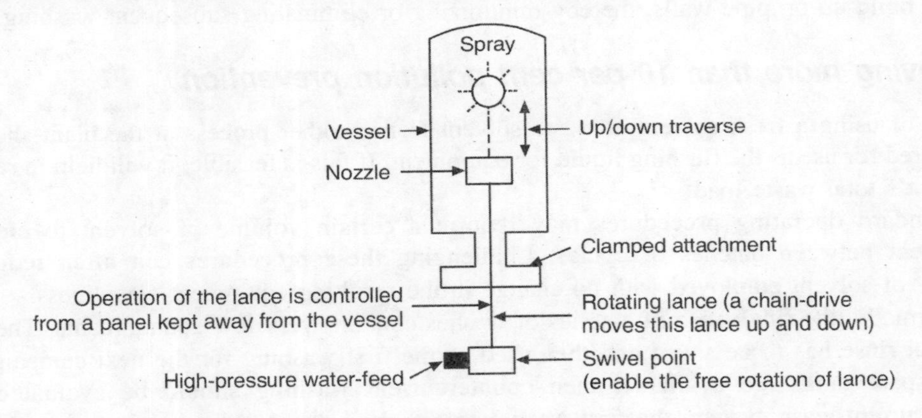

Figure 10.8 High-pressure water wash system.

Adapted from Kenneth, L. Mulholland and James, A. Dyer, *Chemical Engineering Progress*, May 1999.

Media blasting—several media, including plastic beads, baking soda and wheat starch can be employed to strip the coatings away by a combination of impact and abrasion. The plastic bead process employs low-pressure air or centrifugal wheels to project non-toxic plastic media at the coated surface. The beads are collected, cleaned and recycled. As the beads break down, they are discarded and replaced. The discarded material must be disposed of by using an appropriate method.

In the baking soda process, sodium bicarbonate is delivered by the wet blast system to get rid of the coating by impaction. As a consequence, the coating is shattered into a very fine particulate. The sodium bicarbonate medium is employed only once, and is not recycled.

Wheat starch blasting employs low pressure air to propel the particles at the coated surface. The coating is stripped away by a combination of impact and abrasion. The starch can be recycled and replaced as the particles break down. The fine particles and coating material must be disposed of properly.

Aqueous cleaning

Water with the appropriate pH or additives (such as detergents) can be employed to remove soft coatings. The water solution is collected and recycled until it becomes saturated with the material being removed. However, considerations must be given to the possible impact of the detergents and other materials on the effluent treatment facility.

Ultrasonic cleaning makes use of cavitation in an aqueous solution for achieving greater cleaning effectiveness. The high-frequency sound waves generate cavitation zones that exert enormous pressures (in the order of 10,000 psi) and temperatures (approximately 20,000 "F on a microscopic scale). These pressures and temperatures help in loosening contaminants and perform the actual scrubbing of the surface. Intricate surfaces can be cleaned effectively by the ultrasonic cleaning process.

Pipe cleaning device; these devices are pipe cleaning mechanisms made of any number of materials. They are activated by high-pressure water, product or air. These devices remove residue build-up on pipe walls, thereby minimizing or eliminating subsequent washing.

Achieving more than 10 per cent pollution prevention

Instead of using a fresh solvent, a waste solvent from another process at the plant should be considered for use as the flushing liquid for equipment. If this is feasible, it will help in reducing the plant's total waste load.

Standard operating procedures may require a certain volume of solvent to clean the equipment between batches or cycles. Challenging these procedures can often reduce the quantity of solvent employed with no change in the resulting equipment cleanliness.

Normally, the final wash of a series of washes contains very few contaminants. Therefore, this final rinse has to be saved and then used as the first washing for the next campaign.

If spare tanks are available, then countercurrent washing should be evaluated. In a countercurrent wash system, the first wash employs the solvent from the second wash of the previous campaign, the second wash employs the solvent from the third wash of the previous campaign, and so on. The solvent from the first wash of the present campaign is the only wash sent to recovery or waste treatment and disposal. As a consequence, solvent usage can be reduced to that needed for a single flush.

Achieving less than 10 per cent pollution prevention

The cleaning solvent can be replaced with another solvent that has less of a negative impact on the environment or that is not regulated.

10.11.3 Creating Awareness in Employees

Another important aspect of equipment cleaning is creating an awareness in the employees. Plant operators and maintenance personnel are in contact with the equipment every day. Organizing and conducting training programmes that help them to understand the sources and causes of waste generation and its impact on the business and employee safety, will definitely make a difference.

10.11.4 Research and Development in the Field of Cleaning

Research efforts are being made to understand better, the process of adhesion of contaminants to vessel surfaces and to identify aqueous based cleaning solvents that can replace traditional organic-based cleaning solvents. The research studies span the range from fundamental surface studies to characterize surface/organic interactions that promote surface fouling, to the practical applications of designing effective aqueous cleaning solutions and developing a robotic waterjet system for batch vessel cleaning.

One research study has shown that the adhesion of organic materials to vessel surfaces is mainly caused by acid/base interactions between the organic material and the vessel wall. In many situations, the residue material adhering to the vessel surface is different than the starting material and has different functional groups. Organic acids and bases will interact with hydroxyl groups on the metal surface. For this reason, the pH of the cleaning solution becomes critical and must exceed the isoelectric point (point of zero charge) for the metal surface. Above the isoelectric point, the metal surface will be negatively charged. The objective, then is to establish a net repulsion between the charged metal surface and the organic residue. For example, the isoelectric point for stainless steel is about pH 8.5, therefore a pH greater than 8.5 will be needed to clean most organic compounds from stainless steel surfaces. In the case of organic bases, very high pH values are sometimes required.

In another research programme, the scientists have developed an articulated robot arm to manipulate high-pressure water jets to achieve complete coverage of internal areas where fouling is known to take place and defined the jet properties (i.e. impingement conditions, composition, temperature, pressure, etc.) that ensure the desired cleaning efficiency. For instance, it was observed that cleaning efficiency is a strong function of travel speed and the distance of the water-jet nozzle from the vessel surface. The objective here was to maximize travel speed, to minimize cleaning time and waste water generation.

The use of hydrogen peroxide (an oxidant) as a cleaning agent in aqueous cleaning solutions has been examined. Hydrogen peroxide, when safely handled and employed, can help in the dissolution of organic acid residues from vessel surfaces.

Examples of successful pollution prevention effects

The successful pollution prevention efforts in cleaning operations is highlighted by the following four examples:

Minimizing the need for cleaning: A site manufactured two types of chlorinated aromatic compounds that are sold as three different products—pure product one, pure product two and a mixing of the two products. Between campaigns, large quantities of residual product remained that had to be removed to prevent contamination of the product in the next campaign.

In the past, this was carried out by flushing the equipment with the help of an organic solvent. However, using an organic solvent had several disadvantages; a large waste stream was generated; use of solvent contributed to long turnaround times between campaigns; a large amount of off-spec product was made due to contamination of the initial product with the previous campaigns residue; and the solvent wash contained a significant quantity of valuable product.

A pollution prevention team implemented an equipment drainage system to eliminate the need for the organic solvent. Valves were placed at the lowest elevation of the process and the residual product was collected in a special container at the end of the campaign, so that it could be reused in the next campaign of that product (Figure 10.9).

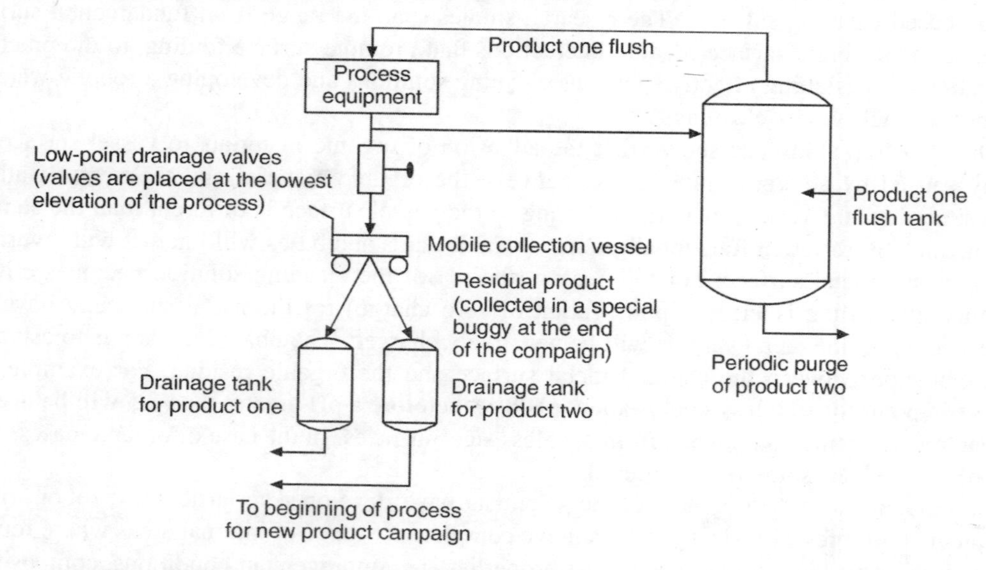

Figure 10.9 Low-point drainage system with portable collection buggy.
Adapted from Kenneth, L. Mulholland and James, A. Dyer, *Chemical Engineering Progress*, May 1999.

In making decision as to whether or not to implement the system, the pollution-prevention team had to consider the effect on product quality. With the new system, after the equipment is drained, a small quantity of the product residues remains. In the case of product two, the purity specifications are loose enough that equipment flushing before the next product campaign is not required. On the other hand, purity specifications for product one, are strict enough that flushing with a reserve amount of product one is required. The flush tank shown in Figure 10.9 was designed with this objective in mind. The amount of product two, within the flush tank is held constant by periodically drawing off small quantities of the flush for recycle back to the process or to make the mixed product.

By adopting the aforementioned technique, a 100 per cent reduction in waste generation can be achieved.

Making procedural modifications and relaxing product specifications: At a chemical manufacturing unit, a series of distillation columns was employed to purify different product crudes in separate campaigns. At the conclusion of each campaign, a portion of the product crude was used to wash out the equipment when the crude became too contaminated, it was sent for destruction in a hazardous waste incinerator.

First, an analysis of the procedure employed to wash a decant tank indicated that only one-tenth of the product crude wash material was really needed to effect cleaning. Secondly, a dedicated pipeline for each crude was installed, which eliminated the need to perform flushing of the line between campaigns. Thirdly, an extended and improved drainage procedure was developed for a large packed bed distillation column. Finally, the product specifications were relaxed, which resulted in a reduction in the number of washes that were required to maintain product specifications.

This organization concerned saw a reduction of 78 per cent reduction in waste generation as a result of taking the aforementioned steps.

Employing high-pressure washing technique: At one chemical manufacturing plant, cleaning of kettles needed 800 hours of labour and 110 tons of solvent per year. The cleaning process sometimes involved employees entering the kettles, to scrap the walls—a process that took each worker's 3 hours. Switching to a high-pressure rotating spray head effected a reduction in the quantity of solvent requirement, thereby reducing the time to be invested for cleaning. In addition to this, worker safety was improved by eliminating the necessity for workers to enter the vessels. By incorporating this change the management had a substantial savings.

In another manufacturing unit, several types of polymers were made in an agitated vessel that had to be cleaned periodically to maintain product quality. The vessel was cleaned by washing with a flammable solvent. To eliminate the use of the solvent a special high-pressure water jet was installed to clean the reaction vessels.

This step effected a reduction of 98 per cent in waste generation and was not very expensive.

In the manufacture of PolyVinyl Chloride (PVC), polymer build-up on reactor surfaces, agitators, brackets and other parts required cleaning after every batch. The manual cleaning operation was replaced by a high pressure water jet. To minimize the need for water jet cleaning, an additive held under patent was employed to suppress the formation of polymer build-up on the walls. After each batch, a low-pressure water rinse removed sufficient material to prevent contamination of the next batch. Only after 500 batches were over, there was a requirement for high-pressure water jet wash.

Flushing equipment with waste solvent: In the manufacture of cast type product, methylene chloride was employed as the process flush and cleaning for over two decades. Two significant changes were incorporated into the cleaning procedure. Firstly, methylene chloride was replaced with a dibasic ester waste stream from another process. Secondly, the waste load was further reduced by modifying the flushing procedure. Methylene chloride emissions were reduced by 97 per cent and a proposed expensive project to control methylene chloride emissions into the air was avoided.

10.12 MINIMIZATION OF WASTES AT OPERATING PLANTS

It has been generally accepted that treatment alone, however, will not resolve hazardous waste problems. It is very important that the generation and the subsequent need for treatment, storage and disposal of hazardous wastes be minimized, if not completely eliminated. The first step towards this end is waste minimization.

Many methods of waste minimization are available which include inventory management, raw material substitution, process design and operation, volume reduction, recycling and chemical alteration of the waste. Here it is outlined as to what each method involves and can serve as a check list of actions reported by Robert Katin that an operating plant can implement.

10.12.1 Inventory Management

It would be desirable to make an inventory of all raw materials and only purchase the minimum quantity required. As new plants are constructed and as processes undergo a variation, the store room is often the last to find out. It is likely that the original raw materials required and the quantity required change over time. There is no need to buy chemicals that are no longer necessary for the process. It would be advantageous to purchase materials from suppliers who will refill used containers. This eliminates the generation of waste water for say, triple rinsing an empty container. It also eliminates disposing of an unwanted container that may have adequate residue to be treated as a hazardous waste.

10.12.2 Raw Material Substitution

It is better to concentrate our efforts on one or two streams that have the largest volume or account for the greatest degree of hazard. Waste streams with the highest costs for treatment and disposal should naturally receive priority. It usually turns out that the largest single waste stream is often the most economical to modify. Substituting raw materials can be a very expensive proposition—a significant amount of labour will be expended, process design variations may be required and equipment may have to be modified or replaced. Therefore, it is of paramount importance to assess the waste streams and prioritize them.

It would be desirable to buy fewer materials that are hazardous; instead, one can go in for either non-hazardous or less hazardous materials. For instance, electronic industry has been using 1,1,1-trichloroethane (TCA) instead of trichloroethylene (TCE). While TCA is less stable and more corrosive than TCE, it has a higher Threshold Limit Value (TLV) for human exposure and is not considered a Volatile Organic Compound (VOC) responsible for generating atmospheric ozone. Likewise phosphate-based corrosion inhibitor chemicals can be used in cooling towers instead of chromates. The end products may be made less hazardous, by redesign or reformulation. To prevent or minimize the formation of hazardous by-products, more pure raw materials can be purchased. Use of plastic media blasting in place of solvent stripping can be considered for removing paint. Detergents (such as alkaline salts, surfactants, and emulsions) or treatment by means of chemical reactions (such as reacting chelating agents with soils to form soluble complexes) may be substituted for industrial solvents. Aliphatic hydrocarbons (naphtha, kerosene, diesel), aromatic hydrocarbons (benzene, toluene, xylene), non-flammable solvents [TCA, TCE, perchloroethylene (PCE)], chlorofluorocarbons (CFCs) and polar solvents (ketones, alcohols, esters, ethers, amines) are among the solvents that can be frequently replaced.

Lower volatility materials may be used to clean metal parts. For cleaning pre-assembly and sub-assembly aircraft components prior to surface bonding, less harmful materials like trichlorotrifluoroethane and TCA instead of high volatility hydrocarbons such as Methyl Ethyl Ketone (MEK) can be used.

Heavy metals may be removed from liquid streams by precipitation before discharging the liquids to the plant's liquid effluent treatment system. For instance, a number of streams containing chromic acid are employed in metal surface finishing. The heavy metals are precipitated as sludges using either sodium hydroxide, lime (calcium hydroxide) or ferrous sulphate. A ferrous sulphate-sodium sulphate process, tested in a company, has achieved sludge weight reduction of up to 90 per cent over lime precipitation. An organization has started using a terpene-water emulsion as a substitute for ozone-damaging chlorofluorocarbons in some electronic degreasing operations.

10.12.3 Process Design and Operation

Equipment that generates minimal or no waste may be installed and modification can be made on the equipment to enhance recovery or recycling operations. Equipment or production may be redesigned to produce less waste. The generation of paint waste can be minimized by employing high-volume low-pressure paint guns or electrostatic paint equipment. Employing proportional paint-mixing equipment to prepare two or three component paints can also bring about a reduction in the generation of hazardous paint waste. The operating efficiency of equipment can be improved by maintaining a strict preventive maintenance programme. It is well known that regularly scheduled maintenance of motors, pumps and machines can minimize oil, process fluid and hydraulic fluid losses. The reactions may be optimized, to the extent possible.

10.12.4 Effecting Volume Reduction

Wastes can be segregated by type for easier recovery. Users of large quantities of solvents should segregate, not mix, solvents so that they can be recycled. The sources of leaks and spills can be eliminated. Hazardous wastes can be separated from non-hazardous ones. Good house-keeping can be maintained by employing shovels and brooms instead of washing down equipment pads and the like, with water.

Undesired waste streams may be concentrated and the volume of waste that must be treated as hazardous, may be reduced by techniques like: reverse osmosis, evaporation, ion-exchange, precipitation, centrifugation, gravity separation and distillation.

10.12.5 The Advantages of Recycling

Common industrial solvents can be recycled by distillation. In an organization using 30,000 gal/yr of 1,1,1-trichloroethane to degrease parts after forming and plating, a distillation system was installed and it resulted in lots of savings for the plant. The still bottoms were subjected to incineration and this step resulted in reduction in fuel requirements for the process furnace.

Sour water in a refinery can be recycled into sulphur and ammonia. Steel industry's pickling liquor can also be recycled into ferric and ferrous chloride.

In petroleum refineries, it may be a desirable, to recover waste oil from their API separators, and dissolved air floatation units and feed the recovered oil to the coker to take advantage of the oil's heating value instead of paying for disposal of the waste oil.

Relatively pure materials can often be sent off-site to a recycler.

Small quantities of solvent that would require to be laboratory packed (Laboratory packing in this context refers to the packaging of small, laboratory sized containers in a larger drum), for disposal, can, sometimes be bulked in a mixing vessel for subsequent recycling.

Spent petroleum refinery alkylation acid can be burned in a combustion chamber and regenerated into fresh sulphuric acid. The hydrocarbons present in the spent acid merely add fuel value in the sulphuric acid plant combustion chamber. Likewise soap manufacturing industry spent acid can also be regenerated; because it does not contain hydrocarbons, however, fuel must be added to regenerate it.

Spent caustic from refinery alkylation units contains enough caustic that can be sold to paper manufacturing companies. Similarly, spent caustic from aerospace processes can meet aluminium producer's feed requirements.

10.12.6 Chemical Alteration

Hazardous waste can be encapsulated so that it is sealed with a material that makes the exterior of the waste non-hazardous. An example of this is the encapsulation of nuclear waste in borosilicate glass.

Waste oil can be chemically treated and converted into a saleable aromatic distillate.

Adding lime to a spent catalyst that contains clay and phosphoric acid enables the materials to be sold as concrete for parking lots and pavements.

Waste hydrochloric acid that had been disposed of through deep-well injection is now converted into calcium chloride and employed in the manufacture of polyamide fibre.

It is possible to treat the calcium sulphate waste from the production of hydrochloric acid and use it as a soil stabilizer.

As can be observed from the aforementioned examples, there are a number of ways by which waste minimization can be achieved in chemical process industries.

10.13 AIR EMISSIONS INVENTORY FOR ENVIRONMENTAL PROTECTION

A manufacturing facility can keep track of its emissions and compliance status efficiently and in a changing work environment, by performing a thorough, technical emissions inventory. This can be carried out by developing process-specific factors to determine air emissions from all the processes, preferably linked to easy-to-obtain production data.

There are many techniques that can be employed to estimate emissions from the manufacturing processes. Many of these methods outlined by Marc Karell may be used in any specific manufacturing unit. These are presented here.

10.13.1 Emission Factors for Different Processes

Air pollutants' emission factors are based on information obtained by the Environmental Protection Agency/Pollution Control Boards over many years, including measurements performed on actual operating equipment in the plant. Most of these factors are normalized in terms of pounds or grams of pollutants emitted per usage (for instance, grams per thousand litres of fuel oil burnt). The emission factors are offered in both metric and english units. In addition to the EPA—published factors, in some countries, industry associations and manufacturers publish their own emission factors based on data they have gathered.

The US Environmental Protection Agency (EPA), maintains a compendium (a summary) of emission factors for many different processes called **AP–42**. An advantage of using AP–42 or other published emission factors is the simplicity of the method. Engineering calculations and testing are not required and it is not necessary to hire an engineering consultant. One has to look up the appropriate emission factors and multiply them by the usage in the period of time of interest to make an estimation of emissions. Because most manufacturing facilities maintain usage records such as the quantity of fuel combusted, this turns out to be simple and verifiable. However, there are several major disadvantages in using AP–42 emission factors.

Overall, emission factors are most advantageous when they are specific to the equipment, the plant in question is employing. For instance, many manufacturers of combustion equipment provide emission factors for different pollutants for the specific or related models of equipment for sale. Owing to the fact that the factors are based on testing of that specific model, there is a good chance that emissions of that unit in any plant is likely to be similar.

10.13.2 Material Balance for Estimating Emissions

Another frequently employed technique for estimating emissions is a material balance, in which the fate of each compound is quantified throughout its life cycle in a manufacturing plant. If a plant is in a position to estimate the quantity of compound entering the plant (purchased or used in its processes), the quantity consumed and the quantity disposed of in solid waste or in its liquid effluent and lost in any spills, then the difference between these quantities can be a reasonable estimate of losses by other means, which would mainly be due to evaporation (air emissions).

Many of these quantities can be estimated using Standard Operating Procedures (SOPs) and purchase batch and waste disposal records. In this case a material balance could reasonably represent a relatively inexpensive method to estimate air emissions.

However, employing material balances has several potential disadvantages. Because the fraction of material not accounted for and therefore, considered emitted is generally very small, any error made in measurement or calculation of any parameter will have a major percentage impact on the emission estimate. For instance, if one considers a plant that uses 100,000 lbs/month of a solvent to facilitate a chemical reaction, It is estimated that 98,000 lbs of the solvent is disposed of in waste based on measuring the contents of selected waste drums and waste water samples. Therefore, by striking a material balance, we observe that the plant in question emits into the air 1 ton/month of that solvent. However, if the error in estimating the contents of the various waste drums and waste water was only 2 per cent then the total amount of the solvent emitted could have been closer to 4,000 lbs, two times that of the original estimate. For a complex material balance with many fates of the compound in question, even larger calculation errors are likely to be common. Of greater significance is the fact that material balances may possibly underestimate emissions, potentially giving rise to compliance issues for the operating plant.

Therefore, it is generally recommended that material balances be used only to estimate emissions for processes, the chemicals of which have a known, simple fate. For instance, material balances could be useful in estimating emissions from coating operations. The solvent that carries the pigment or resin to the substrate goes into the atmosphere as emissions. Solvent emissions can, thus, be simply and accurately determined as equal to the solvent fraction of the quantity of coating used.

10.13.3 Direct Measurement of Emissions

Direct measurement of emissions also known as *stack testing*, to develop emission rates represents the *actual* emission estimation technique, since a part of the actual exhaust is sampled during the operation and subjected to analysis. The standard procedure involves the insertion of a probe in the exhaust to pull out a representative sample, which is taken to a laboratory for analysis or for immediate analysis in a continuous emission monitor. The EPA has published techniques for sampling and analyzing that can be followed.

An advantage of stack testing is its acceptance in many situations. If done in accordance with the protocol, the results essentially are unquestionable and taken as an accurate representation of emissions from that process under those conditions.

A major disadvantage is the expense part of this technique. However, the stack tests, are most useful in the determination of the emissions from a small number of specific sources and steps.

10.13.4 Engineering Equations

Emissions may be estimated on the basis of equations that are themselves based on the fate of the compounds during the physical actions that they are subjected to during process steps. The driving force of the physical action and the chemical properties of the components, mainly the relative volatilities, have an effect on the emission rate. The equations derived are based on the ideal gas law and those that may be used to estimate emissions are available for the following common industrial process steps.

Equipment filling

When a volume of material is added to a vessel, such as a reactor or a tank, an equal volume of vapour is displaced and emitted from the vessel, laden with the volatile compounds from existing compounds, and any other substance being added. The rate at which the emission takes place may be calculated on the basis of the pollutants' relative volatilities and the rate of displacement of the vapour. The equations compute the vapour mole fractions and emissions of various compounds in a multi-component system.

Gas sweep

When an equipment (such as a container or vessel partially filled with liquids) is purged with an inert gas (such as nitrogen), volatile compounds are swept into the purge gas and get emitted. The emission rate is determined and the computation is based on the rate of the sweep, the pressure of the air space in the vessel and the vapour pressures of the pollutants.

Evacuation

The emission rate for the contents of a vessel emitted after, if has been evacuated, is calculated based upon the factors such as free space in the vessel, duration of evacuation, differential system pressures and vapour pressures of volatile components.

Heating of the contents of reactor or vessel

When the contents of a reactor or tank are heated, thermal expansion brings about a volume of vapour to be displaced at a relatively high temperature. Emissions are computed on the basis of the change in temperature of the components, the exit temperature of the vessel, the pressure of the system, the headspace volume and the vapour pressures of the volatile components concerned.

Gas evolution

There is a possibility of new compounds getting formed and volatilized during a reaction. The rate of evolution of the gas and its molecular weight are needed to evaluate the vapour mole fraction from which volatile emission rate may be computed.

Vacuum distillation process

Emissions from the distillation process may be estimated based upon the component's relative volatilities. The equations take into account the condensation of the exhaust stream to recover solvent. The EPA equation for emissions is based on a driving force (air leaking into the system) and the relative volatilities of the components.

Equations are available also for vacuum drying, evaporation and other operations, plant personnel can use actual operating parameters into the appropriate equations to estimate emissions from each batch step of a process.

The EPA has fully accepted engineering equations as a valid method to estimate emissions in many applications. For instance, emission model (i.e., the use of process-specific equations) is the preferred method for estimating volatile organic compounds and hazardous air pollutants emission from the following:

- Mixing operations such as material loading, heat-up losses and surface evaporation
- Product filling
- Vessel cleaning
- Liquid effluent treatment operations
- Material storage
- Spills that occur

Using engineering equations as an emissions inventory technique has several advantages. While many of the equations are based on theoretical relationships, this approach may be superior in many application to emission factors and material balance owing to the fact that it is based on actual process conditions and in many situations will be more accurate and hence reliable. Another advantage of the engineering equations method is its efficiency, as the same equations can be used repetitively, and consistently for several operations.

10.14 PROCESS CHANGES AND PRODUCT CHANGES FOR WASTE MINIMIZATION AND POLLUTION REDUCTION

Bringing about an improvement in the efficiency of a manufacturing process is very likely to minimize to a very large extent, the quantities of the pollutants generated. These are commonly very effective and economical modes of attack in the pollution prevention domain. Such modifications may include adopting more advanced process technologies, changing over to less polluting reagents, switching cleaning processes and chemicals, employing catalysts to enhance the efficiency of the reaction, segregating effluents and process stream and improving operating and maintenance procedures. More often than not, more than one of these approaches will be employed in an integrated fashion to accomplish optimum production while minimizing the generation of waste.

10.14.1 Using Latest Developments in Process Technologies

First of all, we should investigate the operation and control of the reaction processes. We should find out whether anything can be done to enhance both the effectiveness of the reaction in order to reduce the quantity of process chemicals required and the conversion efficiency to product. Understandably any improvements in this, will give rise to reductions in the quantities of unreacted chemicals or by-products requiring further processing. In many situations, fairly simple changes in operating conditions, such as reaction temperatures or pressure can greatly enhance operating efficiency. It has often been observed that reagents are used in great excess over that, actually required because the personnel concerned have got used to it. Production personnel routinely state that a particular reaction mix has been employed for a long time and they are not to effect any change to that practice. However, that particular mix may have been developed long before there was a concern for the wastes or effluent streams produced or when the economics was such that industry would not mind wasting chemicals to ensure the quality of the product. With the advent of easy-to-use process control equipment and a better understanding of the chemistry of many of these processes in the recent past, it is often now feasible to effect modification in the process in order to have less pollution without any adverse effect on the quality of the product under consideration.

An illustration of how process changes can be of immense benefit, is the rapidly developing application of power coating in place of traditional paint. Solvent-based paints have commonly been applied to the surface of parts, generally by liquid spray with an idea of having corrosion protection; surface protection; identification and aesthetic appeal. An organic solvent is commonly employed as a carrier for the paint. Painting generates two significant sources of hazardous wastes. Paint sludges result from paint overspray. Only about 50 per cent of paint sprayed actually adheres to the piece being painted, while the remaining 50 per cent of paint is invariably removed from the air in a water scrubber, resulting in a hazardous sludge deposit. The solvents in the paint evaporate into the air, giving rise to another hazardous waste source. Solvents are also used to clean painting equipment after the painting is done, again causing pollution. Water-based paints are becoming more common, but their properties are often not as good as solvent-based paints.

Powder coating also known as *dry powder painting* is now changing the scenario. This technique works on the principle of depositing specially formulated thermoplastic or thermosetting heat fusible powders on the metal parts concerned. In most cases, the dry paint powder is applied electrostatically by ionizing the air employed to spray the paint, which in turn charges the dry power particles. The surface to be coated carries the opposite charge and the powder gets electrostatically attracted to the surface. The coating is then fused to the surface and cured in conventional ovens. No solvents are needed for cleanup. The final product is often superior to that of the paints manufactured by adopting conventional processes. The only shortcoming is that powder coating can be applied only to parts that can withstand the curing temperatures of about 350° F required. This implies that it cannot be used on most plastics or aluminium. Since no solvents are employed in this technique, the problems that are encountered in the case of solvent-based paints, do not exist.

By effecting a better control over the features such as temperature and pressure in a reactor through enhanced feedback control mechanisms, improved metering of chemicals to the reactors, use of sensors to monitor the reaction and the creation of product, and improved automation of the process. The efficiency of reagent conversion can be greatly enhanced and waste production can be minimized.

During the conceptual design stage for a process, several strategic steps should be taken into account in order to minimize pollution (Butner, R.S.). These include the following:

1. To avoid adsorptive separations where adsorbent beds cannot be readily regenerated.
2. To make provision for separate reactors for recycle streams, to enable the optimization of reactions.
3. To consider low-temperature distillation columns when dealing with thermally unstable process streams.
4. To consider high-efficiency packing in place of conventional tray type columns thereby reducing pressure drop and decreasing reboiler temperatures.
5. To consider continuous processing when black cleaning wastes are likely to be appreciable, e.g. with highly viscous, water-insoluble and adherent materials.
6. To consider scraped-wall exchangers and evaporators with viscous materials to avoid thermal degradation of product.

10.14.2 Improvement in Reactor Design

It is well known that the reactor and the conditions under which it is operated are crucial to source reduction. Modifications in stirred-tank reactors can be incorporated in order to improve the degree of agitation achieved and thus the mass transfer efficiency between reactants through use of more efficient mixers and baffling systems. These can also help to improve the heat transfer, making the contents of the reactors more homogeneous and augmenting the conversion efficiency. Insulating the reactors will entail improvements in temperature control within the reactor. Adoption of high efficiency heat exchangers, if they are not currently being employed, can often result in reduction in energy costs.

The operation of reactors in a continuous flow mode is often less polluting than employing the batch mode. Because most reactions never go to completion, some unreacted process chemicals usually remain at the end of the reaction; these must be separated and disposed of. The cleaning of batch reactor between each run can create a significant waste load. Besides, the process control is much easier in a continuous flow reactor, where the reactor contents and environmental conditions are identical throughout, than in a batch reactor where the reaction may be causing environmental conditions to continually alter. Reduction of the quantity of pollution resulting from a chemical reaction may also be accomplished by employing something besides the common stirred-tank continuous flow reactor. For instance, plug flow reactors allow for staging, with different variables, such as temperature, closely controlled in each stage to optimize output. Switching to a fluidized bed reactor from a stirred-tank reactor may also achieve appreciable effluent reductions in some situations.

Other reactor design considerations include ensuring easy access in order to simplify cleaning of the interior of the reactor and implementing a design that enables complete draining of the vessel. The piping system should be designed such that piping run length and the number of valves and flanges are minimized. Drains, vents and relief lines should be routed to recovery or treatment equipments.

In some situations, radical changes to a production process may be advantageous, but the improvements in efficiency must be balanced against the costs involved. For instance, steel industry has found that new electric arc furnaces are lesser polluting than the more common open hearth and basic oxygen processes, but the cost of converting these plants to electric arc furnaces is almost prohibitive. These changes will come slowly as older units need to be replaced, but the significant improvement that would be realized through immediate use of this new technology cannot be accomplished overnight.

Other process modifications that are given below can have a significant impact on water use and can also have an impact on source reduction.

1. Augmenting the number of stages in extraction processes that use water
2. Using spray balls as a scouring agent to remove caked-on solids for more effective internal vessel cleaning
3. Changing over from wet cooling towers to air coolers
4. Improving control of cooling tower blow down
5. Attaching triggers to hoses to prevent unattended running
6. Improving energy efficiency to reduce steam demand and hence, reducing the waste water generated by the steam system through boiler blow down, aqueous waste from boiler feed water equipment and condensate loss
7. Increasing condensate returns from steam lines to reduce boiler blow down and aqueous waste from boiler feed water equipment
8. Improving control of boiler blow down

10.14.3 Improvements in the Control of Reactors

In many reaction systems, improving reactor environment controls is likely to have a substantial impact. Many a time when reactors are controlled by people, who may not be property attentive, accidents may take place. Close reaction control may turn out to be difficult to regulate and off-specification product may result. This may be brought down by installing automated monitoring and control devices for such variables as temperature, pressure, feed rates and duration of reaction. Robots employed for welding and numerically controlled cutting tools are examples of successful changes. However, to be successful, these devices need to be very accurate and reliable and should have a low failure rate. For instance, people can sometimes perform better by eyes, of detecting defective compounds than an optical sensor that does not perceive slight changes in a component. The use of automated control can be a great benefit to quality control and waste minimization, but they must be employed carefully. Control system efficiency can be attributed to a combination of the characteristics listed out hereinafter.

1. Measurement accuracy, stability and reproduceability
2. Sensor locations
3. Controller response actions (Proportional, integral, derivative, cascade, feed forward and stepped)
4. Process dynamics
5. Final control elements (Valves, dampers, relays, etc.)
6. Characteristics and location
7. Overall system reliability

Application of process controls for prevention of pollution can be divided into various categories including reduction of waste production by adopting process efficiency improvements, pre-treatment of pollutant containing effluents through chemical reactions, capturing and recycling of pollutant containing waste by-products and collection and storage of pollutant containing waste by-products (Doun, R.D.). It is likely that this involves the use of process controllers that continuously compare measured variables to desired values and regulate such variables as reaction temperature and pressure and reaction time or that regulate pumps and valves to recycle flow streams or to send off-specification materials to waste or to a reprocessing step.

10.14.4 Improvements in Separation Process

Many industrial processes depend on separation units to recover products from a mixture of unreacted chemicals and undesirable by-products. These processes may involve separation of gases, liquids or solids from gas, liquid or solid streams. A variety of separation processes are available, some of which, are more appropriate for a particular application than another. Some potentially useful separation processes for different applications are listed in Table 10.5.

Table 10.5 Potentially Useful Separation Processes

Separation process	*Recover from gas*			*Recover from liquid*			*Recover from Solid*		
	Gas	*Liquid*	*Solid*	*Gas*	*Liquid*	*Solid*	*Gas*	*Liquid*	*Solid*
Absorption	⊗	⊗							
Adsorption	⊗				⊗				
Centrifuge					⊗	⊗		⊗	⊗
Chemical Precipitation						⊗			
Coalescence and Settling		⊗		⊗	⊗	⊗			
Condensation		⊗							
Crystallization						⊗			
Distillation					⊗				
Drying/Evaporation						⊗		⊗	
Electromagnetic separation									⊗
Electrostatic precipitation		⊗	⊗						
Elutriation									⊗
Filtration			⊗			⊗		⊗	
Gravity sedimentation		⊗	⊗		⊗	⊗		⊗	
Ion-exchange					⊗				
Membrane separation process	⊗				⊗	⊗			
Screening									⊗
Scrubbing process	⊗	⊗	⊗						
Solvent extraction					⊗			⊗	
Sorting technique									⊗
Stripping		⊗		⊗	⊗		⊗	⊗	

Long, R.B., suggests several general rules of thumb for decreasing energy consumption in separation processes and techniques.

1. Carrying out mechanical separations first if more than one phase exists in the feed
2. Avoiding over design and using designs that operate efficiently over a range of conditions, simple processes may be favoured
3. Preferring processes transferring the minor rather than the major component between phases
4. Favouring high-separation factors
5. Recognizing value differences of energy in different forms and of heat and cold at different temperature levels
6. Investigating the use of heat pump, vapour compression or multiple effects for separation with small temperature ranges
7. Using staging or countercurrent flow where applicable
8. For similar separation factors, preferring energy-driven processes to mass separating agent processes
9. In the case of energy-driven processes favouring these with lower heat of phase change.

Despite the fact that this list is aimed at energy reduction, these strategies often reduce the quantity of waste generated. For instance, extracting the minor by-products from the process stream to another phase, rather than extracting the product itself, usually results in a much smaller volume of material to be reprocessed. Choosing processes that operate over a wide range of conditions allows for more sturdy operations and also allows for operation, with less operator interference and thus less chance of accidents or improper control settings.

10.14.5 Improvements in Cleaning/Degreasing Processes

Cleaning and degreasing operations are commonly employed by industry to remove dirt, oil and grease from both process input materials and finished products. In the case of metal finishing industry, cleaning normally follows machining and comes before other surface finishing steps, such as anodizing or metal plating. Circuit boards and other electronic equipments must be scrupulously clean, in order to be effective. In paper and textile industries, cleaning of equipment to remove inks, dyes, oils and other materials is frequently carried out. Solvent cleaning is the major process in the dry cleaning industry. Needless to mention all industrial production processes require cleaning or degreasing to a certain extent.

In some situations, water is suitable for cleaning of components, but in many situations an organic solvent must be used. The most commonly used solvents are trichloroethylene (TCE) and perchloroethylene (PCE) owing to the merits they possess like the ability to dissolve a wide range of organic contaminants, their low flammability (flash point) and their high vapour pressure, which allows them to readily evaporate from coated surfaces. The relevant properties and characteristics of some of the typical solvents are shown in Table 10.6.

Table 10.6 Properties and Characteristics of Typical Solvents used in Cleaning

Property	*Water*	*Trichloroethane*	*Trichloroethylene (TCE)*	*Perchloroethylene (PCE)*	*Ethylene chloride*
Molecular weight	18.0	133.5	131.4	165.9	84.9
Boiling point °C	100	72.9	86–88	120–122	40
Density (gms/c.c)	1.0	1.34	1.46	1.62	1.33
Vapour pressure (mm of mercury)	17.5	100	58	14	350
Flash point (°C)	None	None	None	None	None

A density greater than water allows the solvent to be gravity separated from water after use, and a boiling point difference from that of water, enables its separation by the process of distillation. All of the solvents listed in Table 10.6 have these properties; however, there is an inherent problem; all are halogenated solvents that are major environmental contaminants, posing significant potential public health threats. Thus, all efforts must be expended to bring about a reduction in the loss of these materials to the environment or to substitute them with less polluting solvents.

Contaminants generated during cleaning and degreasing may include (1) liquid waste solvent and degreasing compounds containing unwanted film material, (2) air emissions consisting of volatile solvents, (3) solvent contaminated waste water from vapour degreaser—

water separators or subsequent rinsing operations, and (4) solid wastes from distillation systems, comprising oil, grease, soil particles and other film materials removed from manufactured parts (Thom, J. and Higgins, T.).

In some situations, incorporating modifications in the manufacturing process may result in a reduction in the cleaning requirements for a particular process. For instance, changing a piece of equipment to minimize carryover of material from one production step to another, may minimize or even totally eliminate the need for intermediate cleaning. These possible modifications should be looked into before contemplating changes at the cleaning process itself.

In a commonly employed cleaning system which usually uses water as the solvent, the component or part to be cleaned is placed in a rinse tank or consecutively in several rinse tanks containing either stagnant or flowing water. The objective here is to transfer the contaminants on the surface of the manufactured part to the water by the process of dissolution and dilution. This is the typical process employed to remove excess plating solution dragged out of a plating bath. It is only useful, though, for removing materials that are water soluble. A single stagnant rinse tank can be used as a cleaning device, but it is not very effective and cleaning ability decreases very significantly with time as the level of contaminants in the rinse tank builds up. Large quantities of water are required to effect cleaning. A series of stagnant rinse tanks brings about improvement in the situation since subsequent rinse tanks in the progression contain fewer contaminants than the previous one, but they are still not very efficient. By employing a flowing tank or series of tanks, cleaning efficiency can be appreciably increased and water consumption decreased. Running the rinse water through the rinse tanks counter current to the flow of the pieces of components being cleaned, provides the best and most efficient cleaning possible. Each additional cleaning step in the series contains cleaner water and improves the cleaning operation, but there is a limit to this improvement. By and large little benefit is achieved by operating more than three countercurrent tanks in series.

Some industrial units employ aqueous cleaning with the addition of various detergents. The parts or components to be cleaned are either tumbled in an open tilted vessel that rotates and that contains the cleaning solution or they are dipped into a tank containing the cleaning solution. In either case, the parts must be rinsed thoroughly in a series of rinse tanks or under a continuously flowing stream of clean water to get rid of the cleaning solution and contaminants. In the recent past, automated washers with low volume, high pressure water jets or water knives have been introduced to reduce the quantity of water needed for cleaning and rinsing. Aqueous cleaning processes generally need large quantities of water and thus generate large volumes of dilute waste water.

Many of the cleaning operations for removal of dirt or organic compounds, such as oils and greases are dependent on organic solvents. Common solvent cleaning operations include cold cleaning and vapour degreasing. Cold cleaning is the simplest, most economical and most widely employed solvent cleaning process. Typically this is carried out in open top dip tanks. The piece to be cleaned is dipped into a tank containing a liquid mineral spirit solvent. After transferring contaminants from the piece to the solvent, the piece is removed and solvent evaporates from it leaving a clean surface. Unless collected and condensed, these vapours can turn out to be a source of major air pollution. Some industrial establishments employ hoods over the cleaning bath and allow the cleaned pieces to dry within the hood, so that any volatized solvent can be collected, condensed and placed back into the cleaning tank. Eventually the solvent bath, which accumulates all of the dirt and grease that was sticking on the part, must be cleaned.

In the process of vapour-Trichloroethane, Perchloroethylene, Trichloroethylene degreasing, chlorinated hydrocarbons, such as or methylene chloride in the vapour phase, are used to clean surfaces. Vapour greasing takes place at an elevated temperature necessary to produce the solvent vapour. A tank, filled to about one tenth of its depth with liquid solvent, contains steam coils or some other heating mechanism to heat the solvent and produce the necessary vapours. The vapours are heavier than air and hence remain in the tank. Components to be cleaned are placed in the vapour region of the tank. Owing to the lower temperature of the components being cleaned, solvent vapours immediately condense on the surface of the piece in question, dissolve the contaminants and then drip back to the bottom of the tank where they can be re-vaporized. The cleaning process continues until the part is heated to the same temperature as the vapour at which time no more vapour condensation takes place and the part is removed. In order to prevent loss of solvent vapours from the top of the cleaner, cooling coils are located in the top of the tank to condense any vapours that do condense on the parts returning them to the tank bottom. Solvent loss can also be brought down by increasing the free board in the tank. This enables the retention of more of the solvent vapours within the confines of the tank. Many solvent cleaning baths are now covered to minimize the loss of solvent. Other units are completely enclosed thereby preventing loss of nearly all of the solvents.

The vapours of the solvent are generally pure solvent, but periodically the liquid solvent must be reprocessed to get rid of the contaminants and the dirt that accumulates on the tank bottom must be removed. Some units have the provision for this in the form of built-in recovery stills to clean the solvent. The volume of waste generated by a solvent cleaning system is only a small fraction of that produced by an aqueous cleaning system. However, solvent is generally suited only to cleaning organic contaminants; this method is, for instance, ineffective for removal of water based metal plating drag out.

Ultrasonic cleaning devices are now having wide applications. These use ultrasonic waves in 20–40 kHz range to induce vibrations that cause the rapid formation and collapse of microscopic gas bubbles in the solvent. The process, termed cavitation creates pressures as high as 10,000 psi and temperatures as high as 20,000 degree F on a microscopic scale that loosens contaminants and scrubs the workpiece. The vibrations are created by a transducer placed in the solvent bath. Ultrasonics can be employed with a variety of cleaning solutions and will remove a wide range of organic and inorganic contaminants. The speed of cleaning is increased and often lower concentration of cleaning solution can be used. The workpiece must still be rinsed clean with clean water, generating an aqueous effluent stream, but the quantity generated is generally relatively low. Descriptions of the equipment and its functioning can be found elsewhere (Gavaskar, L.).

Other cleaning devices are also available for use. These include a vacuum furnace to volatilize oils from parts and a laser ablation system (vaporization during high-speed movement through the atmosphere) which employs a laser to rapidly heat and vaporize a thin layer of the part's surface and any contaminants sticking on it.

The entire spectrum of associated costs and environmental impacts need to be taken into consideration while selecting a cleaning system. For instance, vapour degreasing minimizes solvent use and solvent loss to the environment, thus minimizing environmental impacts, but it requires heating of the solvent and has all of the cost and environmental consequences of increased energy consumption.

10.14.6 The Process of Equipment Cleaning

A fundamental method for effecting reduction in the quantity of waste from equipments (such as heat exchangers, chemical reactors, piping, etc.), cleaning operations, is to avoid unnecessary cleaning and to reduce the frequency of cleaning. This can be substantially accomplished by switching from batch reactors to the continuous flow ones as and when possible. Improvements can also be realized by choosing reactors that have easy access for cleaning and that are designed for ease of cleaning that is those with no internal parts that will accumulate contaminants.

Studies should be conducted to determine how frequently cleaning is necessary. In the past, solvents commonly were reused until they were too contaminated to clean effectively. Then they were discarded. Unfortunately, this extended use often made them so contaminated that recovery of the solvent turned out to be infeasible. An improved approach is to install a batch solvent recovery system and process the solvents more frequently on-site. This makes it possible to have limitless use of the solvents, minimizing the cost of solvent purchase and also greatly reduces the quantity of hazardous effluent that must be disposed of. Cost recovery for this system generally takes place within months (Thomas, S.T.).

10.14.7 The Recycling Process

Recycling or reuse of materials in a process effluent is often an attractive way of reducing effluent streams requiring treatment and their associated costs while at the same time reducing the demand for virgin chemicals and their associated costs. Many industrial establishments are observing that recovery and reuse of materials that were earlier considered wastes can be cost-effective. The Environmental Protection Agency (EPA) of the USA does not consider recycling or reuse to be pollution prevention, but both can be cosidered to fall under the broader definition of pollution prevention owing to their significant benefits to industry. The recycling within a process is discussed here. Materials can be recycled in several ways as shown in Figure 10.10.

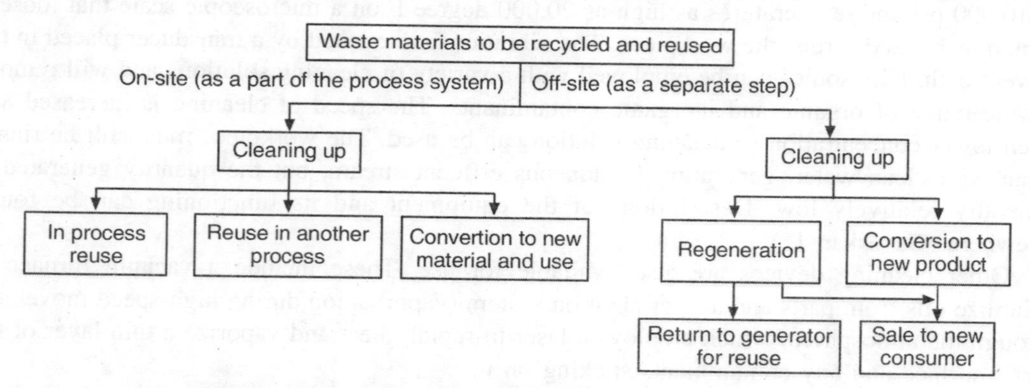

Figure 10.10 Options available for recycling and reuse.

Recovery may be on-site or off-site. It may be carried out on-site as a part of the process system or it may be accomplished in a separate step. The materials may be recycled to the process step where they were produced or to some other process in the facility. There is a possibility of converting them to a new material before reuse.

The effective use of recycle or materials recovery options requires the efficient segregation of waste and process streams. Segregating wastes allows for minimization of contamination of process streams with effluents from other streams, simplifying recycling or reuse of the chemicals and reducing the volume of effluents requiring treatment. Segregation should occur as close to the source as possible. In the past, it was a common practice to dump all effluent streams into a common sewer, where effluents mixed together and flowed from the plant. Following combination with other effluent streams is very difficult, but not economically impossible, to recover materials for reuse. The problem is compounded if a potentially recoverable effluent stream is mixed with a high flow dilute effluent stream. The recoverable material will be diluted further, making recovery difficult while the large volume dilute effluent stream which may have required only minor treatment before discharge, may now need extensive effluent treatment. In the past, hoses employed to wash equipment or fill reactors were run continuously, because water was inexpensive and this practice was found to be more convenient than constantly opening and closing valves. The unused water entered along with all other materials. This signified that all of this clean water get polluted and had to be treated. If the contaminants were toxic, then the whole volume of effluent would be considered to be toxic, despite the fact that much of it was clean water that had never been used. Fortunately most facililties have now have done away with this practice.

In order to make a recycling or reuse option viable, the chemical composition of the recycle stream must be compatible with the process to which it is directed; the expenses involved in regenerating or cleaning up the material or of modifying the process to accommodate recycled stream, must be economically justifiable; and the material must be available in a consistent quality and volume. Any of the techniques followed for process stream separation can be employed to separate clean-up, and concentrate a chemical for recycle or reuse. In some situations, the material may be converted to a new chemical by chemical reaction before being reused in another process.

Recycling or reuse often takes place on site with unused process chemicals being separated from an effluent stream and returned to the same process from where they came. This is commonly occurring in the metal plating industry, where the excess metal solution dragged out from a plating bath and removed from the plated work piece during the rinse steps are fortified, possibly employing ion-exchange or reverse osmosis, and returned to the plating bath. By this technique, a reduction in the quantity of process chemicals being wasted to the sewer is brought about and also a reduction in the quantity of chemicals needed to be purchased for plating, is effected.

- The recovered materials may also be used on-site in a different process or process step than where they originated. Usually, it is impossible to regenerate the chemical stream to as high quality as virgin chemicals. For many applications this is not a very significant factor and the recovered materials may be mixed with the new process chemicals going into the reactor. However, in certain situations, impurities in the recycle stream may be harmful to the process, making the stream's direct recycle to that reactor, ineffective. The recovered materials may be suitable for use in another process step even though, the purity specifications may not be as exacting. For instance, copper recovered from a printed circuit board electroless plating process is likely to contain impurities that would prohibit its being sent back to the electroless process bath where tight quality control is maintained on the bath contents, but the recovered materials may be appropriate for use in decorative electroless plating over plastic components where the specifications are not as critical. Another example is the use of recovered solvents employed in solvent cleaning of parts. The recovered solvent may not be as clean as virgin solvent and may not be suitable for the final cleaning step on a component that must be extremely clean, but it may be all right for cleaning steps earlier in the production process. A detailed discussion of solvent recovery can be found elsewhere (Thomas, J. and Higgins, T.).

A significant commodity for reuse that is often overlooked is water. In the past, waste has been assumed to be a limitless low-cost resource. However, this attitude has changed as industrial organizations have had to deal with not only the increasing cost of water, but also the cost of treating polluted water. Many industrial establishments use enormous quantities of water during processing. For instance, paper mills use anywhere between 1000 and 100,000 gallons of process water per ton of product (averaging about 50,000 gallons per ton of product). A typical pulp and paper facility produces about 1000 tons of paper per day.

Thus, about 50 million gallons of water are used per day in the typical paper plant. By reusing reclaimed water, many plants have reduced their water consumption to the extent of 80 per cent. This lowers not only their costs of process water purchase and treatment, but it also greatly reduces the volume of effluents requiring treatment and concentrates the constituents in the effluent to such an extent that many of them are economically recoverable.

This example concerns one of the largest industrial water users, but essentially all industries can derive major benefits by availing themselves of water minimization steps. Figure 10.11 indicates typical water uses in an industrial chemical process facility.

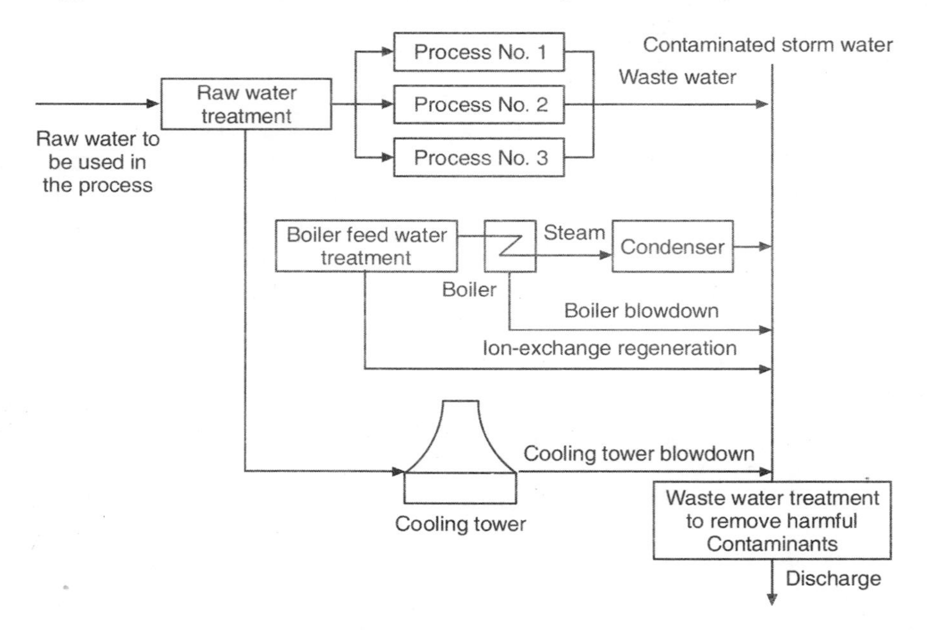

Figure 10.11 Typical water use in an industrial chemical process facility.

Adapted from Smith, R., *Waste Water Minimization in Waste Minimization through Process Design,* A.P. Rossiter (Ed.), McGraw Hill, New York, 1995.

Water consumption can be minimized by (1) incorporating process changes in which less water-demanding equipment is employed, (2) reusing water directly in other operations, provided the level of previous contamination does not interfere with the process, or (3) treating the effluent to remove harmful constituents that are likely to prohibit the water's reuse and reusing the water in the original process or in another process.

It may be mentioned in this context that some industries are now exploring ways to integrate industrial processes in order to minimize water consumption.

10.14.8 Reuse of Recovered Materials

Recycling or reuse of materials is generally desirable from the perspective of conservation of resources, but it may not be so in terms of environmental life cycle assessment. In many circumstances, more energy and resources are needed to recover the material or more wastes are generated during recovery than in the case when generating virgin materials. It may be preferable to dispose of the waste materials and use virgin materials for new products. A life cycle analysis is required to take a correct decision in this regard.

Recovered materials that are useless to the industry generating them, may be of value to other companies; as a matter of fact, a large number of companies have developed a market for recovered chemicals. They may either sell the chemicals to another company or trade them for chemicals that they can use which are recovered by the second company. In some countries waste exchange services have been set up to assist companies in finding markets for their recovered materials for sale or for trade or to locate sources of recovered materials, that a company requires. This may be a very economical transaction because the recovered material can often be acquired for much lower cost than virgin materials. The company disposing of the material needs to charge only for the net cost of recovery after the anticipated cost of treatment and disposal of that material if it was not reused, is factored in.

In many situations, the quantity of effluent generated by an industry is too small to economically justify on-site recovery of materials; or the industrial establishment may not have the expertise or the inclination to operate a recovery system. An establishment may still eliminate the need to treat or dispose of these wastes, with all the costs associated with this, by sending them to an off-site recovery centre. Many such centralized facilities have been established to carry out the process of recovering oils, solvents, scrap metals, process baths, plastic scrap, cardboard and electroplating sludges. By combining effluents from many small industrial establishments, the recovery processes can be made fairly economical. The recovered materials may be sent back to the generator for reuse or may be sold to other consumers.

10.14.9 Making Changes in Product to Reduce Pollution

One of the more effective ways to bring down pollution at the source is to make modifications in the product itself or in the chemicals employed to make the product. For instance, new metal planting technology makes it possible to eliminate much of the cyanide that was required in cadmium plating baths. Water-based paints have replaced many solvent-based paints. Work pieces with fewer turnings and cavities will reduce the quantity of drag out from process tanks.

Replacement of steel automobiles bumpers with aluminium or polymer composite bumpers makes them lighter and sometimes stronger and also improves the automobile's fuel efficiency. However, when evaluating a change in a product to minimize pollution associated with its production, one must bear in mind that the main constraint is that the product, as made, serves its intended purpose in the best possible way. It makes no sense to change over to a new material or product configuration in order to effect a reduction in pollution if the resulting product cannot be economically manufactured or if it is inferior in quality to the original product. A life cycle assessment of the existing and proposed product can help in deciding which will be the most cost-effective and environmentally harmless.

10.14.10 Storage Operations

Materials storage operation can turn out to be a major source of pollution in manufacturing process, but these areas can be designed and operated to bring about minimization of possible emissions. As a matter of fact, many industrial facilities that are used to store hazardous materials or hazardous water are regulated and are expected to meet stringent requirements in respect of harmful releases.

Proper material storage and delivery system should be designed and managed with utmost care to make sure that materials are not lost from storage tanks or piping systems through spills or leaks and to minimize losses during transfer operations. In many situations, precautions based on common sense will be adequate. Some commonly suggested approaches include the following (Hunt, G.E.):

1. A spill prevention, control and counter measures plan must be established.
2. Only properly designed storage tanks and vessels for their intended purpose should be employed.
3. Overflow alarms on all storage tanks should be installed.
4. Secondary containment areas to collect any spilled materials must be installed.
5. All spillage should be documented.
6. Containers need to be spaced to facilitate their inspection for damage and potential leakage.
7. Containers should be stacked properly in order to minimize tipping, breaking or puncturing.
8. Containers need to be raised off the floor to minimize corrosion, from *sweating* concrete.
9. Different hazardous substances need to be separated in order to minimize cross contamination and also to separate incompatible materials.
10. A just in-time order system for process chemicals to minimize the quantity of stored materials need to be used.
11. Procuring reagent chemicals in exact quantities in order to minimize the amount of stored chemicals/materials.
12. Establishing an inventory control programme to trace chemicals from cradle to grave.

13. Rotating chemical stock to prevent chemicals from turning out to be outdated, necessitating disposal.
14. Validation of shelf life of chemical expiry dates, elimination of shelf-life requirements for stable materials and testing of the effectiveness of outdated materials.
15. All materials and containers need to be labelled properly with material identification health hazards and first aid recommendations made available.
16. Switching to less hazardous raw materials.
17. Switching to materials packaging and storage containers that are less susceptible to corrosion or leakage.
18. Using large containers where possible, to minimize the tank's surface-to-volume ratio, thereby reducing the area that has to be cleaned.
19. Using rinsable/resusable containers.
20. Drums and containers need to be emptied thoroughly before cleaning or disposing.
21. Elimination of storm drains in liquid storage and transfer areas.

Table 10.7 presents the results of industrial chemical storage facilities (Lagrega, M.D.; Buchingham, P.L. and Evans, J.C.).

Table 10.7 Most Common Deficiencies Encountered in Storing Hazardous Wastes and Hazardous Materials

Deficiency	*Audited facilities having this deficiency, %*
Inadequate containment/tank storage area	68%
Inadequacy in tank integrity testing	65%
Inadequate inspection programme for storage	48%
Improper labelling	44%
Inadequate training programme to the staff	40%
Inadequate spill prevention, control and contingency plan	34%
Incomplete waste water analysis	24%
Inadequate hazardous waste contingency plan	18%
Inadequate handling of materials containing PCBs (Polychlorobiphenyls)	18%
Storm drains in liquid storage and transfer areas	18%
Inadequate handling of empty drums	17%

As can be seen from Table 10.7 that many storage facilities had deficiencies. The most common problems were the existence of storm drains in liquid storage and transfer areas that could contaminate the relatively clean storm water, drains in storage areas that connected to waste water sewers and could hence contaminate the waste water with process chemicals, improper handling of empty drums, use of containers that were in poor condition and inadequate secondary containment facililties.

Many of the aforementioned problems could be eliminated if the industry instituted and enforced a Spill Prevention, Control and Countermeasures (SPCC) plan. This is a plan developed by each industry to make sure that spills or accidental releases from storage areas will not take place, and to outline corrective measures that will be taken, if they do. An SPCC plan is required for industrial oil storage facilities under the Oil Pollution Act, and should also be developed for other chemicals or toxic wastes stored on-site as well. The SPCC plan should address the three areas hereinafter.

- Operating procedures that prevent spills and leaks from occurring.
- Control measures installed to prevent the occurrence of a spill or leak reaching soil or water.
- Countermeasures to contain, clean up and mitigate the effects of the spill or leak.

The plan should also address all of the items listed earlier for precautions to be taken in order to prevent spills and leaks.

The industry should also work in close coordination with the local community committee responsible for the local emergency plan, to ensure that the community is prepared to handle any leaks or spills that do take place.

Invariably many industrial establishments generate hazardous effluents during manufacturing process. These effluents are usually stored for at least sometime before being subjected to treatment or being shipped off-site for disposal. All facilities that store hazardous wastes need to be regulated. Any facility that stores more than 1000 kgs of wastes for more than 90 days is expected to have a storage permit and meet very stringent requirements (facilities that store only 100 to 1000 kgs can store for up to 120 days before they need a permit). Storage facililties are typically areas where drums or tanks of hazardous waste are kept until they are processed or shipped; but they also include pits, ponds, lagoons, tanks, piles and vaults.

In general, storage of hazardous waste should be done such that it minimizes accidental releases. All containers need to be labelled properly, appropriate containers need to be used, a hazard communication plan should be in place and the workers trained to implement it and a detailed inventory of stored wastes should be maintained. Facilities have to be so designed as to minimize the possibility of fire, explosion or any unplanned sudden or non-sudden release of hazardous waste to the air, soil or water. Materials need to be stored at least 50 feet from property lines and away form any traffic including forklift and foot traffic. An impermeable base should underlie all storage containers to hold any leaked or spilled material or accumulated rainfall. In some states, epoxy-coated secondary containment may be required to contain leaks. Adequate ventilation needs to be provided to prevent build-up of gases. Drums should not be stacked more than two in a vertical row, and stacking of flammable liquids should be avoided. Incompatible wastes should not be stored together and should be separated by barriers, such as well or berms. A drainage system should be installed to avoid spilled wastes or precipitation remaining in contact with the containers. Storm water run-off into the containment area must be prevented. A more detailed description of storage requirement can be found elsewhere (Kuhre, W.L.).

None of the aforementioned actions to minimize leaks and spills during materials storage or preparations for emergency response to spills that do take place will be of use though, unless they are meticulously carried out, their implementation is enforced strictly and all employees are well trained and informed of the need for close attention to the contents of the plan. Needless to add, proper education and training are of utmost importance.

10.14.11 Management's Commitment to Pollution Prevention Programme

It is well known that no pollution prevention programme will be successful unless the management is fully committed to its concepts. Employees can be expected to comply with a plan only if they feel that it is important to the organization concerned and the management provides proper support for it to be properly carried out. Among the factors that are critical to success are a well-conceived and functional preventive maintenance programme, proper employee training and good record keeping.

Housekeeping

It is generally well known that the pollution prevention activity with the greatest potential impact and the one that should be addressed first, is improvement of housekeeping practices. This low-cost and sometimes no-cost action usually results in significant reductions in effluent of process chemicals, water and energy and it usually improves worker health and safety. Many industrial establishments needlessly generate waste without even realizing it until they closely examine their housekeeping activities. A metal plating company that allows rinse waters to continue to flow through a rinse tank when no workpieces are in the tank may be sending a very large quantity of clean water to a sewer where it mixes with wastes, becomes contaminated and must be treated. A chemical processing company that hoses down floors to remove spilled solid chemicals, rather than sweeping them up dry, is also increasing its waste water pollution load and the resulting treatment costs. Inadequate equipment maintenance can many a time lead to leakage of pumps and valves that give out process chemicals or oil; thereby wasting them and necessitating waste treatment, improper metering by chemical feed equipment; leading to wastage of materials or possible production of off-specification products that must be disposed of, and cross contamination of chemicals. Improper storage of materials, such as incorrect stacking or crowding of the storage area, can possibly give leaks and spills. Lax labelling practices more often than not lead to materials being used improperly or being discarded unnecessarily owing to the reason that the correct contents of a container are not known.

Many a time, potentially polluting or harmful practices are performed solely because the employees are not aware of the consequences. For instance, flushing out the solid residue in the bottom of a storage container with water before cleaning the container may make sense to the operator; moreover, this long-standing practice had been taught to the operator who therefore is concerned only with ensuring that the tank is clean and that any residues will not contaminate the next batch. The employee is probably not aware of the impact of flushing the dry chemical into the sewer. The remaining material in the container could probably be vacuumed up and either reclaimed or disposed of in a dry form, but if the operator does not understand why this is preferable, the expedient practice of water flushing will likely to continue to be employed. It is desirable all employees be educated about the impacts of their practices on the overall operation of the facility and on the environment, and should be briefed about ways to minimize harmful impacts. Many improper and incorrect practices could be avoided solely through education.

Simple housekeeping practices which are often overlooked are given here.

1. Closing or covering solvent containers when they are not in use
2. Isolating liquid wastes from solid wastes
3. Turning off equipment and lights when not in use
4. Eliminating leaks, drips and other fugitive emissions from the units of the plant
5. Controlling and cleaning up all spills and leaks as they take place
6. Developing a preventive maintenance schedule and enforcing its use meticulously
7. Scheduling production runs to minimize frequency of cleaning
8. Improving lubrication of equipment
9. Keeping machinery running at optimum efficiency
10. Dry sweeping of floors whenever possible
11. Not allowing materials to mix in common floor drains
12. Insisting on proper labelling of all containers
13. Briefing all employees on the need for proper housekeeping practices
14. Including housekeeping reviews in all process inspections
15. Adopting a total quality management philosophy

It is well known that preventive maintenance plays a very important role in good housekeeping practices. Regular inspection and maintenance of all equipments, including replacement of worn or broken parts before the equipment actually fails will result in desirable developments such as increased efficiency and longevity of equipment, fewer disruptions owing to equipment failure and less waste from leaking equipment or off-specification products. In particular, seals and gaskets in hydraulic equipment should be examined frequently and replaced as often as possible to prevent leaks. A properly designed and implemented preventive maintenance programme can often eliminate or minimize production shutdown and prevent many potential sources of waste, turning out to be a problem. Although these programmes are quite expensive, they usually pay for themselves many times over by avoiding production delays and unnecessary effluent treatment. Needless to add good housekeeping practices are by and large based on common sense. It is essential though, that all employees be trained to adopt proper procedures and taught why they are very important. Their attention to good housekeeping practices must be monitored regularly; human nature being what it is very easy to lapse into a mode of doing what is most convenient rather than what is right.

Training programme for employees

It is very important that once the operating and maintenance procedures are established, they must be fully documented and made part of the employee training programme. Proper and continuous employee training is of vital importance to the success of any pollution prevention programme; without it, the chances of failure of any plan are very high. The objective of the training programme should be to make every employee; from the operators to the executive officers, aware of sources of waste generation in the facility, the impact of waste on the organization and the environment and, ways to reduce the occurrence of pollution. Training should be an ongoing programme with frequent review updates.

Employee training should focus on spill prevention, response and reporting procedures, good housekeeping practices; material management practices and proper fuel and storage procedures (Shen, T.T.).

The pollution prevention programme should set goals for substantial waste reduction. The training programme is designed to provide the information necessary for the personnel to achieve those objectives. A variety of training techniques should be used. Seminars, written materials and in-plant training can be included in the programme. All plant personnel should be active participants in the training in order to ensure that everyone feels they are making a collective effort and attacking the pollution problem as a team. A major role in training preparation is expected of the operators, as they have the most knowledge about the processes they are working on.

A pollution prevention training programme should be designed to accomplish the following objectives (Osantowski, R.A.; Liello, J.C. and Applegate, C.S.):

- Promoting awareness of waste reduction initiatives
- Establishing quantifiable waste water reduction objectives
- Giving training to personnel in waste reduction techniques and water use minimization practices

4. Implementing plant wide waste reduction techniques and water use minimization practices
5. Monitoring progress towards attaining waste reduction objectives and to readjust objectives if they prove to be unattainable
6. Attaining waste reduction objectives

It is well known that pollution prevention and worker safety go hand in hand. An additional benefit of a well-conceived pollution training programme is improved safety of workers.

Record keeping

Effective record keeping can significantly help in both the development and the implementation of a sound pollution prevention programme. Documentation of process procedures, process control parameters, chemical specifications, chemical use, energy use, waste generation and spill frequencies and causes will help to pay attention on areas in which source reduction activities will be very useful. Inventory control, for both process chemicals and waste materials, will lead to reduction in the volumes of these materials that are kept on hand and thus reduce costs and potential losses to the environment from leaks or spills. It will also reduce the necessity to dispose of contaminated, off-specification or obsolete reagents. Good record keeping can also be used to ensure compliance with environmental regulations while showing the public that the organization has a true concern for environmental quality.

10.15 A SYSTEMATIC AUDITING PROCEDURE FOR WASTE MINIMIZATION

Many plants are trying to identify some good ways to minimize the quantity of wastes leaving their sites. There are three basic techniques for waste minimization. They are source reduction, recycling and waste treatment. Invariably, reducing or avoiding waste generation is the most desirable objective and hence, should be given top priority and explored first. Then only comes recycling followed by treatment. Figure 10.12 details basic contents of these approaches.

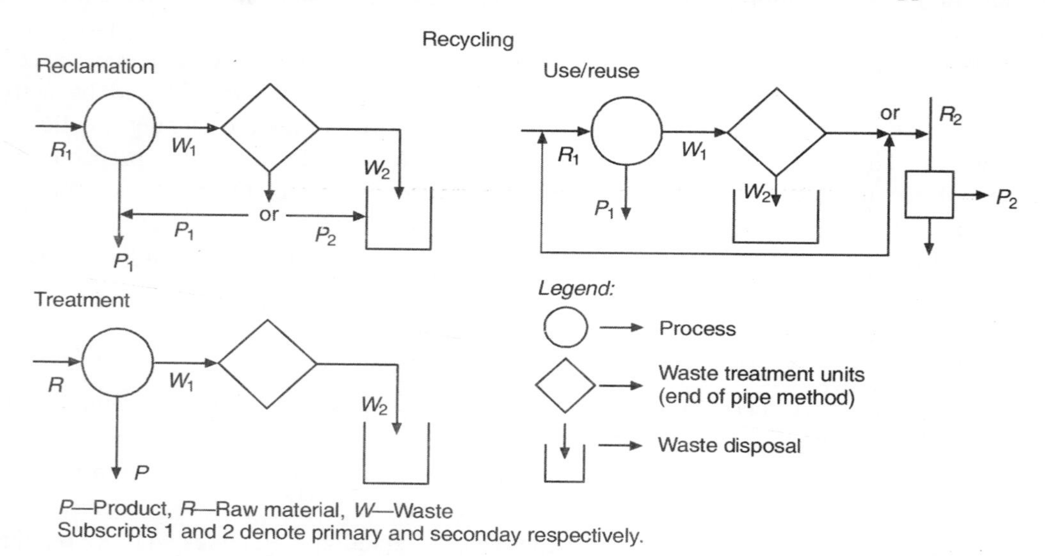

Figure 10.12 Waste minimization approaches through source reduction.

Table 10.8 Source Reduction Approach to Waste Minimization

Alteration of product		
Product substitution	Conservation of product	Alteration of the composition of the product
Source control		
Alteration in input material	*Change in Technology*	*Procedural and institutional* modifications
Purification of material	Process modifications	Procedural steps
Substitution of material	Equipment, piping or layout modifications	Prevention of loss
	Additional automation of process	Personnel practices
	Modifications in operational settings	Segregation of waste stream
	Conservation of energy	Improvements in the handling of materials
	Conservation of water	

Adapted from Carl. H. Tromm, et al, *Chemical Engineering*, September 1987.

It may be noted that $W_2 < W_1$ in respect of both volume and toxicity of the waste stream.

Identifying all waste minimization possibilities and then selecting the best ones can turn out to be a real challenge. An auditing procedure suggested by Carl Tromm, et al. can greatly help to uncover the most workable options.

Employing such auditing, a waste minimization programme involves six stages as shown in Table 10.9.

Table 10.9 Waste Minimization Programme—Stage Wise

S.no	*Step*	*Elements*
1.	Initiation of the programme	Secure commitment and authority, define goals, establish organization that is needed
2.	Pre auditing	Prepare *needs* list and inspection agenda, conducting pre audit meeting and inspection, compile data and select target waste streams
3.	Auditing	Inspect plant, generate a comprehensive list of waste minimization options, evaluate these options (separately by audit team and plant personnel), select options for further analysis of the problem
4.	Post auditing	Conduct technical and economic feasibility studies
5.	Recommendation	Suggest preferred options
6.	Implementation of the programme	Design, procure and construct, start up, monitor performance.

10.15.1 Attending to Essential Preliminaries

Getting started with a successful waste minimization programme demands top-management commitment, allocation of adequate financial and technical resources, an appropriate organization and definition of goals and planning.

From the onset, it is imperative to realize that the programme may possibly have to overcome such barriers as a lack of awareness of the benefits of waste minimization steps; limitations of technical staff; concern over tampering with an adequately running process; and organizational inertia and politics. Thus, constructive approaches may have to be employed to deal with these factors effectively.

The auditing procedure can be carried out by an internal taskforce with or without an outside consultant. The leader of the team should have solid technical credentials as well as proven problem solving abilities. It is better if the team includes at least one outsider, e.g. a person from another plant or production unit; that person will help spot and avoid inbred plant prejudices.

The audit team should be empowered with enough authority so that it can gain access both to all required technical documentation (process flow diagrams, inventory and operations logs, etc.) and to all the technical personnel at the plant.

Now the team can start the pre audit stage. Review of pertinent documentation and references should result in a well-defined list of information needs or a plant inspection programme designed to fill in the gaps.

A visit to the manufacturing site including a guided tour of the facility, then should provide first-hand information about the plant, particularly with an eye to filling information needs and gaps in essential data.

At this point of time, the views of plant operating personnel on the focus and function of the audit should be sought. It is certain that, the visit would establish a good working relationship with the site personnel. The initial point of contact (for instance, the plant manager or environmental coordinator) should be enlisted to support the programme.

Information pertaining to all the waste streams leaving the plant should be compiled. Reviews of the list of wastes, process flow diagrams, piping and instrumentation diagrams and the heat and material balances are especially useful. Each diagram has to be analyzed from all points where waste generation could occur.

Waste resulting from the cleaning of equipment should also be taken into consideration. As a matter of fact, in batch processes, large quantities of wastes may be associated with operation, such as cleaning a reactor or mixing vessel. Yet these incidental wastes rarely appear on process flow diagrams or are quantified in process descriptions hence, discussions with key operation personnel are essential to define such wastes.

Waste stream should be quantified on a uniform basis. If a waste stream is intermittent, it should be represented as a pseudo-continuous one. Once all waste streams are quantified, each should be expressed as percentage of the total quantity of waste leaving the operating plant.

Equipped with this information, the team can select the waste stream to be targeted for immediate attention. Since this establishes a focus on the entire audit activity, a great deal of thought has to be given to this targeting. Usual criteria for this include (1) method of disposal and cost of disposal, (2) composition of the waste, (3) quantity—present and future, (4) degree of hazards like toxicity, flammability, etc., (5) potential for minimization, and (6) compliance status—current and future.

Once this analysis is carried out, the results should be summarized in a written report, which should include the following; the location of the facility and its size, a description of the process operations that are of concern, with diagrams necessary to detail relevant aspects of waste generation; details about the waste streams—including generation rates, compositions, disposal and raw material costs-focussing on sources and current methods of management, the rationale for selection of that waste stream.

10.15.2 Conducting the Audit

With the objective of the audit now focused, the audit team should carefully carry out an inspection to develop an understanding of how the targeted waste is generated. This inspection would also furnish any additional information needed to decide about minimization options.

The suggestions that may be helpful are listed here.

1. Having an agenda ready.
2. Planning to observe the operation of different points of time during a shift. Sometimes, monitoring of all shifts may turn out to be absolutely necessary—especially when waste generation is highly dependent on human involvement (as is typical of many discrete manual operations, such as painting or parts cleaning).

3. Obtaining permission to directly interview the plant operators and supervisors. Listening attentively and not hesitating to interview more than one person on the subject. Evaluating their awareness of the waste generation aspects of the operation.
4. Obtaining approval to photograph the facility, photographs are particularly valuable in the absence of lay out drawings and can capture many useful details that will help in the later phases of the audit.
5. Taking a look at the housekeeping check for signs of leaks and spills. Visiting the maintenance department and getting to know its problems in maintaining the equipments leak free.
6. Evaluating the organizational structure and level of coordination of environmental activities between various departments.

Besides the aforementioned suggestions, it may prove advantageous during the preparation and conduct of the audit inspection to mentally *walk the line* from the suspected source of waste generation to the point of exit.

At this point of time, the team should be in a position to generate a comprehensive set of waste minimization options. A brainstorming session involving the entire audit team is one way of doing this, or each member can develop lists on an independent basis.

In exploring options, it has to be re-emphasized that source reduction should be considered before recycling or treatment.

It is important to prepare as comprehensive a list of options as possible. The check list appearing in Table 10.10, may be helpful in thinking about alternatives.

The waste minimization measures presently taken are also worth listing, since they may lead to additional approaches. At this point, one need not worry about the viability of the suggestions. A written rationale, backed by literature references or recorded discussions with equipment and materials, suppliers or consultants, should be adequate.

Once a thorough list has been developed, the team's attention can turn to evaluating the various options. Alternate options that do not merit further attention can be dropped and the remaining choices ranked in the order of desirability.

Ranking can be carried out by taking each option in turn and looking at how it stacks up on key criteria such as:

1. Effectiveness for reducing waste
2. Technical risk involved
3. Extent of current use of the option in question in the facility
4. Industrial precedent
5. Capital and operating costs incurred
6. Effect on the quality of product
7. Possible impact on plant operations
8. Required time for implementation of the option under consideration
9. Other aspects important in the particular situation

Table 10.10 Check List for Considering Alternatives and Adding to the Existing Waste Reduction Efforts

Sources of waste generation	*Options of waste reduction efforts*	*Sources of waste generation*	*Options of waste reduction efforts*
Materials	Buy higher purity raw materials	Filtration and washing	Employ efficient washing and rinsing methods
	Switch to less toxic raw materials		Eliminate the use of filter aids, to the extent possible
	Use non-corrosive raw materials		Adopt counter current washing technique
			Recycle spent wash water
Operating practices	Tighten equipment inspection and maintenance	Leaks and spills	Employ bellows sealed valves
	Improve operator training		Go in for pumps with double mechanical seals or canned (seal less) ones
	Provide closer supervision of the operating personnel		Maximize use of welded, instead of flanged, pipe joints
	Practice better housekeeping		Cut down on water use for spill cleanup
	Employ better process monitoring systems		
	Improve quality control		
Materials handling	Segregate containers by prior contents	Equipment cleaning	Use corrosion resistant materials
	Use washable/recyclable drums		Convert from batch to continuous processing
	Buy materials in bulk or larger containers		Modify production schedule to minimize the number of cleanings
	Purchase materials in preweighed packages		Increase equipment- drainage time
	Employ pipelines for intermediate transfer		Agitate storage tanks as often as possible
Reaction/ Processing	Optimize the reactor design and reaction conditions		Use mechanical wipers on mix tanks
	Optimize the method of addition of the reactant		Clean mix tanks immediately after use
	Eliminate the use of toxic catalysts		Install a high-pressure spray wash system
			Adopt a counter-current rinse sequence
			Re-examine the need for chemical cleaning
			Use in-process cleaning devices
			Recycle spent rinse water
			Blanket with nitrogen to reduce oxidation

Adapted from M. Froeman, et al., *Chemical Engineering,* September 1987.

Assigning a numerical value (say, from 1 to 10, with 10 being best) for the weight of each of these criteria and then, rating each option with respect to each criterion on the same scale and adding the products of the weight and rating will be desirable.

Such evaluations should be conducted independently by the audit team and facility personnel. This provides a useful framework for identifying and resolving, differences of opinion. The ratings should be compared and discussed at a joint meeting between the audit team and the facility personnel so as to evolve mutually acceptable ratings with the audit team leader in firm control of discussions.

It is of significance to know as to how some options are to be introduced—the audit team should be particularly sensitive to the way in which the housekeeping measures are presented. For instance, a recommendation to keep cover on degreasers may appear trivial and even a little insulting. However, if in this case, a rough estimate of (the savings in the cost due to the avoidance solvent-replacement are made known) the suggestion will likely be better understood and not dismissed as trivial.

This reconciliation session should lead to a list of waste minimization options with revised ratings.

Choice of options for feasibility analysis should be based on the revised weighted-sum values. The number of alternatives considered further has a bearing on the time, budget and other resources available for such study.

10.15.3 Finalizing the Best Option

The waste minimization approaches chosen, should be explored during the post audit stage in adequate detail to derive study-grade estimates of capital and operating costs. Capital and operating costs should then be subjected to standard profitability analysis, which must be taken into consideration avoided costs for disposal and materials.

Technical feasibility based on industrial precedent or additional bench scale or trial production run testing should be evaluated. Close attention should be paid to effects on product quality, especially in cases of material substitution.

Such an approach leads to rational basis for selection. The findings of the analysis should then be summarized in a final report. These should be clear recommendations for the course of action to be taken. It should also document the basics and methodology employed to derive the expected results.

This completes the audit portion of a waste minimization project. If management decides to go ahead with the waste minimization project, the implementation stage then proceeds along a well established route of preliminary design, final design, procurement construction and start-up. In an alternative approach, this phase may be preceded by additional research and development.

However, many waste minimization options uncovered during the audit can be put into practice immediately as they do not involve any cost. These options usually fall into the areas of good operating practices or housekeeping, for instance, better operator training or waste segregation.

The success of a waste minimization programme can be ascertained by determining the relative waste generation rate, which is based on the quantity (kgs) of waste constituents of concerned per unit production. Alternatively, one can use the ratio of input materials mass to unit production.

Needless to add although this systematic approach should prove quite useful in identifying and evaluating waste minimization options and in conducting effective audits, its success requires cooperation between the audit team and plant operating personnel, as well as a good understanding of the process and its underlying principles.

10.16 DESIGNING OF MANUFACTURING PROCESSES FOR WASTE MINIMIZATION

It is well known that incorporating waste elimination considerations into the design of a manufacturing process can substantially reduce the generation of waste streams during plant operations. Two stages of a chemical manufacturing operation present opportunities for waste elimination as outlined by Richard Jacob. The first stage is while original design and development of a process is carried out. It is during this time that raw materials, process conditions and equipment selection are the most flexible. The second stage is when the process is already in operation. The actual job of elimination of waste involves operating parameter changes and process modifications that need to be balanced against economic costs and other potentially detrimental effects on the process.

Before strategies for waste reduction are taken up for discussion however, a definition of waste generation is required. In the broadest sense a waste may be considered to be any material or energy input into a process that does not become incorporated into the desired final product. A more useful definition of waste generation takes into consideration only materials that enter and leave the chemical process. As per this definition, sources of waste can be categorized as: (1) unrecovered raw materials, (2) unrecovered products, (3) useful by-products, (4) useless by-products, (5) impurities in raw materials, (6) spent process material, and (7) packaging and container wastes.

Chemical engineers can easily deal with these sources on the familiar grounds of process optimization and yield improvement. The engineering tools, however, must be viewed with a new outlook. The objective is to make products not only of the highest quality and lowest cost, but also to make them while generating the negligible quantity of waste. This implies that waste elimination needs to assume equal importance in the chemical engineer's thinking, as for safety, product quality and cost efficiency.

In the methodology of implementation of a waste elimination strategy, process design can be divided into four phases. Each of them presents different opportunities for implementing waste reduction measures.

1. Product conception
2. Laboratory research
3. Process development (based on pilot plant studies)
4. Mechanical design

10.16.1 Conception of a Product

A product is usually conceived to satisfy a specific market need with little thought given to parameters (involved in the manufacturing process). At this stage of consideration, it may turn out to be possible to avoid some significant waste generation problems in future operations by answering a few simple questions.

1. What are the raw materials employed to manufacture the product? Are there any new chemicals for the organization to handle, that will require the application of new operating and environmental procedures?
2. Are there any toxic or hazardous chemicals likely to be generated during the manufacturing process? Special consideration must be given to by-product and waste stream in the regard.
3. What performance specifications is the new product expected to meet? Is extreme purity needed? If so, more separations will probably be required. Can specifications be widened to include impurities in the raw materials?
4. How reliable will the production process be? Are all steps commercially proven? Does the organization have experience with the unit operations required for the production process?
5. What types of wastes are likely to be generated? What is their physical form? Are they hazardous? Is the organization currently managing these wastes?

By putting these questions to management and knowledgeable technical staff in the organization, significant waste issues can be identified at any early stage in the new product decision process. These issues can bring about a change in the conception of the process design or they may even help elimination of the product from further consideration. In case process design does move forward, the engineers will have a firm understanding of what waste issues are likely to arise.

10.16.2 Laboratory Research Efforts

Once the organization has taken a decision to market a new product; the research and development staff commences laboratory investigations to define the key process parameters and operating conditions necessary for optimum production. Typically, chemical engineers pay attention on mass balances and yield calculations to determine how much of the raw materials end up in the final product.

The strategy for waste minimization or waste elimination, however, demands that an equal amount of attention be focussed on how much of raw materials does not end up in the final product. The questions such as what happens to them? and are they, vented or lost through fugitive emissions come up? Even 99 per cent 'yields are not good enough if the resulting losses give rise to toxic emissions or by-products that cannot be accounted for. Even a speck in the bottom of the laboratory flask could become thousands of kilograms of solid waste each year in full-scale manufacture. Meticulous attention to material balances with complete accounting for all process inputs is of paramount significance to waste identification and elimination in the laboratory.

Another key waste reduction opportunity that should be explored in the laboratory is simplification of process in question. A multistep batch process with intermediate separations and purifications is likely to generate more waste than a continuous process. The quantity of product that will be manufactured is an important consideration in evaluating the importance of the rate of waste generation per unit of a product.

In the determination of the rate of waste generation, consideration of the contribution of all those activities that are necessary for the process, but that are not directly associated with production is of importance. This includes following:

1. Start-up and shutdown losses
2. Reactor washings between operations
3. Sampling and analytical losses
4. Catalyst usage in the production process and its losses
5. Incidental losses from spills, equipment cleanings, etc.
6. Packaging requirements for raw materials and products

In a complex multistep process, situations may arise such that these losses can overshadow the losses associated with product yields. The underlying message is that a simpler process gives less waste than a complex one.

10.16.3 Development of Process

Once the process has been defined in the laboratory and the expected yields are known, the waste streams that are most likely to be generated can be quantified. The next step is to verify these data on a scale that can be used to design the commercial manufacturing process. This is done by building a pilot plant and conducting relevant studies employing it.

Besides verifying the chemistry proven in the laboratory, an effective waste elimination strategy warrants that the pilot plant be employed to quantify those non-production activities mentioned earlier and to commence the assessment of the impact of the process on the public. Important parameters to assess during pilot plant studies include the following:

1. Flexibility in the selection of raw materials to minimize the volume of effluent stream and its toxicity.
2. Methods of improving process reliability to keep losses, spills and off-specification products at a minimum.
3. Ability to keep track of and exercise control all waste streams.
4. Potential impacts of the process on the public including odour generation, visible emissions, fear generated by the handling of toxic materials, emergency considerations, and so on.

In the selection of raw materials, the key waste generation factor, after yield to product is the quantity and types of impurities present in them. Any impurity that cannot be passed through to the product, naturally ends up as a waste. Thus, it may be more economical and advantageous to purchase a higher purity raw material than to manage the waste associated with a less pure feedstock. This economic tradeoff is hard to evaluate owing to the fact that many of the costs associated with waste management are difficult to quantify. Disposal costs are generally known, but the time and labour required to comply with regulations and permits, the production losses associated with waste and the future environmental liability costs are not readily available.

Process reliability is generally considered from the health and safety viewpoint. Reliability also affects waste generation by preventing situations that might give rise to releases or the production of off-specification product. Sequencing operations to reduce equipment cleaning and reactor washing between steps also effect a reduction in the quantity of non-production wastes associated with the process.

A major problem that clashes with the ability of many operating plants to minimize their waste is a shortfall of adequate process measurements. Many waste streams come out of an operating unit unnoticed because they get mixed with other streams prior to being fed to a centralized treatment system that serves the entire complex. Owing to the fact that the waste stream is not measured at the point of generation, no one assumes responsibility for controlling the quantity of its generation. The waste is simply treated to meet the discharge permit requirements for the complex.

During the pilot plant stage, one has a good opportunity to evaluate all of those real world possibilities for process upset.

10.16.4 Mechanical Design

With the waste elimination strategies fully implemented during the product conception, laboratory research and pilot plant stages of the new product development, the task during the course of mechanical design is to suitably incorporate all of the waste elimination steps previously identified into the final plant design. Extra attention should be given to process improvements in the following area:

1. Carrying out the reduction and/or control of fugitive emissions (spill control, leak less valves, closed-loop sampling, etc.).
2. Making the measurements necessary to quantify and control wastes which may imply putting flow meters in unusual places.
3. Concentration of waste stream to the maximum extent possible by adopting the methods, such as thin-film evaporation, ultra-filtration, etc.
4. Carrying out the reduction of interim storage and container management to the extent possible (containers must be washed or disposed of, and these activities generate wastes).
5. Utilization of waste treatments that enable recovery and reuse of raw material and intermediates through methods like liquid ion-exchange vs filtration, acid stripping vs neutral stripping, recovery of HCl from incinerators.

To make a plant implement a waste elimination strategy in an operating facility the important steps to be followed are:

1. Establishing definitions and objectives
2. Conducting on inventory of all waste streams
3. Establishing a prioritization, reporting and tracking systems for wastes
4. Systematically studying each waste to determine how it can be managed, and
5. Implementing specific projects or actions for each waste

This information is valuable as a knowledge base from which a waste minimization programme can be initiated.

Each operating plant would know best as to how to define and prioritize its waste streams and that within broad corporate guidelines that provided some consistency. Decisions regarding wastes that would be addressed first should be made by the operating plant. Periodically, representatives of all plants can gather and share their ideas and experiences in the waste reduction exercises. This would enable the plants to learn from each other and allow the organization to move towards a common understanding of its waste generation universe.

10.17 IDENTIFICATION OF PROCESS IMPROVEMENT OPTIONS AND APPLICATIONS OF PROCESS INTEGRATION TO MINIMIZE EMISSIONS AND WASTE GENERATION

The development of process improvements, such as new types of equipment and process changes to effect a reduction in emissions has been achieved on an ad hoc basis by creative engineers. Several worthwhile advances have been made in this way. As a definite advance over purely ad hoc methods, some efforts have been made to categorize methods for reducing wastes and emissions and to provide check lists for application in pollution control and waste reduction projects.

However, even more rapid progress can be made if a systematic technique is employed to identify process improvement opportunities. This kind of an approach would enable engineers to identify improved emissions reduction opportunities more easily and more rapidly than they would otherwise. In addition, the process changes identified could not only reduce emissions, but also bring about energy savings and improved raw materials efficiency. Rather than focussing on localized effects, the procedure would provide an overall process perspective, enabling benefits and costs to be evaluated in terms of their impact on the process considered as a whole.

Here the salient points of such a systematic method for identifying process modifications to minimize waste generation discussed by Alan Rosster et al. are presented. This method is based on the hierarchical decision procedure, which provides a framework for identifying process improvement options and evaluating heat and mass integration opportunities. This method can also be employed to assess the incentives for developing alternative technologies to solve specific problems.

10.17.1 Hierarchical Review of the Process

Over the past two decades, several process synthesis techniques have been developed to provide a systematic framework for various types of process design and improvement problems. These techniques include the following:

1. *Pinch analysis* started as a methodology for designing heat exchanger networks. Later it has developed into a comprehensive package of tools for improving processes and total sites.
2. Hierarchical decision procedures which provide a basis for developing good designs from preliminary process concepts.
3. Cost modelling methods which use simplified mathematical representation of process cost elements to define optimum process structures and equipment sizes.

Process integration work has generally focussed on minimization of cost, with the emphasis on the tradeoffs between capital and operating costs. However, essentially, the same methodologies can be employed to explore the three-way tradeoffs among the three factors viz., capital costs, operating costs and environmental impacts. The following section deals specifically with an adaptation of the hierarchical decision approach for use in pollution abatement applications.

The basic premise of hierarchical review is that process designs are developed through a series of decisions which provide successive levels of details. By identifying the key decisions and implementing them in the correct order it is possible to identify good design options with the minimum effort and re-work. Alternative designs for evaluation may also be generated by this procedure. At each level, an economic evaluation of the process can be done, providing a screening tool to eliminate non-workable processes without having to use up excessive design effort.

The hierarchical approach has been employed in the design of various types of plants and has been adapted for retrofit applications. In retrofit application, rather than providing a logical sequence for making design decisions, the procedure provided a systematic framework for reviewing and improving existing installations. In either case, a similar overall hierarchical structure can be developed, although the detailed questions that must be addressed at each decision-making stage, have a bearing on the type of problem being analyzed.

It is important to note that the procedure cannot ensure optimum designs and it does not guarantee minimum achievable emissions. What it provides is an objective structure for generating, screening and comparing various options, and defining realistically attainable targets for emission levels using available technologies.

10.17.2 The Methodology of Making Decisions

The hierarchy of decisions given here can be used to analyze new and existing plant designs with the objective of reducing emissions.

Stage I Processing mode, batch versus continuous processing

Stage II Input-output structure of the flow sheet

Stage III Recycle structure and product formation considerations
Stage IV Separation systems
Stage V Product drying
Stage VI Energy systems
Stage VII Equipment and pipe work specifications

Options for the design of new processes and possible improvements for existing ones are considered sequentially in the order of the decision hierarchy, employing a combination of heuristics, questions and numerical calculations. Rough economics for each option are evaluated as the option is identified to avoid wasting effort on options that are clearly not feasible.

Heat and material balances are required as a basis for these evaluations. It is generally difficult to obtain accurate information on the quantities of materials released by processes and conventional flow sheets often show these streams as *trace* flow. In practice, the hierarchical approach can be applied even in case where only very approximate data are available.

Use of the procedure in new plant design generally gives rise in one or more good design (s) for the process. In retrofit situations, it generates a list of potential process modifications to reduce emissions economically.

Stage I: Processing mode

Criteria influencing the fundamental choice between continuous and batch operations are per used and implications for waste generation and disposal are evaluated. The two key differences in respect of waste generation between batch and continuous processes are:

- Waste streams from batch processes are generally intermittent, whereas those from continuous processes are continuous.
- Compositions and flow rates of waste streams leaving a batch process typically vary. The waste streams from continuous processes are nearly constant.

The greater variability of the wastes from batch processes tends to give rise to more difficult waste management problems. For instance, if a given total volume of waste must be handled, the instantaneous maximum flow rate is higher in a batch plant necessitating the need for, larger equipment items to handle these wastes. Moreover, waste generation rates are often high during startup and shutdown periods, and these occur most frequently in batch units. It hence follows that waste reduction factors generally favour continuous rather than batch processing.

The main factors and heuristics that are usually considered in deciding between batch and continuous operations are given here.

Production rate: Under 500,000 tons/year batch processing is almost invariably used; between 500,000 and 5,000,000 batch processing is common, at higher rates of production continuous processing is usually preferred.

Life of product: Batch plants are better suited to products with short life spans; where a rapid response to the market is required.

Multiproduct capabilities: If the unit is required to produce several similar products using the same equipment, batch processing is usually preferable.

Process reasons: A number of process related factors may lead to batch processing, being preferred, for instance, cleaning requirements that necessitate frequent shutdowns, difficulty in scaling up laboratory data, or complicated process recipes.

The aforementioned environmental considerations suggest that another heuristic should be added to this list namely, if potentially serious environmental problems are expected in a particular process, this favours the selection of a continuous unit.

In practice, some of the other factors mentioned above may still dictate that a batch operation is preferred. In this case, consideration should be given to smoothing intermittent or variable flow streams, for example, by adding buffer storage capacity to make processing and recovery of waste materials simple.

Stage II: Input-output structure of a process

The input-output structure of the process is to be considered next. Figure 10.13 shows one possible form of this structure.

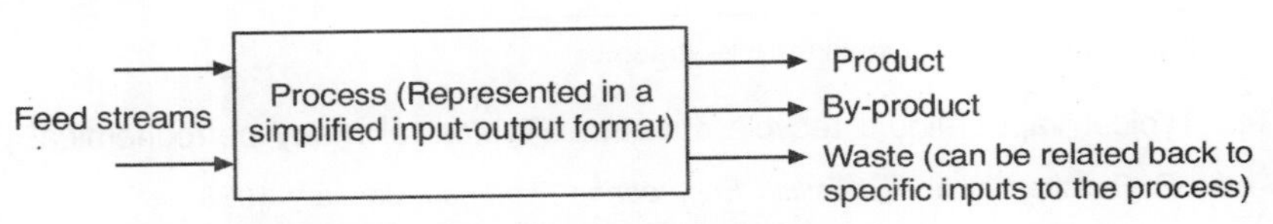

Figure 10.13 Typical input-output structure of a process.

Adapted from Alan P. Rossiter, et al., *Chemical Engineering Progress*, January 1993.

Representing the process in a simplified input-output format gives a useful overview of the material transformations taking place and helps to identify the individual waste streams that are generated from the process.

The wastes can often be related back to specific inputs to the process. For instance, impurities in a feed stream may have to be removed in the form of a process effluent, and it stands to common sense that using a purer feed material may eliminate the need for the effluent streams. Another example can be cited in many separation processes where an additional feed is provided for the sole purpose of assisting in the separation. Perhaps the most common form of this is live steam injection, where the resulting steam condensate generally turns out to be an impure waste stream that has to be eliminated. This waste generation can sometimes be avoided by using a reboiler rather than live steam. A third example is the situation in which one of the feeds is a diluent—in this situation it may be possible to recycle this within the process rather than supply it as a feed.

The points to be considered in the Stage II analysis include the following:

- Whether the feed quality can be improved at low cost.
- Whether the impurities should be removed before processing or allowed to pass through the process.

- Whether any of the input streams can be eliminated.
- Whether any *waste* output streams can be employed advantageously within the plant or sold to an external customer.

Stage III: *Recycle structure and formation of product*

Here the process is broken down into major component sections, typically reaction and separation, together with interlinking streams and recycle streams. A configuration typical of many petrochemical processes is shown in Figure 10.14.

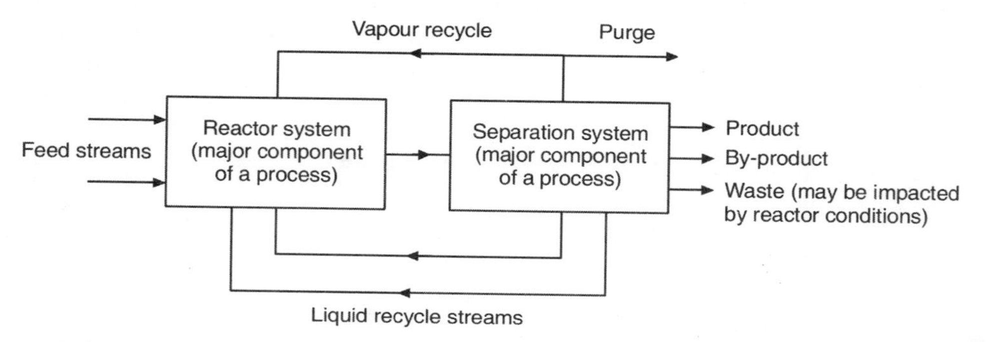

Figure 10.14 Typical vapour-liquid recycle structure (Typical of many petrochemical processes). Adapted from Alan P. Rossiter, et al., *Chemical Engineering Progress*, January 1993.

Where data are available, it is possible to evaluate the impact of reactor conditions on waste formation and to evaluate some of the major tradeoffs in the process, such as reactor conversion, cost of separation/recycle and waste generation due to by-product reactions.

The points to be considered in Stage III include the following:

1. Whether any output streams contain feed or product material that could be recovered and recycled.
2. Whether reaction conditions can be changed to minimize the formation of waste by-products.
3. Whether waste by-products can be recycled to extinction.

Stage IV: *Separation systems*

Stage IV provides a more detailed analysis of separation options. A typical separation system structure for a vapour-liquid system is indicated in Figure 10.15.

At this stage (Stage IV), the impact of the separation technology on waste generation is considered. Possible means of reducing emissions by employing alternative separation techniques or by rearranging the existing ones are evaluated. Interactions of the separation system with the reactor and recycle systems are also taken into consideration.

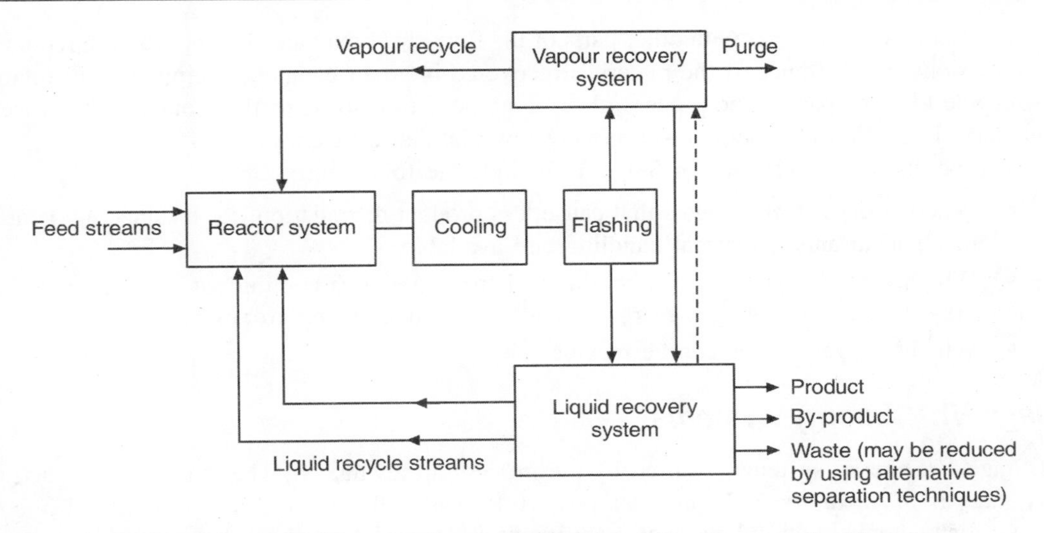

Figure 10.15 Typical separation system structure for a vapour-liquid system.

Adapted from Alan P. Rossiter, et al., *Chemical Engineering Progress*, January 1993.

The following points are to be addressed at Stage IV:

- Whether any waste streams are the result or outcome of poor or inappropriate separation processes.
- Whether any wastes, especially hazardous ones, can be removed from process effluents by adding new separation processes. Typically this will involve additional separation steps to recover the undesirable material from effluent, preferable for recycle.
- Whether there are alternative separation technologies that could replace or supplement the existing separation methods and reduce releases? Most industries have preferred separation technologies—for instance, distillation is the most sought-after unit operation of petroleum refining and petrochemical industries. Sometimes, it may be possible to improve the quality, if process effluents by employing different separation equipments for instance, membrane systems are showing a lot of promise in a variety of applications.

Stage V: Product drying

Stage V is specific to solid-liquid processes, and is mainly concerned with product drying, although some of the same issues are also relevant to other operations that use solid catalysts such as fluid catalytic cracking. Possible contributions to waste generation due to materials degradation in the dryer are evaluated together with the control of material losses from the dryer.

Many a time, the mode of operation of the dryer is a direct cause of material losses for example, high air and particle velocities in flash dryers can give rise to attrition of solid products and thus to the creation of fines, which are difficult to remove from the air stream leaving the dryer. Selection of the most appropriate drying equipment is therefore important, as it makes the cleanup of the air stream much easier.

Interactions of the dryer with other parts of the flow sheet may also be important in reducing net emissions. For instance, if the materials recovered in an air cleaning equipment are suitable for recycle to some part of the process, there is no need to dispose of them outside the process boundary. Recycling also helps in improving raw material efficiency.

The points to be addressed in Stage V, include the following:

- Whether dryer technologies that cause less product degradation, for instance, by gentler mechanical and/or thermal handling, are available.
- Whether waste materials can be removed from the gases coming out of dryer, and thus prevented from leaving the process and getting into the environment.
- Whether dryer wastes can be recycled.

Stage VI: *Energy systems*

At Stage VI, energy systems required by the process are evaluated. The major tool for use at this stage is *pinch analysis*. It provides powerful insights for process integration, and is best known for its application in heat exchanger network design. However, it has other far-reaching applications in decreasing the environment impact of industrial processes.

For instance, pinch analysis can integrate *end of pipe* treatment steps into the rest of the process to reduce capital and operating costs. It can reduce fuel firing and steam consumption which, in turn reduces production of greenhouse gases and also lessens the use of water treatment chemicals. It can help in the selection of appropriate utility systems to maximum cogeneration. It can also help to minimize the required temperature levels for utility heating, and lower temperature utilities are generally less expensive and generate less net pollution compared to higher temperature utilities. Finally, pinch analysis can provide insights into possible modifications to reaction and separation processes for reducing energy consumption, and emissions, and these insights can then be used to explore trade offs.

The points to be addressed during Stage VI analysis include the following:

- Whether the energy consumption of the process be economically reduced to a large extent.
- Whether the temperature levels at which heat is delivered to the process can be lowered.
- What fuels are used to provide heat for the process? Are cleaner fuels available, and if so at what cost?

Stage VII: *Equipment and pipe work specifications*

Stage VII provides an opportunity to assess and also minimize fugitive emissions from individual equipment items, valves and pipe work. The number of potential sources of fugitive emissions is minimized by *designing out* problematic equipment to the extent possible. Emission characteristics are considered in the selection of critical equipment items.

The points to be considered in the analysis at Stage VII are the following:

- Whether the total number of equipment items and connections can be brought down.
- In situations where there are alternative types of equipments for a particular service, which one offers the most favourable emission characteristics? Is the incremental cost justified or offset by the reduced emissions?
- Whether welded pipe can be used in place of flanged pipe.

Heuristics

Heuristics or rules of thumb, provide guidelines for the development of good designs or retrofit options. They are typically based on accumulated experience from a large number of similar applications and represent the erudition of many engineers.

In cases where heuristics are available, it is often possible to go a long way towards defining good design options with only limited calculation and experimentation, thereby saving much work. If heuristics are not available, the burden of calculation and test work is much greater.

The following are proposed for use in working through the seven stages of the hierarchy:

- By changing set points or tightening control variations of key variables, rapid low cost reductions in waste generation can often be achieved. Modifications to single equipment items (e.g. changing column internals) can also result in significant improvements with little capital expenditure.
- Wherever possible, waste materials need to be eliminated at their source. One of the ways of achieving this is by switching raw materials.
- Waste materials can be recycled within the process. In situations in which this is not possible, the possibility of using these materials within other processes, preferably within the same operating facility, can be explored.
- In case waste by-products are formed reversibly within a reaction process, they should be recycled to extinction. Where this is possible, it can help much in completely eliminating a class of waste materials.
- For all heating duties that require utilities, the one with the lowest practical temperature can be employed. This offers several potential benefits, including reduced heat exchanger fouling occurrences and lower utility costs.
- The total number of main equipment items in the process especially in areas that handle toxic materials, may be minimized. The number of pipework connections to and from equipment items may also be minimized.
- Pollution prevention and control is generally more difficult and expensive in batch operations than in continuous ones. Therefore, where feasible, continuous processes are to be preferred.

It may be noted that the inherent flexibility of batch plants often makes raw material and product substitutions simpler in these processes. In this respect, pollution prevention is easier in batch processes, which is a direct contradiction of the seventh heuristic.

This is indicative of an underlying problem with heuristics. They are simply guidelines based on a body of experience and cannot be guaranteed to yield the right answers in every case. They are nevertheless, a very helpful way of getting rapid insights and generating options for evaluation.

10.17.3 The Hierarchical Analysis of a Refinery

The hierarchical analysis procedure was applied to the crude unit, the Fluid Catalytic Creaking (FCC) unit and the sour waster system at a refinery in the U.S. This study resulted in a large number of ideas for reducing releases, some of which are summarized in Table 10.11.

Table 10.11 Selection of Process Improvement Ideas for the Petroleum Refinery Studied

S.no.	*Description of the project*	*Objectives*
1.	Replacement of steam ejectors with vacuum pumps	To reduce sour water flow and steam use
2.	Two stage desalting	To reduce brine flow and oil losses
3.	Recovering and recycling oil in desalter brine	To reduce hydrocarbon losses
4.	Cooling desalter brine	To reduce hydrocarbon flashing
5.	Replacing live steam with reboilers	To reduce sour water generation
6.	Using stripped sour water for line washing	To reduce water consumption
7.	Generating stripped steam from stripped sour water	To reduce sour water surplus; to reduce water consumption
8.	Replacing pump packings	To eliminate seal water losses
9.	Recycling polygasoline water	To reduce feed to sour water systems
10.	Integrating heat	To save upto 10 per cent fuel in crude unit; to generate up to 30 per cent more steam in FCC unit, to reduce NO_x, SO_x and CO_2 emissions
11.	To minimize losses to flare	To improve raw materials efficiency and to eliminate flare emissions
12.	Recovery of flare gas as fuel	To improve raw material efficiency and to eliminate flare emissions—same effect as in minimizing losses to flare
13.	New solid recovery system for FCC unit refrigeration	To reduce fines losses from regenerator
14.	Recirculation of stored decanter oil	To reduce sedimentation

The procedure based on known technologies, and available equipment enabled these ideas to be generated very quickly with only limited process data. Here it is discussed that how this procedure was applied to the FCC unit.

The FCC unit discharges considerable quantities of aqueous effluents. In addition, catalyst fines are lost from the regenerator as airborne dust.

There are three major forms of water entry into the system in addition to the small amount of water in the feed material; stripping/fluidizing steam to the reactor and regenerator, wash waster to vapour lines and steam stripping in light cycle oil and decanter oil stripper. Virtually all the water is removed as condensate within the distillation system; this goes to sour water stripping. A very small quantity of water is fed forward in product streams.

The hierarchical analysis is carried out as follows.

Stage I: *Processing model*

The process is continuous. there is no motive to convert to batch operation.

Stage II: *Input-output structure*

The following points are raised.

- Whether stripping steam can be eliminated in columns by using alternative separation technologies (Stage IV may be referred to)
- Whether the quality of the feed can be improved to eliminate or reduce the need for the condensate washing system
- Whether steam consumption in the reactor can be reduced so that less condensate will have to be removed from the distillation system. This would warrant changing the reactor design and would probably not be a viable retrofit option.
- Within the regeneration system, the loss of fines appears to be a function of air input rate (Stage II) and cyclone design (Stage IV). At Stage II, a reduction in air identified as a possible means of reducing the discharge of fines. However, preliminary evaluation showed this to be less attractive than the other options.

Stage III: *Recycle structure*

The requirement for steam in the reactor is met by using fresh steam. If this could be replaced with steam generated from process water, the liquid effluent from the FCC unit could be greatly reduced. This would enable the condensate recovered within the distillation section (or other contaminated condensate) to be revaporized and recycled to the reactor section. A large purge would probably be required, but the total volume of liquid effluent would be much less than in the existing plant, and a large part of the hydrocarbon content would be recovered. Regeneration also consumes a considerable quantity of steam. It may also be possible to meet this requirement with *dirty steam* since the hydrocarbon content would be incinerated with the coke in the regenerator.

Used wash water is collected as several points and then purged from the process. Instead if it could be recovered and recycled, water entry could be reduced.

Stage IV: *Separation system*

Three options could be studied

- Replacement of live steam stripping with reboilers. In practice, this would reduce water entry by less than 2 per cent of the total and so is unlikely to be practicable.
- Placing additional oil/water separators downstream of existing condensate collection point. These separators could be, for example, coalescers or strippers. Hydrocarbons can be recovered and recyled.
- Improving gas/solid separation after the regenerator. This might simply require better cyclone and/or duct work design or electrostatic precipitation.

Stage V: Protect drying

There is no product drying occurring in the FCC unit. However, recovery of fine particles of catalyst downstream of the regenerator (Stage IV, option 3) presents problems similar to those encountered in the drying of solids.

One possibility for reducing fine particle losses, might be to develop a regenerator with less abrasive particle handling, which would produce fewer fine particles. However, this would not be a realistic or economic retrofit option.

Another option that is worth investigating is the use of more attrition-resistant catalysts.

Stage VI: Energy systems

In steady operating conditions, the FCC unit generated adequate heat to be self-sustaining and required no external heating utility. Much of the heat rejected by the unit could be recovered in 150 psig steam. The rest of the heat may be rejected in coolers or allowed to leave in hot gases. The primary objective of the Stage VI analysis is therefore, to maximize the quantity and the value of the recovered heat.

Pinch analysis shows that it would be possible to increase heat recovery to 150 psig steam by more than 100 lakhs Btu/hour (nearly 31%) with realistic temperature driving forces for heat transfer. It would also be possible to increase the steam pressure to nearly 600 psig with only a very small loss in the quantity of heat that would be recovered. This enhancement in pressure would allow the steam to be passed through a backpressure steam turbine for power generation, without the steam losing its ability to meet heating demands at the 150 psig level. However, it is doubtful whether this would be economical in a retrofit situation.

Stage VII: Equipment and pipe work

The data available to conduct this level of analysis are inadequate.

The FCC unit ideas generated could be further evaluated employing conventional engineering methods. It is likely that in some cases, further evaluation of the concepts may occur and the final form, of the project may differ from the original. Several of these concepts appear, in various guises, in the short list of process improvement options shown in Table 10.11, together with the ideas generated in related studies of the crude unit and the sour water system.

All of these efforts result in savings in raw materials and/or utilities as well as providing environmental benefits which include the following:

- Elimination of the surplus water in the sour water system bringing about a reduction in potential source of odours.
- A 30 per cent reduction in desalter brine flow.
- Recovery of a very large quantity of raw material per year of the raw material with an equivalent reduction in loading in the water treatment plant.
- Savings of more than 300 lakhs Btu/hour in fuel firing. Recovery of an additional 200 lakhs Btu/hour in fuel gas.

10.18 MINIMIZING THE EMISSION OF AIR TOXICS THROUGH PROCESS CHANGES

Pollution prevention may require a change in philosophy so that the generation of emissions and wastes is viewed as an shortcoming or inefficiency in the production process rather than as an inevitable environmental problem. As in the case with effective safety and quality programme, the commitment on the part of management of an industrial organization is crucial in the implementation of a successful pollution prevention programme.

In the following section, some of the opportunities discussed by Nick Chadha, et al. for minimizing emissions of hazardous air pollutants that can be uncovered by looking at manufacturing processes very closely and analyzing situations relating to generation of emissions, are presented. Many of the common sense approaches presented are simple to carry out do not require the development of breakthrough process technologies and could be applied to a chemical process industry (CPI) facility of any size. The technical strategies discussed here are based on experience with the Synthetic Organic Chemical Manufacturing Industry (SOCMI), the pharamaceutical industry and other CPI segments that employ Volatile Organic Compounds (VOCs) as raw materials or process solvents. The approaches outlined here in respect of two specific industries are applicable to most facilities that use or manufacture chemicals.

Needless to mention, a significant advantage of minimizing air toxic emissions is reduced capital and operating costs for air pollution control equipment. Other benefits that could be derived through this approach include cleaner and safe working condition as well as the reduction of potential environmental liabilities and risks. In situations where it is not possible to totally eliminate emissions through process changes, capturing of emissions and reusing in a process can be advantageous, especially in the case of process solvents. This enables recovery of the chemical and/or energy value from streams that did not get converted (in the process) to products or useful by-products.

10.18.1 Sources of Emission

The Synthetic Organic Chemical Manufacturing Industry plants manufacture a large number of intermediate and finished organic products starting from a small number of basic feedstock materials which in turn, are derived from crude oil, natural gas or coal. Major operations involve conversion processes such as air oxidation, and separation processes such as distillation.

In synthetic organic chemical manufacturing industrial units air oxidation processes, one or more chemicals are reacted with oxygen which supplied as air or oxygen-enriched air. Vents from these processes let out large volumes of inert gases containing VOCs because the nitrogen in the oxidation air, passes through unreacted components and carries VOCs out with it. The nature of emissions of VOCs from distillation process vents depends on the operating conditions and various other factors.

Owing to the diversity of operations emissions from synthetic organic chemical manufacturing industry plants differ widely in the types and concentrations of VOCs emitted.

In respect of pharmaceutical industry, the pharmaceutical synthesis operations are typically carried out in batch processes and involve one or more chemical reactions followed by a series of product concentration and purification steps. The equipment of a pharmaceutical plant include reactors, process vessels, centrifuges, dryers, crystallizers and distillation columns. Different products are made in the same equipment at different times and hence the product purity turns out to be critical.

In pharmaceutical plants nitrogen is used abundantly. For instance, for pressure transfer of vessel contents purging of lines and vessels direct-contact cooling of vessel contents and other applications nitrogen is employed. Volatile organic solvents, including many flammable ones are also commonly employed. The nitrogen and solvents typically do not participate in the synthesis reactions.

The liquid nitrogen and organic solvents are usually much cheaper compared to the high-value pharmaceutical products and by-products manufactured. Thus, many situations, there is limited economic incentive for solvent recovery (especially if multiple solvents are involved in the same process). Therefore, the waste solvents are often sent to an on-site hazardous waste incinerator for disposal. Solvent losses cause most of the air toxic emissions from a pharmaceutical plant.

Even though each synthetic organic chemical manufacturing plant and pharmaceutical plant is unique in terms of its process operations and the types of products manufactured, a few similarities can be seen when these processes are closely examined. As shown in Figure 10.16, emissions and wastes from any CPI facility, represent materials that did not get converted to products and/or useful by-products in the process.

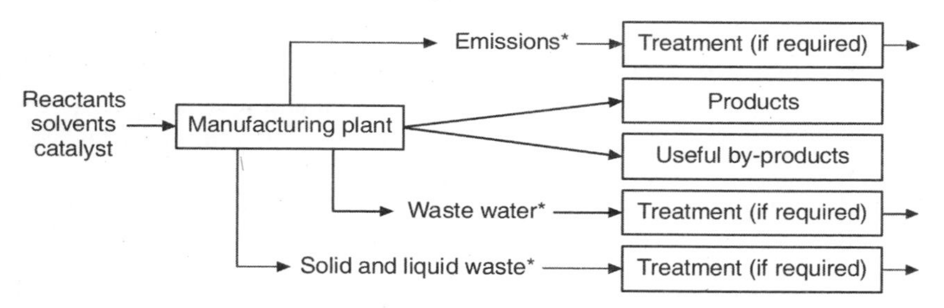

Contaminated by reactants, impurities in reactants, products, usable by-products, non-usable by-products, solvents and catalysts.

Figure 10.16 The wastes or emissions from a typical chemical plant.

Adapted from Nick Chadha, et al., *Chemical Engineering Progress*, January 1993.

These emissions and wastes comprise either raw materials, impurities in raw materials, solvents, process catalysts, products, by-products (usable or non-usable) or products of combustion. Emissions from combustion sources typically contain more inorganic contaminants than organic contaminants and are not discussed further.

10.18.2 Categorization of Emission

Emissions from any Chemical Processing Industry (CPI) can be classified into four major categories in order to provide a structural, approach to minimizing emissions. These are: (1) storage and handling emissions, (2) process emissions, (3) fugitive emissions, and (4) secondary emissions. These four categories are briefly given here.

Storage and handling emissions

These emissions have a bearing on the construction and size of the storage tank, the vapour pressure of stored organic liquid and ambient conditions at place where the tank is located. For fixed roof tanks, emissions are the sum of working losses (due to filling and emptying of tanks) and breathing losses (due to variations in ambient temperature or pressure). Handling losses take place from railcar, tank truck, at marine stations used for loading/unloading and transfer of volatile organic liquids.

Process emissions

Stacks and vents from process reactors and recovery and control equipments such as vent condensers, carbon adsorbers, and absorbers or scrubbers are some of the sources for process emissions. Process emissions also include losses from downstream product separation and purification and separation equipments such as filters, centrifuges and distillation columns. For instance, if distillation technique is employed for fractionation of potential emission sources include condensers, product receiver vessels, vaccum pumps and steam ejectors, associated with the distillation unit.

Fugitive emissions

Fugitive emissions emanate from sources such as leaking pumps, valves, flanges, open-ended lines, agitator seals, instrument connections, sample connections, and so on. Fugitive emissions from leaking equipment depend on the nature of the chemical process, the level of preventive maintenance being carried out and other site specific variables. Despite the fact that fugitive emissions are listed as a separate category, they can emanate from many plant sources, including storage tanks and process operations. Fugitive emission sources from process include covers for manhole and access parts, drains from process vessels, and troughs used for collecting discharges from steam ejectors and barometric condensers.

Secondary emissions

Secondary emissions refer to organic emissions from waste water collection and treatment source such as trenches, sumps, surface impoundments and aeration basins. Organic emissions from solvent recovery operations, liquid waste incineration, accident spills and leaks and other miscellaneous sources also fall under this category.

Emissions from storage and handling, fugitive, and secondary sources have been addressed by a number of technical publications. The following section focuses specifically on emissions from process sources, which represent the major component of a chemical plant's total air toxic emissions.

10.18.3 Generic Ways of Causing Process Emissions

To make the identification of process emission sources easier, it is helpful to review the generic ways in which emissions are caused. In general, emissions are generated when a non-condensible material, such as air or nitrogen is introduced into a process containing volatile organic or when any uncondensed material leaves a process.

The relative importance of the various ways of causing emissions can be predicted to some extent, for instance, the introduction of air into a vessel will likely give rise to more emissions than displacement due to liquid transfer. However, the exact emission generating potential will be governed by many operating parameters and will need to be estimated on the basis of a specific site.

The following are some common generic ways and methods in which process emissions are generated:

1. Introducing air (oxygen) into reactors for oxidation of volatile organic reactors during the process of chemical synthesis.
2. Introducing air into reactors when manhole covers are opened for charging solid powders or other raw matereials that is deliberate ventilation designed to provide an adequate face velocity at the reactor opening.
3. Existence of leakage of air into any process equipment containing volatile organics and operating under vacuum that is incidental ventilation owing to poor equipment maintenance.
4. Employing liquid nitrogen or dry ice to provide direct contact cooling in vessels containing volatile organics.
5. Employing nitrogen (or other non-condensable materials) for pressure transfer of volatile organics from an vessel to another or for blowing lines to clear residual liquids.
6. Employing nitrogen for breaking vacuum or providing an inert atmosphere (by constant purging) in process equipment containing volatile organics.
7. Generating non-condensable gases (such as carbon dioxide or hydrogen), a by-product or a product of the reaction.
8. Generating VOCs with vapour pressures greater than atmospheric pressure at process temperatures (such as ethane or isobutylene) as a product or by-product of the reaction.
9. Evacuating of vessels containing volatile organics that is venting a vessel to vacuum equipment to reduce its operating pressure.
10. Heating of vessels containing volatile organics so as to cause expansion and increase in organic vapour pressures.
11. Stripping of volatile organics from reaction mixtures during the process vacuum distillation.
12. Boiling of pure solvents in vessels to clean them between batches or different product campaigns.
13. Charging of liquids into a vessel containing volatile organics or bringing about level equalization between two vessels so as to cause volume displacement.
14. Exhaustion of uncondensed organic vapours from vent condensers and others recovery or control equipment that handles VOCs (this could happen due to several reasons).

10.18.4 Commencing with an Accurate Emissions Inventory

Experience in processing plants indicates that 80 per cent of the emissions will be generated by 20 per cent of the sources. Therefore, it is imperative that one focuses on identifying and quantifying the major contributors.

Typically, engineering calculations are employed to quantifying emissions from process sources. However, this approach has some hidden difficulties related to the various simplifying assumptions that must be made while estimating process emissions.

As a matter of fact engineering estimates can hide potential opportunities for minimizing emissions in some situations. For instance, in a pharamaceutical plant, centrifuge exhausts could be a major (unidentified) source of solvent emissions owing to air, in leakage rates being several orders of magnitude higher than what would have been assumed in the engineering calculations. This in turn may be due to high equipment vibration missing gaskets, age or even obsolescence of the machines and other maintenance factors.

10.18.5 Unidentified Emissions and Fugitive Emissions

The total fugitive emissions from leaking process equipment and other sources typically consist of 20 per cent to 50 per cent of the total emissions from a processing plant. The balance is made up of stack emissions. However, it is often difficult to close the material balance adequately for batch processes in spite of the fact that the emission estimates developed by engineering calculations are usually conservative. All unidentified emissions are then incorrectly accounted as fugitive emissions in the baseline emissions inventory.

It is very likely that many of these emissions, however, are actually unidentified point-sources emissions from some not so obvious process operations and sources. These sources must be identified by addressing the generic emission generation mechanisms that have been discussed earlier.

10.18.6 Reviewing the Available Emission Data

This systematic approach to minimizing emissions commences with a pollution prevention audit to review available emission data and identify data gaps. When the auditing is being carried out, the generic emission categories and the generic ways of causing emission are taken up for analysis so that any previously unknown point source emission sources can be uncovered. During problem definition one has to go beyond traditional development of baseline emissions inventory to ask not only *what* and *how much* but also *why*, *how* and *when* for all possible emissions from each source.

The generation of emissions normally follows repeating patterns that are independent of the specific industry. This is because the true source of air toxic emissions is either the process chemistry or the way in which the process is engineered, operated or maintained. For instance, if emissions are caused owing to generation of a volatile organic product or by-product during a chemical reaction, then the true source of emissions can be interpreted as the process chemistry and modifying the engineering design will not solve this problem. The following sections present some generic strategies for minimizing air toxic emissions and provide specific examples to illustrate those strategies.

10.18.7 Modifications in Process Chemistry

In some situations emissions are generated by reasons related to the process chemistry, such as the reaction stoichiometry, the kinetics or the conversion or product yield. Emissions generation may be minimized by strategies varying from simply adjusting the order in which reactants are added to major modifications that call for significant process development work and capital expenditures.

10.18.8 Changing the Order of Reactant Additions

A pharmaceutical plant made process chemistry modification in order to minimize the emissions of an undesirable by-product, isobutylene from a mature synthesis process. As indicated in Figure 10.17, the process consists of four batch operations. Emissions of isobutylene were brought down by identifying the process conditions that gives rise to its formation in the third step of the process.

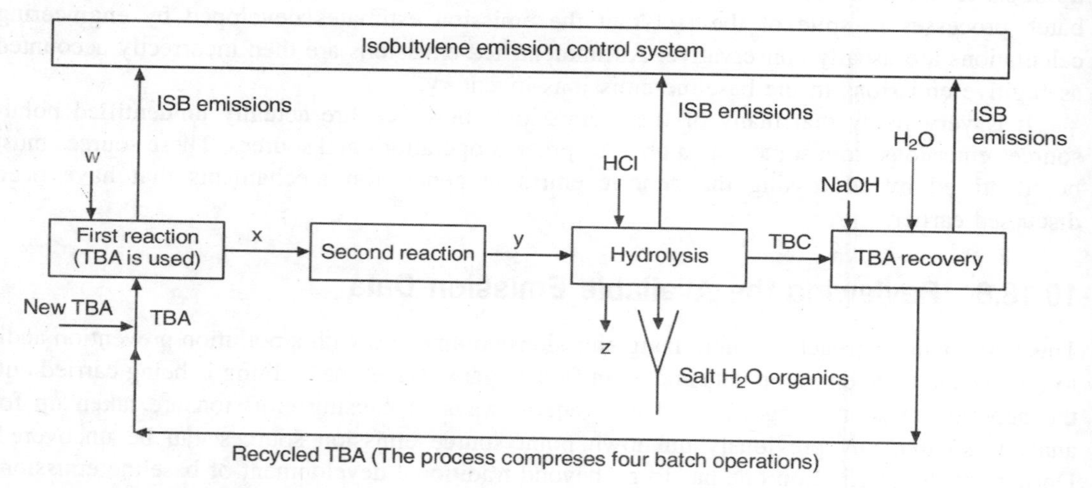

Figure 10.17 Reduction of emissions through process chemistry changes.
Adapted from Nick Chadha, et al., *Chemical Engineering Progress*, January 1993.

In the first reaction Tertiary Butyl Alcohol (TBA) was used to temporarily block a reactive site on the primary molecule. After the second reaction gets to completion TBA is removed as Tertiary Butyl Chloride (TBC) by hydrolysis with hydrochloric acid. To improve process economics the final step involved recovery of TBA by reacting TBC with sodium hydroxide. TBA recovery could be incomplete because isobutylene (a chemical related to both TBA and TBC) may be unintentionally formed during the TBA recovery step.

Investigation of the actual or true source of these emissions indicated that isobutylene formation was not inevitable. The addition of excess NaOH gave rise to alkaline conditions in the reactor, accidently resulting in conditions that favoured the formation of isobutylene over TBA. When the order of addition of the sodium hydroxide and TBC reactants was reversed and the NaOH addition rate was controlled so as to maintain the pH low say between 1 and 2, isobutylene formation was almost completely eliminated. As a result of these process changes, the installation of add-on emission controls turned out to be unnecessary and the only capital expense incurred was for the installation of a pH control loop.

10.18.9 Changing the Process Chemistry

In one chemical plant, odorous emissions had been observed for many years near a drum dryer line used for volatilizing an organic solvent from a reaction mixture. Although there were two dryers/product lines, the odours were noticed in the vicinity of only one line.

Analysis and field-testing showed that the chemical compounds causing the odours were actually generated in upstream unit operations due to the hydrolysis. The hydrolysis products were stripped out of solution by the process solvent and appeared in the form of odours fumes at the dryer. Conditions for hydrolysis were favourable at upstream locations because of temperature and acidity conditions and the residence time available in the process and the water needed for hydrolysis was provided by another water-based chemical additive employed in the dryer line that had the odour problem.

Because the actual cause of the odourous emissions was the process chemistry, the plant needed to evaluate ways and means to minimize hydrolysis and the consequent formation of undesirable odourous by-products. Ventilation modifications to lessen the odours level would have served only to transfer fumes from the drum dryers to locations outside the process building and definitely not provided a long-term solution to the odour problem.

10.18.10 Modifications in Engineering Design

Emissions may be generated due to such factors as equipment operating above its rated capacity, pressure and temperature conditions, improper process controls or faulty instrumentation. Strategies to minimize emission, can vary from troubleshooting and debottlenecking existing equipment to designing and installing hardware.

10.18.11 Vent Condensers

Vent condensers are commonly used in synthetic organic chemical manufacturing processes and pharamaceutical process for the recovery and/or control of organic contaminants. In a number of plants vent condensers were significant emission sources owing to one or more of the conditions given here.

1. Field modifications bypassed the vent condenser, but the associated piping changes were not documented in the engineering drawings.
2. The vent stream was too dilute to condense owing to changes in process conditions.

3. The condenser was overloaded (that is the heat-transfer area was not adequate) owing to gradual increase in prodution capacity over the years.
4. The overall heat transfer efficient was much lower than the design value due either to gradual fouling by dirty components or to condenser flooding with large quantities of non-condensible nitrogen gas.
5. The condenser cooling capacity was limited by improper control schemes. In one case, the coolant supply was unnecessarily throttled because the coolant supply value was adjusted by a flow controller that was equipped only with a proportional band control (that is no reset control). In another situation only the coolant return temperatures were controlled (that is, process stream outlet temperatures were not monitored).
6. In each of the aforementioned situations, design modifications were needed to reduce emissions.

10.18.12 Nitrogen Usage

In some large pharmaceutical plants nitrogen consumption can be several lakh cubic feet per year. In many plants, nitrogen flows for purging of reactors and other process equipment are not monitored or controlled. Pressure transfer and line blowing operations are conducted manually and the duration of the transfer operation depends very much on the operator. It is likely that the operator may not be present to shut-off the nitrogen supply value when the operation is complete.

Identifying ways to reduce nitrogen usage will help in the minimization of solvent emissions from a process. For instance, every 1000 cubic feet of nitrogen (if it is assumed to be completely saturated) will vent approximately 970 lbs of methylene chloride with it at 20°C and 130 lbs of methylene chloride with it at 10°C. The problem gets aggravated if fine mists or aerosols are generated due to pressure transfer or entrainment and the nitrogen becomes supersaturated with the solvent.

Some operating plants for instance, could monitor and reduce nitrogen consumption by installing flow meters in the nitrogen supply lines to each process building. Within each building simple engineering modifications such as installation of flow rotameters, programmable timers and automatic shut-off valves could help in minimizing solvent emissions.

10.18.13 Modifications in Operational Methods

Operational factors that can influence or impact emissions include the operating rate, scheduling of product campaigns and the plant's standard operating procedures. Implementation of operational modifications often requires the least capital compared to the other strategies presented in this section.

One synthetic organic chemical manufacturing plant wanted to bring down emissions of cyclohexane solvent from storage and loading/unloading operations. The tank arrangement had organic liquid storage tanks with both fixed-roof and floating roof construction. The major source of cyclohexane emissions was observed to be liquid displacement due to periodic filling of fixed roof storage tanks. Standard operating procedures for cyclohexane loading were modified so that the fixed-roof tanks were always kept full and the cyclohexane liquid volume varied only in floating-roof tanks. This relatively simple operational modification effected reduction of cyclohexane emissions from the tank by more than 20 tons/year.

A pharmaceutial manufacturing company wanted to reduce emissions of methylene chloride solvent from a process comprising a batch reaction step followed by vacuum distillation to strip-off the solvent. The batch distillation was conducted by piping the reactor to a receiver vessel evacuated through a vacuum pump. The changes listed out here were incorporated in existing operating procedures to minimize emissions.

1. The initial methylene chloride charge was added at a reactor temperature of 10°C rather than at room temperature. Provision of cooling on the reactor jacket lowered the methylene chloride vapour pressure and minimized its losses when the reactor hatch was opened for charging solid reactants later in the batch cycle.
2. The nitrogen purge to the reactor vessel was shut-off during the vacuum distillation step. The continous purge had been overloading the downstream vacuum pump system and it was found to be unnecessary because methylene chloride is not flammable. This modification brought about a reduction in losses due to stripping of methylene chloride from the reactor mix.
3. The temperature of the evacuated receiving vessel was lowered during the vacuum distillation step. Provision of maximum cooling on the receiving vessel helped minimization of methylene chloride losses owing to revaporization at the lower pressure of the receiving vessel.

10.18.14 Modifications in Maintenance Pattern

Strategies to minimize emissions in any manufacturing facility should also be linked to prevented maintenance practices. Emissions are generated by leaking equipment (inadequate sealing or corrosion), infrequent and irregular preventive maintenance, lack of instrument calibration and the maintenance of chemicals, and procedures employed.

Maintenance modification to minimize emissions are often obvious once the specific sources and causes of the emissions are identified.

For instance, if there is a leak in a pump seal, fixing the packing materials or mechanical seal is the first strategy that is usually taken up for consideration. For many synthetic chemical manufacturing plants and pharamaceutical plants, simply expanding the existing preventive maintenance programme to include centrifuge dryers and other process equipment that employ volatile organics has resulted in substantial reductions in emissions.

The experience gained in the successful minimization of emissions of the toxics through process changes indicated the following:

1. Generation of air toxics emissions can be minimized even for mature manufacturing processes.
2. Generation of emissions follow repeating patterns that are quite independent of the industry.
3. Defining the true source of emissions, is crucial for quickly focusing on long-term solutions for minimizing emissions.

4. An accurate emissions inventory is quite helpful in uncovering potential opportunities for minimizing emissions.
5. Unidentified emissions in a plant's emissions inventory may not always be fugitive emissions.
6. Evaluating generic ways of generating emissions from a manufacturing process can help identify unexpected opportunities for minimizing emissions.
7. Not all pollution prevention strategies require process development and/or significant capital investments.

EXERCISES

10.1 List out the steps involved in the waste stream analysis and process analysis which will help in the implementation of waste minimization.

10.2 Write an explanatory note on the waste management priority hierarchy.

10.3 How are the pollution prevention options identified in waste minimization programme?

10.4 Describe the process of demulsification in the treatment of waste streams.

10.5 What is the methodology followed in introducing environmental, health and safety considerations into a new product or a new manufacturing process?

10.6 Describe the methods of minimization of wastes at operating plants of a manufacturing facility.

10.7 What are the techniques available for estimating the emission of air pollutants from a manufacturing process?

10.8 Explain with examples how modification in the product itself would help to reduce pollution.

10.9 Describe as to how incorporating waste elimination considerations into the design of a process can greatly reduce waste generation during plant operations.

10.10 Describe the techniques that can be used to estimate air emissions from the manufacturing processes.

10.11 Explain as to how a systematic auditing procedure helps in minimizing wastes.

10.12 Describe how the identification of process improvement options and application of process integration bring about the minimization of emissions and waste generation.

11

Planning Process for Prevention of Pollution

It is needless to mention that successful minimization of waste generation by industry calls for careful planning to make sure that the best pollution prevention activities are carried out. In the following section, ways to set up and implement effective pollution prevention planning programmes, are briefly presented.

It is well known that pollution prevention planning requires a detailed understanding of how an industrial organization does business and how it manufactures its products. The resulting plan should provide a mechanism for a comprehensive and continuous review of the company's activities as they are very relevant to environmental issues. This task can appear discouraging at the beginning, especially to management personnel, who are likely to be unfamiliar with many of the concepts involved in the planning process. The extent of information and data needed for an effective pollution prevention plan may also give many pauses. However, the task is not likely to be as burdensome as it may appear at first sight. Many of the procedures employed are actually similar to those commonly used to conduct a business activity; the nomenclature may be different. In view of the fact that pollution prevention activities can have a major impact on the day-to-day operation of a business, pollution prevention planning should be carried out in parallel manner with other business planning.

11.1 STRUCTURE OF THE POLLUTION PREVENTION PLANNING PROCESS

The major steps involved in developing a pollution prevention programme are outlined in Figure 11.1.

The major elements of this programme include building support for pollution prevention throughout the organization, organizing the programme, setting goals and objectives, performing a preliminary evaluation of pollution opportunities and identifying problems and solutions.

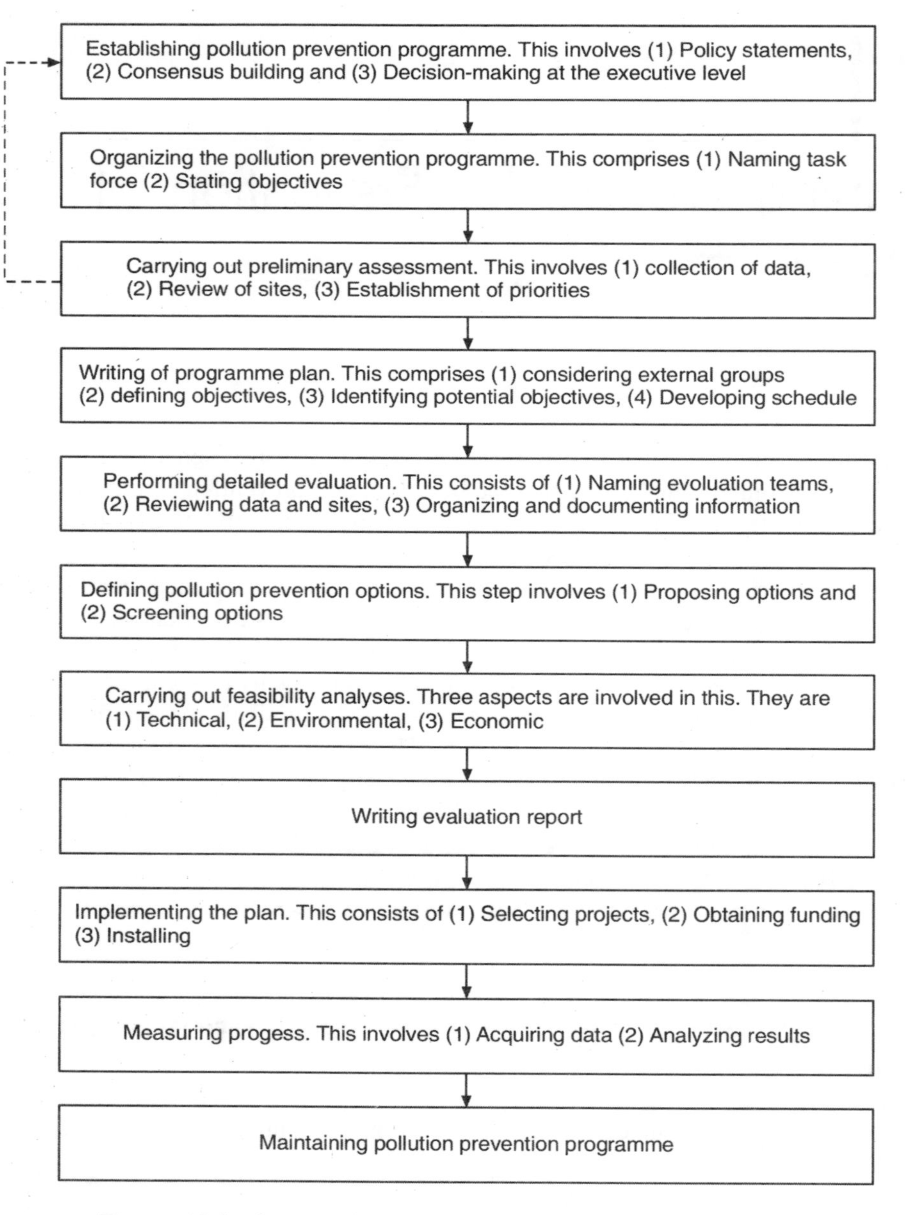

Figure 11.1 An overview of pollution prevention programme.
(Adapted from US, EPA,1992).

11.1.1 Organizing the Pollution Prevention Programme

The initiative to develop a company pollution prevention programme may be forthcoming from the upper management level, but more typically it begins in middle management at which cadre the employees are closer to the production line and can more easily see the potential benefits of pollution prevention or with the environmental control executives in the organization. Wherever it commences, it is imperative to impress upon the company executives about its merits at an early stage. This may imply having to perform an initial sketchy evaluation to indicate where potential cost savings could occur. Invariably it is the executives who make the final decisions on where to invest capital.

Once the senior management accepts the concept of pollution prevention and decides to set up a pollution prevention programme, they should transmit this message to all employees in the organization to make sure of its full compliance. This step may take the form of a formal policy statement. Why the organization is keen on implementing the new programme, what all will be done to implement it, and which staff will be responsible, should all be specified. Whatever method is employed, it is important that all employees recognize that the pollution prevention programme will not just remain on paper, but a real programme that will be enforced. The active cooperation of all employees will be absolutely necessary for the programme to prove to be a success. This is particularly true for the production-level employees, who may have to modify their transitional routines. Upper management must be vigilant to extend support for the programme on a continuous basis.

Care should be exercised while constituting the team that will develop and direct the implementation of the pollution prevention programme. These representatives should collectively be knowledgeable in all aspects of the business, both technical and business view points, and represent all levels of employment. This team may include executives, environmental engineers and plant process engineers and supervisors, experienced production line workers, purchasing agents and personnel in the quality control division. The programme leader should be someone with authority in the organization who enjoys broad-based support and the influence to make sure that the programme development and implementation stay on right track.

The objectives of the programme should be established at an early stage in order to give direction to the programme. The goals should reflect the stated policies of the organization, be clearly spelt out so that everyone understands their meaning and be challenging enough to motivate the employees without being unreasonable or impractical. Some organizations begin with a goal of zero discharge, which is probably impractical and then modify this goal as more information becomes available. The goals should be quite flexible and easily adaptable (Adapted from US, EPA, 1992).

11.1.2 Preliminary Evaluation of the Pollution Prevention Programme

The main objective of this phase of the pollution prevention plan development is to review and evaluate existing data and establish priorities and procedures for detailed assessments. Much of the data required for this phase can usually be got as a part of normal plant operations or in response to existing regulatory reporting requirements. The things needed include such documents as plant purchasing, accounting and inventory records; shipping manifests for hazardous wastes shipped from the plant; plant design documents and equipment operating manuals. There are sources available from which one can get information on the volume, concentration and degree of toxicity of effluent discharged. Guides providing a number of generic and industry-specific worksheets that can be used for gathering these data, are also available.

Two primary factors need to be taken into account while conducting this preliminary evaluation. The first factor is that a multimedia approach (air, water and solid waste) to pollution prevention should be the objective. This would include consideration of all energy and effluent streams involved. The second is that the objective of the analysis is pollution prevention and not extensive and expensive data acquisition. The amount of data required should be governed by the needs of the study. Detailed data beyond what is absolutely necessary will only add complexity and cost to the exercise.

The evaluation team should first get familiarized with the targeted processes. This can be accomplished by going into the facility and conducting a detailed study of the process. This study should be carried out while the process is in operation and if possible during a shutdown/clean-out/start-up period in order to identify the materials used and effluents generated by this procedure. The evaluation team should conduct a detailed discussion with the line personnel comprising equipment operators, maintenance and housekeeping staff and foremen, who by virtue of their close association, possess the best working knowledge of the processes being assessed. While making a study of the process, the team should note all pollution prevention opportunities and should pay particular attention to the aspects listed hereinafter.

- Procedure of operation adopted by line workers
- Quantities and concentrations of materials (especially effluents)
- Collection and handling of effluents (if the effluents are mixed, it should be noted)
- Record keeping
- Flow diagram (follow through the actual production or any other process)
- Leaking lines or poorly operated equipment
- Any spill residue
- Damaged containers and vessels
- Physical and chemical characteristics of the effluent or release.

The assessment team should also go into details of supplemental operations such as shipping/receiving, purchasing, inventory, vehicle maintenance, effluent handling/storage, laboratories, power houses/boilers, cooling towers and maintenance.

Most pollution prevention plans will provide a mechanism for investigating potential pollution prevention initiatives. The studies that are carried out must be very detailed and highly focussed. The level of detail required at the preliminary evaluation stage for the programme development is much less. The objective here is to assign priorities to processes, operations and materials that need to be addressed later in detail when the plan is being actually executed. Table 11.1 lists typical considerations employed when prioritizing effluent streams for further investigation.

Table 11.1 Typical Considerations for Prioritizing Effluent Streams for Further Investigation

- Compliance with current and anticipated regulations
- Costs of effluent stream management (pollution control treatment and disposal)
- Potential environmental and safety liability;
- Quantity of waste generated
- Hazardous properties of the effluent (including factors such as toxicity, flammability, corrosivity and reactivity)
- Other safety hazards to employees
- Possibilities for pollution prevention
- Possibilities for removing bottlenecks in production or effluent treatment.
- Possibilities of recovery of valuable by-products
- Available budgetary provision for the pollution prevention assessment programme and projects.
- Minimizing effluent discharges.
- Reduction in energy use

Adapted from US, EPA, 1992.

Table 11.2 given here presents the components of a pollution prevention plan.

Table 11.2 Components of a Pollution Prevention Plan

- Corporate policy statement of support for the pollution prevention programme
- Description of the pollution prevention team composition, authority and responsibility.
- A detailed account of how all the groups (production, laboratory, maintenance, shipping, marketing, engineering and others) will work together to effect reduction in waste production and energy consumption)
- Plan for publicizing and gaining organizationwide support for the pollution prevention programme.
- Plan for effectively communicating the success and failures of pollution prevention programmes within the organization.
- Description of processes that generate, use or release hazardous or toxic materials, including a clear definition of the quantity and types of substances, materials and products that are being considered.
- List of treatment, disposal, and recycling facilities and transporters presently used. Preliminary review of the expenses to be incurred in pollution control and effluent disposal.
- A detailed account of current and past pollution prevention activities at the manufacturing facility.
- Evaluation of the effectiveness of past and on going pollution prevention measures.
- Criteria for prioritizing the facilities, processes and streams for pollution prevention projects.

Adapted from US, EPA, 1992.

11.1.3 Development of Pollution Prevention Programme Plan

The pollution prevention plan task force employs the information gathered during the preliminary evaluation to prepare the detailed pollution prevention plan. The plan should define the pollution prevention programme objectives, identify possible obstacles and solutions and define the data collection and analysis procedures that will be employed in detailed pollution prevention investigations that would be conducted later. Table 11.2 is a detailed list of items that should be included in the plan. It may be noted that this list does not include recommendations for pollution prevention projects. Rather, it gives a detailed account of current practices and a methodology for evaluating proposed projects.

Many a time it is quite useful to consult people outside the organization concerned; government officials, local community representatives or people from similar commercial ventures, would belong to that category of people. They can provide a broader perspective to the pollution prevention planning process and can help to provide credibility and a sense of community involvement to this endeavour.

A final, crucial part of a successful pollution prevention plan is the development of a schedule for implementation. Significant developments within each phase of the plan should be detailed and realistic; target dates should be assigned. A mechanism for monitoring and ensuring compliance with the implementation schedule should also be included. On some occasions many well-conceived plans turn out to be failure owing to lack of proper follow-up, to confirm that the plan is carried out as designed.

11.1.4 Developing and Implementing Pollution Prevention Projects

Preparing the pollution prevention plan is only the first step towards accomplishing a successful pollution prevention programme. If continued, further action to carry out the provisions of the plan is not performed, the plan will turn out be to useless.

The first step towards the implementation of the plan is to conduct detailed evaluations of the potential areas of opportunity identified when the preliminary assessment is made. Assessment teams assigned to each operational area of any processing or manufacturing facility should review all existing information related to that area and develop more detailed lists of potential waste reduction or energy-saving projects. These may require acquisition of additional data. A lot of useful information can often come from the interviews conducted with the production line workers. Analysis should also include preparation of process flow diagrams and materials, and energy balances in order to determine pollution sources and opportunities for getting rid of them.

Once the sources and nature of effluent streams generated are determined, the evaluation team should propose and then do the screening of the pollution prevention options. Their goal should be to generate a comprehensive, prioritized set of options for doing a detailed feasibility evaluation. The general categories of pollution prevention actions that should be taken into account by each group, in approximate order of increasing complexity, include

(1) Housekeeping improvements
(2) Effluent segregation
(3) Material substitution
(4) Process modification
(5) Recycling and reuse of material
(6) Additional effluent treatment

Some options will be found to have no cost or risk attached; these can be implemented immediately. A few others which are likely to be found to have marginal value or to be impractical will be dropped from further consideration. The options that are remaining will generally be found to require feasibility evaluation.

A number of methods are available for prioritizing pollution prevention projects, The priority for a project will have a bearing on a number of factors and are likely to vary from one facility to another, depending upon the pollution prevention objectives established during the planning process. One prioritizing procedure that is employed to a large extent is the option rating weighted-sum method developed by the Environmental Protection Agency (EPA, 1992) of the US. This method provides a means of quantifying the important criteria that affect effluent management in a particular facility. The first step in this procedure is to determine the important criteria in terms of the objectives of the pollution prevention programme.

Some of the criteria are:

1. Reduction in effluent quantity
2. Reduction in effluent hazard
3. Reduction in effluent treatment/disposal costs
4. Reduction in raw materials costs
5. Previous successful use within the organization
6. Reduction in insurance and liability costs
7. Previous successful use within the industry
8. Not harmful to product quality
9. Low capital cost
10. Low operating and maintenance costs
11. Short implementation period involving minimal disruption of plant operations

The weights (on a scale of 0 to 10, with 0 being of no importance and 10 being highly important) are determined for each criterion (in pollution prevention) in relation to its importance. Thus, if reduction in effluent treatment and disposal costs is very important, it may be given a weight of 10, and in case previous successful use within the organization is of only minor importance, it may be given a lower weighting of only 1 to 2. Criteria that do not have any significance are dropped from the analysis. Each option is then rated on each criterion, again employing a scale of 1 to 10. Finally, the rating of each option for a particular criterion is multiplied by the weight of the criterion. An option's overall rating is the sum of these products.

The options chosen for further analysis should now be examined for technical, environmental and economic feasibility. The option must first be evaluated from a technical standpoint to make sure that it will actually work for a specific application. This may call for extensive laboratory or pilot-scale testing and evaluation by a variety of people—possibly including representatives from many departments such as production, maintenance, quality assurance (or control) and purchasing. It may also need input from the purchasers of the product (consumers) to verify that any specification variations are acceptable to them. The option must also be evaluated from the environmental viewpoint to make sure that waste is reduced in an eco-friendly condition. This should include a total life cycle evaluation of the product and materials and processes employed. In this analysis energy consumption must also be included. Finally, an economic evaluation of the option must be carried out. No pollution prevention option will be implemented if it is not economically justifiable. Projects that require significant capital costs will need a detailed cost analysis, to enable one to decide whether to go ahead with them or not.

11.1.5 Implementation of the Pollution Prevention Plan

Once the viable pollution prevention options have been assessed, prioritized and tested for worthiness, it is time to sell the best ones to the management of the company. Often this turns out to be the most difficult part of the process. Much of the work to this point has dealt with technical issues and most of it has probably been carried out by technical staff. Now these options must be sold to the management side of the business, which is likely to have limited understanding and comprehension of the environmental issues. It is likely that these proposed projects may well be competing with other revenue-generating projects for limited resources. Accurate life cycle costing is necessary for this reason. Showing that in the long-term the pollution prevention project will save the organization's a lot of money, even in case it is through reduced liability or reduced treatment costs rather than through increased sales, should convince management about the merits of the project. A properly structured pollution prevention plan, in which the budgeting decision-making process for pollution prevention is described in detail, will simplify this exercise.

Many pollution prevention projects may need modifications in operating procedures, purchasing methods or materials inventory control. This may also influence employee training procedures. These modifications need to be carried out carefully in order to minimize disruption that is likely to occur in normal business activities. A seamless transfer to pollution prevention should go a long way in allaying the common fear of change.

A pollution prevention plan needs to be a *living document*. It should be continually updated as and when new alternatives are developed. A programme that includes rewards and recognition to employees who come out with pollution prevention ideas or who successfully put those ideas to practice is often helpful in earning the cooperation of the employees, and in making them feel that they are in reality a part of the programme. In other words, this will give them a sense of belonging. The suggestions must be properly evaluated periodically and good ones must be implemented.

EXERCISES

11.1 List out the major steps to be followed in the development of a pollution prevention programme.

11.2 What are the components of a pollution prevention plan?

11.3 Describe as to how one should go about implementing the pollution prevention programme.

12
Strategies for Pollution Prevention

It is well known that pollution prevention encompasses a wide range of activities. Some of these activities require a long-lead time and major investment while others involve procedural and operational variations requiring little investment and giving rise to an immediate financial benefit.

In the following section, some important aspects of the pollution prevention ideas offered by Kenneth Mulholland, et al. that require minimal investment and can bring about a substantial reduction in the cost of manufacture are given. A good number of pollution prevention ideas which may be applied during process or product conceptualization and development, plant design or plant operation, are presented here.

12.1 CONCEPTUALIZATION AND DEVELOPMENT

1. It is advantageous to include a zero waste discharge option in all alternative process evaluations. This enhances waste awareness, which can in turn lead to valuable cost saving ideas.
2. It is worthwhile to consider buying purer raw materials or removing the impurities before they enter the process. Many a time this is best accomplished by the suppliers, because it is likely that they may already have the required infrastructure in place.
3. The reaction kinetics can be improved. In order to reduce by-product and co-product generation, it may be desirable to consider reactions with higher selectivity and lower conversion over reactions with higher conversion and lower selectivity. In the case of reversible reactions, it may be advantageous to consider recycling by-products back to the reactor. Reaction sequences may be modified to bring about a reduction in the quantity of or change the composition of intermediates and by-products.
4. The molar excess for minor ingredients in batch reactions should be carefully perused. The relationship between yield and cost has to be understood. The excess use of inorganic compounds such as mineral acids and bases has to be avoided.
5. The solvent selection has to be considered carefully. In the past it had been the practice to select solvent on the basis of only functionality, availability and purchase cost. Now, solvent selection must take into account other important criteria or factors like hazard

and toxicity, regulatory impact and cost of recycle or destruction. Table 12.1 presents a solvent selection hierarchy (in order of decreasing desirability) based on environmental, toxicity and process hazard concerns.

Table 12.1 The Hierarchy of Solvent Selection (in Order of Decreasing Desirability)

Water	
Non-polar organics	*Polar organics*
Aliphatics (with flash point above 60°C)	Alcohols
Aliphatics (with flash point below 60°C)	Organic acids
Aromatics	Other oxygenates
Halohydrocarbons	Nitrogen-bearing compounds
1. Hydrochloroflurocarbons (HCFCs)	Halogenated organics such as Trichloroethylene and Tetrachloroethylene
2. Chlorocarbons	Chlorinated organics such as chlorophenols hexachlorobenzenes
3. Chloroflurocarbons (CFCs)	

6. Oxygen can be used instead of air for oxidation reactions. This helps in minimizing the introduction of non-condensible, which must be purged from the process and will require treatment.
7. Catalyst selection must be reconsidered. Frequently heterogeneous heavy metal catalysts are more easily retained within the process than homogeneous (soluble) ones. Noble metal catalysts can often be recycled either on-site or by off-site reclaimers.
8. The residual solvent in polymer pellets and films shipped to customers must be minimized, this may solve a customer's waste problem and will result in a fresh product.
9. Solvent emissions from paints have to be lowered. Emissions from solvent-based coatings can be brought down by using extrusion, powder, CO_2 or water based coatings.
10. Solvent less separations processes may be employed. Unit operations or separation technologies, such as membranes, melt crystallizers, and so on that do not require the addition of solvents or other non-reactant chemicals may be assessed.
11. New packaging materials may be developed. The packaging materials may be made out of the end product. For instance, polymer resin can be sold in bags made of that resin so that the customer can grind the bags and use them as a feedstock.
12. The siting of the plant for producing a chemical may be optimized by locating it next to a consumer. This can entail a reduction in the raw material costs, eliminate packaging and maximize product recycle.

12.2 POLLUTION PREVENTION STRATEGIES IN PLANT DESIGN

In this area, the following strategies may be adopted.

1. The streams leaving the flow sheet may be evaluated. It can be ascertained as to what cost these streams represent in terms of replacement solvent or lost product. Also the costs they represent in waste treatment and utilities consumption may be ascertained. In the case of gas streams attempt can be made to find out what can be done to reduce the volume of air going to pollution control devices.

2. The solvents used in the process can be recovered and recycled to the extent possible. Staging the use of wash solvents, that is, mimicking counter flow by recycling used solvent going out of one stage as the input to the previous stage, can help in reducing makeup solvent and waste disposal costs.
3. High-quality water can be put to reuse. The possibility of the recovery of condensate as boiler feed water makeup, process water, or cooling tower makeup can be explored. Condensate often turns out to be economical in place of soft water for application such as diluting caustic soda.
4. The liquid effluent stream that contains Volatile Organic Compounds (VOCs) can be stripped and the organics recovered. This is frequently a more attractive proposition than stripping followed by abatement of the overhead stream for two important reasons—it offers more flexibility in designing the treatment system, and it recovers value from the recycled material.
5. Solvent displacement from solid products and intermediates can be improved. Solvents can be more completely removed by improved solid/liquid separators, such as pressure filters, before they are brought into contact with water. In addition to this, washing can be carried out as a multi-stage operation to reduce dilution and make recovery more economical.
6. Liquid/liquid separation processes can be improved. As a matter of fact, factors like more number of stages, colder operating temperatures and improved interface control can improve decanter performance. Also the possibility of having more stages in distillation and extraction operations can be explored.
7. Hazardous air pollutants can be eliminated to the extent possible. No single approach can be identified to accomplish this. However, constructive and positive thinking can help identify opportunities (for example, in place of regulatory listed hexane, unlisted heptanes can be used).
8. Air pollution should not be allowed to become or turn into water pollution. Water scrubbing as a means to solve air pollution problems should be avoided, if this scrubbing process in turn creates a water pollution problem.
9. The addition of inert materials for blanketing, purging and pressure transfer has to be controlled, for instance by using flow meters or rotameters.
10. Water has to be considered as an expensive solvent and has to be treated as any other solvent. By this statement it is implied that its introduction into the process has to be minimized. It may be reused directly or subjected to treatment and then recycled.
11. Air streams need to be segregated. Process contaminated air has to be kept separate from heating and ventilation air.
12. Dryer operation needs to be optimized. The emissions from indirectly heated convection dryers that use once though air can be minimized by following steps:
    (a) Air can be recycled.
    (b) In addition to recycling air, the amount of leakage can be minimized. This can be achieved by improved baffling, seals, air locks, etc.
    (c) Organics from the purge streams can be recovered.
    (d) Energy value from the purge stream can be recovered.
    (e) Organics from the purge stream can be abated.

13. Use of inerted dryers operating “fuel-rich” with condensation for solvent recovery, can be considered.
14. Use of dense-phasing conveying in lieu of dilute phase conveying can be considered.
15. Contact condensers may be installed to make recovery more economical. The downstream processing and treatment consequences may be considered before installing direct-contact heat exchangers. Contact condensers can under certain circumstances be economically replaced with absorbers (replacing the contacting fluid with a higher boiling or immiscible fluid) or refrigerated condensers. The more concentrated volatile organic is typically cheaper to recover. In addition to this, use of reboilers can be considered in place of live steam injection.
16. Optimum process conditions for phase separation can be determined. For instance, colder decanter temperatures can be attempted.
17. Process waste water can be segregated from storm water run-off by covering areas and draining by a ditch. Roofs and curbs may be less expensive and more environmentally sound than constructing larger treatment facilities to handle the additional flow.
18. Optimization of plant layout to shorten piping runs and to minimize the number of emissions points, particularly fugitive (that is minimizing the number of valves, flanges and pumps) can be carried out. The number of outlets for waste streams (e.g. drains, hoods and tankage) may be minimized).
19. Dedicated shipping containers, bulk packaging and supply can be used. Also shipping by pipeline can be adopted.
20. Use of pressure vessels or floating roof tanks in lieu of traditional tanks with conservation vents can be considered.
21. Pressure or gravity feed can be considered because it is well known that with fewer pumps there are fewer seal leaks.
22. Product degradation can be minimized; staged heating (that is, multiple heat exchangers in series using progressively hotter heating media) can be employed in order to minimize product degradation and undesirable side reactions due to high temperatures at the wall of the tubes.
23. Replacing a solvent and using the actual product or a process intermediate as a carrier for stabilizers and inhibitors instead of introducing additional solvents to the process, can turn out to be beneficial from pollution prevention viewpoint.
24. Selection of mechanical seals over packing can be tried since well maintained mechanical seals have been found to generate less waste than packings with seal flushes.
25. Usage of vacuum pumps instead of steam jets can be considered.

12.3 POLLUTION PREVENTION IDEAS IN PLANT OPERATION

In respect of plant operation, the following steps may be quite beneficial:

1. The wastes need not be over treated. Significant energy can be saved by lowering the temperature of thermal oxidizers, by reducing the pressure drop across scrubbers and by turning off surplus aerators in effluent treatment.

2. In case aerators are run to meet an occasional peak demand, perhaps one or more can be shutdown except during turnarounds or other peak demand periods.
3. It has to be understood whether the waste treatment plant adds alkali or acid or to balance pH wherever possible, the use of pH control chemicals may be reduced. All acid additions into the process have to be monitored. Use of waste acid from another part of the plant or an acid that is generated as a by-product within the process may be considered. This strategy effects savings on money twice through reduced ingredients cost, and through reduced use of an alkali to neutralize the acid.
4. Excessive carbon dioxide scrubbing in caustic scrubbers can be avoided. Scrubbers can be operated at pH less than or equal to 8 and they still remove chlorine, bromine, sulphur dioxide and hydrochloric acid. Operating at higher pH will result in wastage of caustic by scrubbing carbon dioxide. For instance at pH 10 and typical carbon dioxide partial pressures observed in oxidizer off-gases, the increasing caustic consumption for scrubbing carbon dioxide is quite appreciable.
5. Equipment cleaning practices need to be watched closely. High waste loads resulting from equipment cleaning set the peak demand on many effluent treatment plants and represent product down the drain. If the wash fluid is compatible with the process, it may be desirable to recover the first wash and return it to the process during the start up. Attempts may be made to reduce peak loads on the effluent treatment plant. In case cleaning has to be performed often, the last stage of rinsed fluid can be saved and used as the initial rinse for the next shutdown. As an alternate measure the first rinsing can be made with mother liquor. Water washing can be made much more efficient by adjusting pH.
6. Samples can be recycled to the process. This is particularly cost-effective if excess samples are hazardous waste. The dead volume of samples lines should be eliminated and the volume of samples should be reduced to actual analytical requirements. The impact of discarded laboratory samples on waste water effluents should not be under estimated. There have been a number of cases where the dominant sources of heavy metal contamination in the waste water was from laboratory samples. In general, it is advisable to minimize the number and size of samples.
7. It may be worthwhile to consider operating with fewer full tanks in lieu of more partially filled tanks. The tanks can be operated at a constant level. The ingredient losses can be brought down by reducing the number of tanks in service and operating them at constant level.
8. Product purity needs to be controlled. Customers want consistency in quality. Variable purity, including intermittent over purification reduces capacity and increases waste without providing value to the customer. Examples include strict control of near-boilers in refining trains, strict control of filtrate conductivity, avoiding excess solvent wash and measuring the moisture level in front of a drying step. Effective control of product purity can increase capacity for sold-out process.
9. Production schedules can be modified. It can be ascertained whether waste generation costs are to be considered when scheduling multiple products on one production line.

10. The process can be sealed up, and reduction in air exchange between process equipment and the work place can be effected. This brings about savings in money by reducing heating ventilation and air-conditioning costs and also the operating cost of emissions control equipment.
11. The personnel may be advised to routinely monitor for fugitive emissions. Leaks of organics into utility water should be detected and corrective action taken early. Typical sources of leaks include heat exchangers and pump seals. Both the leaks can be economically corrected with proper maintenance.
12. Cleaning practices can be reviewed periodically. The personnel may be advised to use improved cleaning technology, such as pipe cleaning devices, rotating spray heads in tanks, high pressure water jets, anti-stick coatings on vessel walls, better draining equipment, mechanical cleaning and sweeping and multiple small rinses in lieu of filling and draining, to minimize solvent and wash water wastes. Water wash downs can be minimized or even eliminated, and a reduction in the number of water hose stations can be considered.
13. The storage of raw materials, intermediates and products can be brought to the minimum.
14. Start-ups and shutdowns can be minimized. When the optimization of production schedules is carried out, the waste generated during start-ups and shutdowns can be considered.
15. Air leakage into vacuum systems may be minimized. This leakage will eventually become so contaminated with organics or particulates, as to necessitate treatment.
16. Idle equipments such as steam jets, cooling tower water to heat exchangers, water flushes and so on can be turned off or shutdown.
17. Tank and reactor vents during filling, unloading and transfer operations may be equalized.
18. Packaging impurities and unused raw materials may be returned to the suppliers.
19. Efforts can be increased to find markets for by-products and co-products.

The ideal way to cut down waste is to avoid making it in the first place. When this is not practical, a second technique is frequently employed; recycling waste products back into the process. Both these approaches work. Specific improvements involve raw materials, reactors, heat exchangers, pumps, furnaces, distillation columns, piping and control schemes. In the following section, some practical ideas to reduce waste outlined by Nelson which worked for a leading chemical manufacturing company are presented.

12.4 IMPROVEMENTS IN RAW MATERIALS

It is a common practice to buy raw materials from an outside source or transferred from an on-site plant. Each raw material needs to be studied to ascertain how it affects the quantity of wastes generated. The specifications for each raw material entering the operating plant should be closely examined.

Improvement in the quality of feeds

Despite, the percentage of undesirable impurities in a feed stream being low, it can be a major contributor to the total wastes generated in a plant. Reducing the level of impurities may involve working with the supplier of a raw material that is purchased, working with on-site plants that supply feed streams or installing new purification equipment. In some situations the effects are indirect; for instance, water gradually kills the reactor catalyst causing formation of by-products; so a drying bed or column is added.

Use of off-spec materials

Once in a while, a process can use off-spec materials (that would otherwise be burned or land-filled) because the particular quality that makes the material off-spec is not important to the process.

Improvement of the quality of products manufactured

Impurities in the products of an organization may be generating wastes in the customers' plants. In addition to its being expensive, it may make some customers look elsewhere for higher quality raw materials.

Use of inhibitors

Inhibitors help in preventing unwanted side reactions or the formation of polymers. Various types of inhibitors are available in the market. In case inhibitors are already being employed it may be desirable to check with the suppliers for improved formulations and new products.

Changing shipping containers

In case raw materials are being received in containers that cannot be reused and need to be burned or landfilled, it will pay to switch to reusable containers or bulk shipments. Likewise, employing alternative containers for shipping plant products should be discussed with the customers.

Re-examination of the requirement for each raw material

Sometimes the requirements for a particular raw material (one which ultimately ends up as waste) can be brought down or even eliminated by modifying the process or improving the control. For instance, the need for algal inhibitors in a cross flow cooling tower could be reduced by shielding the water distribution decks from sunlight.

12.5 IMPROVEMENTS AND MODIFICATIONS IN REACTORS IN GENERAL

It is well understood that the reactor is the heart of the process and can very much be a primary source of waste products. The quality of mixing in a reactor is very important. Many a time, it has turned out that inadequate design time is spent on scaling up this key parameter and a new production facility has disappointing yields when compared to laboratory or pilot plant date. The first three ideas listed here are related to quality of mixing in reactors.

Improvement in physical mixing taking place in the reactor

It has been shown that modifications to the reactor such as adding or improving baffles, installing a higher rpm motor on the agitators or employing a different mixer blade design (or multiple impellers) can result in improvement in the mixing process. Pumped re-circulation can be added or increased. Two fluids going through a pump however, do not necessarily mix well and an in-line static mixer may be needed to establish good contacting.

Better-feed distribution

This area deserves more attention that it actually gets. The following sketch illustrates the problem. The reactants enter the top of a fixed catalyst bed. Part of the feed short-circuits, down through centre of the reactor so that there is inadequate time for conversion to the desired products. Conversely, the feed closer to the wall remains in the reactor for a too long time and *over reacts* to by-products that eventually turn out to be waste.

Despite the average residence time in the reactor being correct, insufficient feed distribution brings about both poor conversion and poor yield. One effective solution is to add some sort of distributor that makes the feed move uniformly through all parts of the reactor as illustrated in Figure 12.1. Some sort of special collector at the bottom of the reactor may also be necessary to prevent the flow from narrowing down to the outlet.

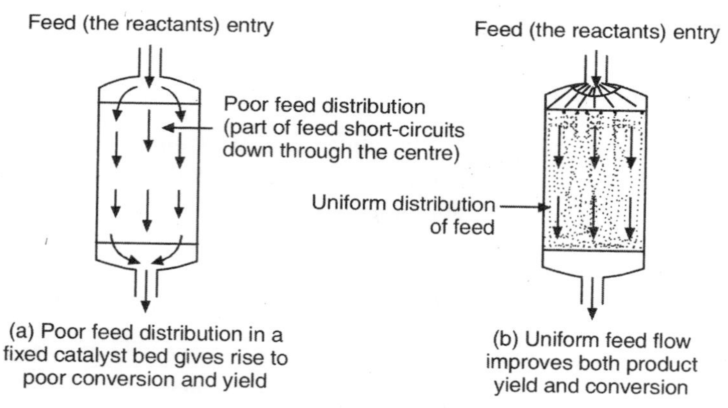

Figure 12.1 Distribution pattern of feed in the reactor.

Adapted from Nelson, K.E., *Hydrocarbon Processing*, March 1990.

Improvement in the method of addition of the reactants

The expectation here is to get closer to ideal reactant concentrations before the feed enters the reactor. This helps in avoiding secondary reactions which lead to the formation of unwanted by-products. The methods of addition of the reactants (both erroneous and right methods) are illustrated in Figures 12.2(a) and 12.2(b).

It is quite unlikely that the ideal concentration exists anywhere in the reactor. A consumable catalyst especially should be diluted in one of the feed streams, the one that does not react in the presence of the catalyst. Figure 12.2(b) illustrates one approach to improve the situation by employing three in-line static mixers.

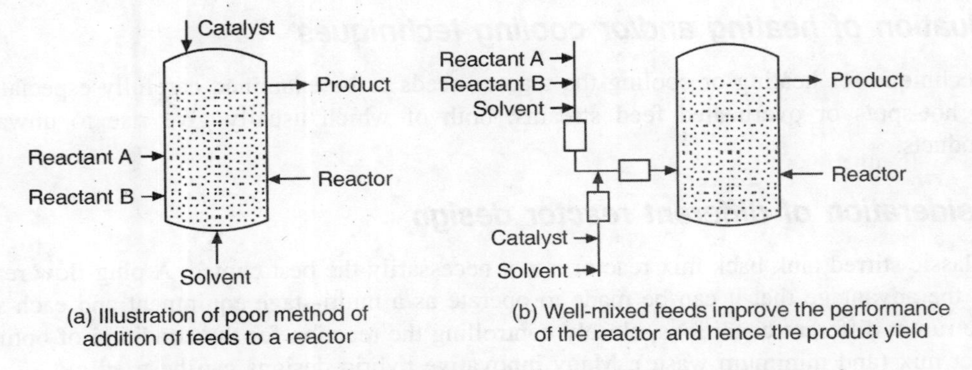

(a) Illustration of poor method of addition of feeds to a reactor

(b) Well-mixed feeds improve the performance of the reactor and hence the product yield

Figure 12.2 Method of addition of the feed to a reaction vessel.

Adapted from Nelson, K.E., *Hydrocarbon Processing*, March 1990.

Improvement in the quality of catalyst

Searching for better catalysts should be a continuous or an ongoing activity owing to the significant effects a catalyst has, on the extent of conversion and product mix. Variations in the chemical makeup of a catalyst, the method by which it is prepared or its physical characteristics (size, shape, porosity, etc.) can give rise to substation improvements in catalyst life and effectiveness.

Provision of separate reactor for recycle streams

In principle, recycling by-product and waste stream is an excellent technique for reducing waste, but frequently the ideal reactor conditions for converting recycle streams back to usable products are considerably different from conditions in the primary reactor. A possible solution, indicated in Figure 12.3, is to provide a separate smaller reactor for handling recycle and waste streams.

The parameters viz. temperatures, pressures and concentrations can be optimized in both reactors to take maximum advantage of reaction kinetics and equilibrium condition.

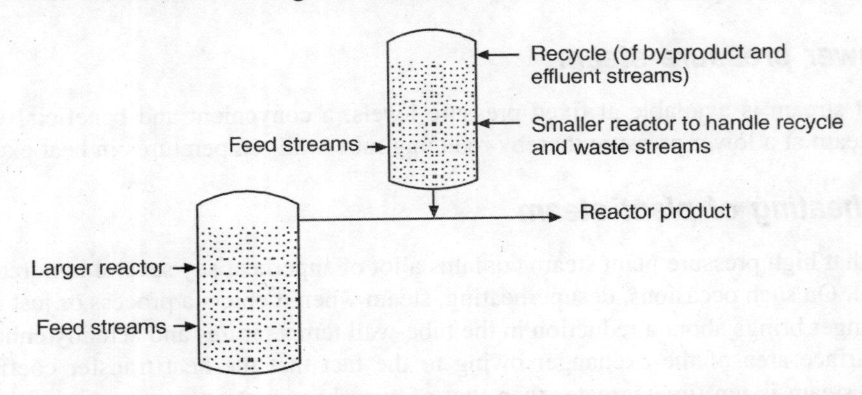

Figure 12.3 Provision of separator reactor to handle recycle waste streams.

Adapted from Nelson, K.E., *Hydrocarbon Processing*, March 1990.

Evaluation of heating and/or cooling techniques

The technique for heating or cooling the reactor needs to be examined carefully especially to avoid hot-spots or overheated feed streams, both of which usually give rise to unwanted by-products.

Consideration of different reactor design

The classic stirred tank back mix reactor is not necessarily the best choice. A plug flow reactor offers the advantage that it can be made to operate as a multi-stage equipment and each stage can be run at different conditions, closely controlling the reaction for the condition of optimum product mix (and minimum waste). Many innovative hybrid designs can be tried out.

Improvement of reactor control

For a given reactor configuration, there is one set of operating conditions that is optimum at any given time. The control system should be aware of those conditions and make them take place with little fluctuation. Such control may be somewhat complex, particularly in the case of batch reactors, but can yield major improvements in the reactor control. In the case of less sophisticated systems, by simply stabilizing reactor operation frequently a reduction in the formation of waste products can be effected.

Advanced computer controlled systems have the capability to respond to process upsets and product changes swiftly and smoothly, producing minimum of unwanted by-products.

12.6 IMPROVEMENT IN THE WORKING OF HEAT EXCHANGERS

It has been observed that heat exchangers can be a source of waste stream especially with the products that are temperature-sensitive. A wide variety of techniques for minimizing the formation of waste products in heat exchangers are available. Many of them are associated with reducing tube-wall temperature.

Using lower pressure steam

When plant stream is available at fixed pressure levels, a convenient and beneficial way is to switch to steam at a lower pressure thereby reducing tube-wall temperatures in heat exchangers.

Desuperheating of plant steam

It is likely that high pressure plant steam contains a lot of superheat say several hundred degrees of superheat. On such occasions, desuperheating, steam when it enters a process or just upstream of an exchanger brings about a reduction in the tube-wall temperatures and actually enhances the effective surface area of the exchanger owing to the fact that the heat transfer coefficient of condensing steam is ten times greater than that of superheated steam.

Installation of a thermo-compressor

Another good method of effecting a reduction in the tube-wall temperature is to install a thermo-compressor. The thermo-compressors are relatively inexpensive and these work on an ejector principle combining high and low pressure steams to produce an intermediate pressure steam. Variable throat models are available that operate like control valves, automatically mixing the correct quantities of high and low pressure steam. Figure 12.4 illustrates the principle of working of a thermo-compressor.

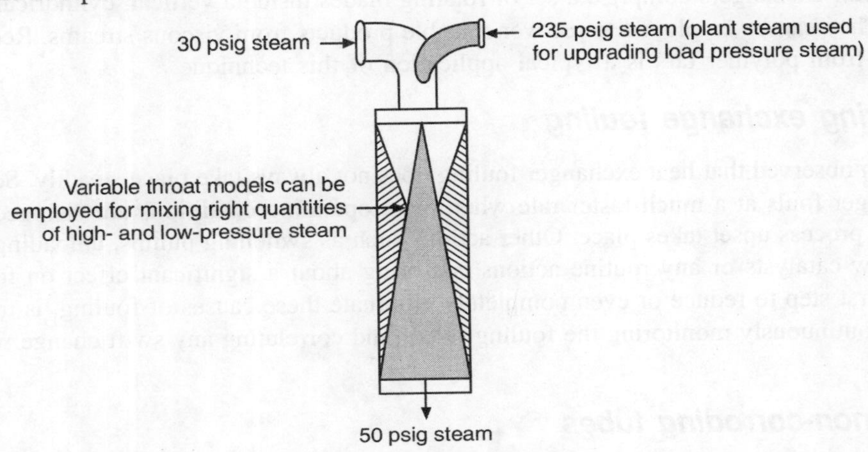

Figure 12.4 Thermo-compressor for upgrading low-pressure steam.

Adapted from Nelson, K.E., *Hydrocarbon Processing*, March 1990.

In the situation depicted by the above sketch plant steam at 235 psig is employed to upgrade 30 psig steam to 50 psig. Without the use of thermo-compressor, only 235 psig steam was used to supply the required heat.

Use of staged heating

If a heat-sensitive fluid has to be heated, staged heating can minimize degradation. One can start, for example, with waste heat and then move on to the use of low-pressure steam finally to desuperheated high pressure steam. Figure 12.5 indicates a typical arrangement.

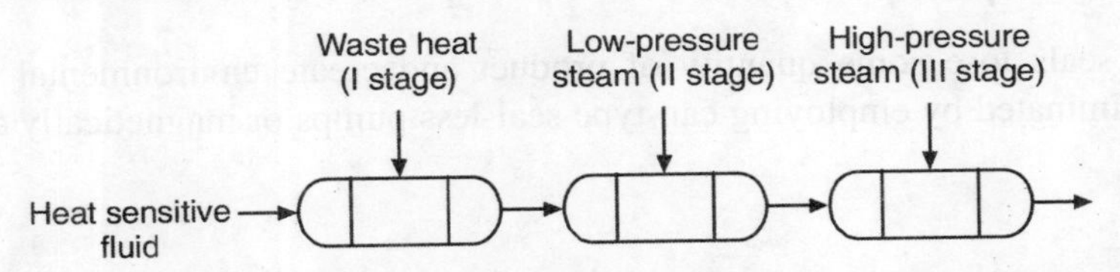

Figure 12.5 Minimization of product degradation by multi-stage heating.

Adapted from Nelson, K.E., *Hydrocarbon Processing*, March 1990.

Use of on-line cleaning techniques

On-line cleaning devices like recirculated sponge balls or reversing brushes have been available in the market for quite a few years. Besides reducing heat exchanger maintenance, they also keep tube surfaces clean making it possible to use lower temperature heat sources.

Use of scraped-wall exchangers

Scraped-wall exchangers comprise a set of rotating blades inside a vertical, cylindrical jacketed column. They can be employed to recover saleable products from viscous streams. Recovery of monomer from polymer tars is a typical application of this technique.

Monitoring exchange fouling

It has been observed that heat exchanger fouling does not always take place steadily. Sometimes an exchanger fouls at a much faster rate when plant operating conditions are changed too fast or when a process upset takes place. Other actions such as switching pumps, unloading tankers, adding new catalysts or any routine actions can bring about a significant effect on fouling.

The first step to reduce or even completely eliminate these causes of fouling, is to identify them by continuously monitoring the fouling factor and correlating any swift change with plant events.

Use of non-corroding tubes

Corroded tubes surfaces are very likely to foul more quickly than non-corroded tube surfaces. Switching to non-corroding tubes can bring about a significant reduction in fouling.

12.7 PUMPS

There is no contribution from pumps to wastes except in two ways.

Recovery of seal flushes and purges

Each pump seal flush and purge is to be examined as a possible source of waste. Most can be recycled to the process with little difficulty.

Use of seal-less pumps

Leaking pump seals lose some quantity of product and create environmental problems. The losses can be eliminated by employing can-type seal-less pumps or magnetically driven seal-less pumps.

12.8 FURNACES

Advances in furnaces technology have been taking place constantly. Furnace manufacturers should be contacted for the latest techniques in introducing optimization in furnaces operations and reducing the formation of tar.

Replacement of coil

In some applications, significant improvements can be made by replacing the existing furnace coil with one having improved design features (e.g. tubes with low residence time or designed for split flow). Even though it may not be practical to replace an undamaged coil, alternative designs should be explored whenever the necessity for replacement comes up.

Replacement of furnace with intermediate exchanger

A desirable option is to eliminate direct heating in a furnace (which necessarily exposes the heated fluid to high tube-wall temperatures) by employing an intermediate heat transfer medium. Figure 12.6 illustrates the principle.

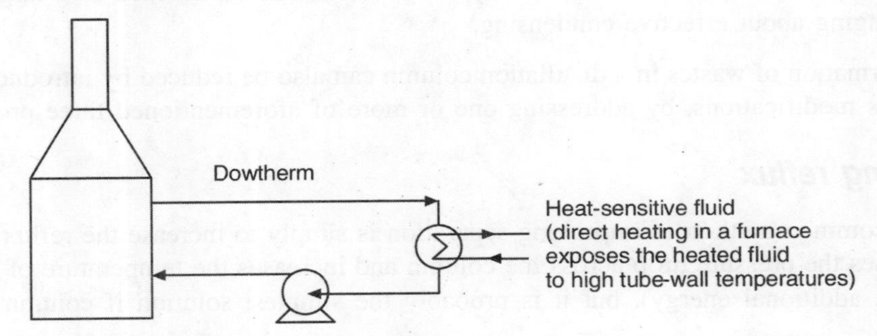

Figure 12.6 Use of an intermediate exchanger to avoid contacting process fluid with hot furnace tube-walls.

Adapted from Nelson, K.E., *Hydrocarbon Processing*, March 1990.

Making use of existing steam superheat

Even if the temperature required for certain processes is above the saturation temperature of the highest pressure of plant steam available, adequate superheat may be available to totally eliminate the need for a furnace.

In Figure 12.7, a process steam is heated to 550° F by employing the superheat available in 475 psig steam. No fired heater is employed so the fluid being heated is not subjected to the condition of hot tube-wall temperatures.

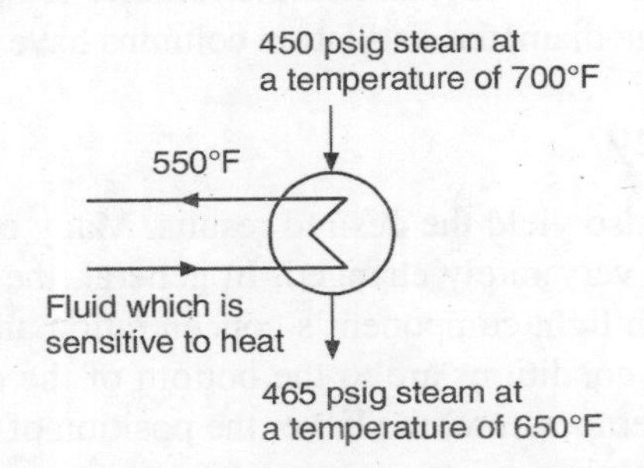

Figure 12.7 Use of steam superheat to meet high temperature needs.

12.9 WASTE REDUCTION IN DISTILLATION COLUMNS

There are three ways by which wastes will be generated by distillation columns.

1. By allowing impurities to remain in a product. The impurities ultimately turn out to be waste. The solution to this problem is to go in for better that is more effective separation. In certain situations, it is desirable to exceed normal product specifications.
2. By forming waste within the distillation column itself, usually because of high reboiler temperatures which brings about polymerization. The solution to this problem is to maintain lower column temperatures.
3. Wastes may also be produced in a distillation column by inadequate condensing which results in vented or flared product. This problem can be avoided to a large extent by bringing about effective condensing.

The formation of wastes in a distillation column can also be reduced by introducing column and process modifications, by addressing one or more of aforementioned three problems.

Increasing reflux

The most common method of improving separation is simply to increase the reflux ratio. This, is turn, raises the pressure drop across the column and increases the temperature of the reboiler (employing additional energy), but it is probably the simplest solution if column capacity is adequate.

Adding new section to column

In case a distillation column is operating close to flooding, a new section can usually be added to increase capacity of the distillation column and improve separation. The new section can have a different diameter and may employ trays, regulator packing or high efficiency packing. The additional new section need not be consistent with the original column.

Retraying or repacking column

Retraying or repacking part of the entire distillation column is another method of increasing separation. Both regular packing and high efficiency packing typically of lower pressure drop through a column, bringing about a decrease in the reboiler temperature. Packing is no longer limited to small columns; large diameter distillation columns have also been successfully packed.

Changing the feed tray

Changing the feed tray may also yield the desired results. Many columns are built with multiple feed trays, but the valving is very rarely changed. In general, the closer the feed conditions are to the top of the column (high light component's concentration and low temperature) the higher the feed tray; the closer feed conditions are to the bottom of the column (high concentration of heavy components and high temperature) the lower the position of the feed tray. Experimentation turns out to be easy if the valving exists.

Insulation

Good insulation is necessary to prevent heat losses; poor insulation needs higher reboiler temperatures and also allows the fluctuation of column conditions with weather conditions.

Improvements in feed distribution

The effectiveness of feed distributors (particularly in packed columns) needs to be analyzed to make sure that distribution irregularities are not lowering overall column efficiency.

Preheating column feed

Preheating the feed to the column should bring about an improvement in column efficiency. Supplying heat in the feed requires lower temperatures than supplying the same quantity of heat to the reboiler, and it reduces reboiler load. More often than not the feed is preheated by cross exchange with other process streams.

Removing overhead product from tray near top of distillation

In case the overhead product contains a light impurity, it may be possible to get a higher purity product from one of the trays closer to the top of the distillation column. A bleed stream from the overhead accumulator can be recycled back to the process to purge the column of light components. Another solution to the problem is to install a second column to remove small amounts of light components from the overheads.

Increasing the size of vapour line

In low pressure or vacuum distillation columns, pressure drop is especially critical. Installation of a larger vapour line brings about a reduction in pressure drop; it also effects a decrease in the reboiler temperature.

Modification in the reboiler design

A conventional thermo-siphon reboiler is not always the best choice, especially in the case of heat sensitive fluids. It may be preferable to employ a falling film reboiler, a pumped recirculation reboiler or high-flux tubes to minimize the degradation of distilled product.

Reduction of reboiler temperature

The same general temperature reduction techniques presented earlier (for heat exchangers) such as desuperheating steam, using lower pressure stream, installing a thermo-compressor, employing an intermediate heat fluid, etc. are applicable also to be reboiler of a distillation column.

Lowering of column pressure

A reduction in the column pressure will also decrease the reboiler temperature and may favourably load the trays or packing as long as the distillation column stays below the flooding conditions. The overhead temperature, however, will also be reduced in this process, which may in turn give rise to a condensing problem.

Improvement in the performance of overhead condenser

If overheads are lost because of an undersized condenser, retubing, replacement of the condenser or addition of a supplementary vent condenser to minimize losses, may be considered. It may also be possible to reroute the vent back to the process (if process pressure is quite stable). In case a refrigerated condenser is employed, it has to be ensured that the tubes are kept above 32°F, if there is any moisture present in the steam.

Improvement in column control

The suggestions given earlier in the section on reactors about closer process control and controlling at the right point apply to distillation columns and also to reactors.

Forwarding of vapour overheads to next column

If the overheads stream is sent to another column to undergo further separation, it may be possible to employ a partial condenser and introduce a vapour stream to the down streams column.

In general, before any equipment modifications are undertaken, it is recommended that a computer simulation is carried out and that a variety of operating conditions be perused. If the column temperature or pressure changes, equipment ratings should also be examined.

12.10 WASTE REDUCTION IN PIPING

Something as seemingly non-harmful as plant piping can sometimes give rise to waste and a simple piping change can result in substantial reduction in waste. Following piping changes to a process may be considered:

Recovery of individual waste streams

In a number of plants, various waste streams are combined and sent to an effluent treatment facility as shown in Figure 12.8.

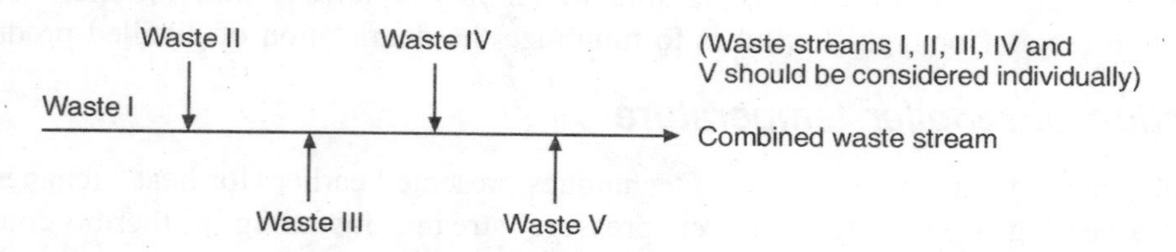

Figure 12.8 Examination of waste stream individually.

Each effluent stream should be considered individually. It is likely that the nature of the impurities makes it possible to recycle or otherwise reuse a particular stream before it is mixed with other effluent streams and becomes unrecoverable. Stripping, filtering, drying or some other type of treatment may turn out to be necessary, before reusing the stream.

Avoding overheating lines

In case a process stream, consists of temperature-sensitive materials both the quantity and temperature level of line and vessel tracing and jacketing needs to be reviewed. In case plant steam levels are too hot, a recircualted warm fluid can be used to prevent the process steam from freezing. It may be desirable to choose one that does not freeze if the system is shutdown in cold climate. Electric tracing is also a worthwhile option.

Avoidance of sending hot material to storage

If a temperature-sensitive material is to be sent to storage it should first be cooled. In case this turns to be uneconomical because the stream from storage needs to be heated when it is used. Simply piping the hot stream directly into the section of the storage tank pump as indicated in Figure 12.9 may solve the problem to a large extent. It has to be ensured that the storage tank pump can handle hot material without cavitating.

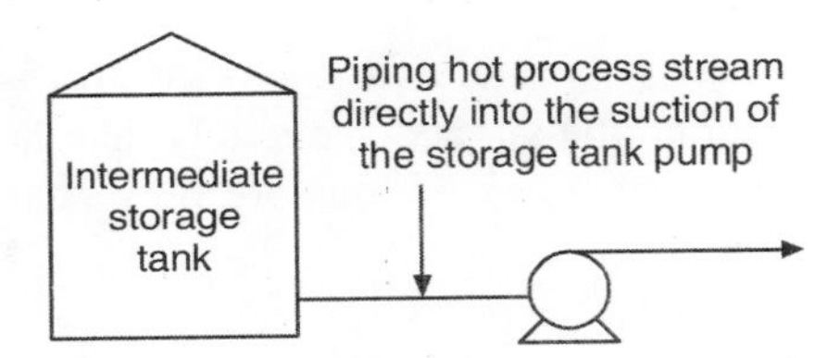

Figure 12.9 Conveying hot stream directly into suction line of storage-tank pump.

Eliminating leaks

It has been observed that leaks can turn out to be a major contributor to a plants overall waste, especially in situation in which the products cannot be seen or smelled. A good method to document leaks is to measure the amount of raw materials that must be bought to replace *lost* streams (e.g. the quantity of refrigerant purchased).

Changing metallurgy

The type of metal employed for vesels or piping may give rise to a colour problem or be acting as a catalyst for the formation of unwanted by-products. If this is the situation, it may be desirable to switch to more inert metals.

Using lined pipes or vessels

Using lined pipes or vessels is often a cheap alternative to using exotic metallurgy. A good number of coatings are available for different applications.

Monitoring major vents and flare system

Flow measurements need not be highly accurate, but should give a reasonable estimate of how much product is lost and when those losses take place. Intermittent losses such as equipment purges can particularly escape detection. Corrective action depends on the specific situation that is encountered; a good number of alternatives exist are available. Frequently, venting or flaring can be reduced or even eliminated.

Recovery of vented product

It turns out to be often worthwhile to install whatever piping is necessary to recover products that are vented or flared and reuse them in the process under consideration (or in another process). Storage tanks, tank cars and tank trucks are common sources of vented product. A condenser or small vent compressor may be all that is needed for recovery of the vented product. Additional purification step may be necessary before the recovered streams can be put to reuse.

12.11 REDUCING WASTE BY HARNESSING PROCESS CONTROL SYSTEMS CAPABILITY

Advances in technology have made it possible for us to install highly sophisticated computer control systems that respond far more quickly and accurately than human beings. This capability can be employed to reduce waste. Some suggestions are presented here.

Improvement in online control

It is well known that good process control reduces waste by minimizing cycling and improving a plant's ability to handle normal variations in flows, temperatures, pressures and compositions. Statistical quality control techniques help plants to analyze process variation and documents improvements. On a few occasions, additional instrumentation or online stream monitors (e.g. gas chromatography) are necessary, but good process control optimizes process conditions and reduces the tripping of the plant—which is a major source of waste.

Optimizing daily operation

If a computer is incorporated into the process control scheme, it can be programmed to analyze the process continuously and optimize operating conditions effectively. If the computer is not an integral part of the control scheme, offline analyses can be performed and employed as a guide for setting process conditions at the optimum level.

Automation of start-ups, shutdowns and product changeovers

Very large quantities of wastes can be generated during plant start-ups, shutdowns and product change-overs, despite such events being well-planned. Programming a computer to control these situations, brings the plant to stable operating conditions at a very short time and minimizes the time spent generating off-spec product. In addition to this, since minimum time is invested in undesirable running modes, equipment fouling and damage are also correspondingly reduced.

Programming the plant to handle unexpected or unforeseen upsets and tripping

Even in the case of the best control systems, upsets and trips take place. It is not possible for us to anticipate, but operators who have been working with the plant for several years may remember most of the important occurrences and know the best way to respond to a particular situation. With the help of computer control, optimum responses can be programmed in advance. Then, when upsets and trips occur, the computer takes over, minimizing incidents such as down time, spills, equipment damage, product loss and waste generation.

12.12 OTHER MISCELLANEOUS IMPROVEMENTS

In addition to the aforementioned ideas, there are a number of other miscellaneous improvements that can be incorporated to bring about a reduction in waste generation.

Avoidance of unexpected trips and shutdown

Minimization of tripping of the plant and unplanned shutdowns can be accomplished by adopting a good preventive maintenance programme and adequate sparing of equipment. Another key is to provide an effective warning system for critical equipment (e.g. vibration monitors). Plant operators can be extremely cooperative and helpful by reporting unusual operating conditions so that minor maintenance problems get rectified before they become major and give rise to a plant trip.

Using waste stream from other plants

Within a chemical complex (a manufacturing facility), each plant's waste streams should be clearly identified. The quantity and quality of these streams should be documented well including the presence of trace quantities of metals, halides or other impurities that render a stream useless as a raw material. The list should be reviewed periodically by all plants in the complex to determine if any are suitable as feedstocks.

Reducing number and quantity of samples

It is well understood that drawing frequent and large samples can generate a surprisingly large amount of waste. Many plants have been observing that the quantity and frequency of sampling can be brought down and that the samples can be returned to the process after the analysis is carried out.

Recovering product from tank cars and tank trucks

It is worthwhile recovering and reusing product drained from a tank car or tank truck especially those dedicated to a single service.

Reclaiming waste material

It is likely that in some situations waste products—not all of which are chemical streams—can be reclaimed. Rather than sending them to a burner or to landfill, chemical plants have devised ways and methods to reuse them. This may involve processes like physical cleaning, special treatment, filtering or other reclamation techniques.

Installation of reusable insulation

When conventional insulation is removed from equipment it is typically scrapped and sent to landfill. A good number of organizations manufacture reusable insulation. Their products are particularly effective on equipments where the insulation is removed regularly with an idea of carrying out the maintenance work (e.g. heat exchanger heads, valves, transmitters, etc.).

Maintaining external painted surfaces

Even in plants which handle corrosive materials, external corrosion turn to be a major cause of pipe deterioration. Piping and vessels should be painted well before the insulation is done and all painted surfaces should be well maintained.

Using more computer communications and less paper

Computers are not the tools for every communications chore, but many routine messages can be sent through computer.

Finding a market for waste products

Converting a waste product to a saleable product may call for additional processing and some creative salesmanship, but it can be an effective means of reducing waste. The converted product should not give rise to a waste problem for the customer.

EXERCISES

12.1 What pollution prevention strategies can be applied at the conceptualization and development stages of a process?

12.2 What steps can be taken to effect pollution prevention during plant design and in the plant operations?

12.3 How can the quality of mixing be improved in reactors?

12.4 Describe the steps by which waste minimization in heat exchangers and distillation columns can be achieved.

12.5 What miscellaneous steps can be adopted in order to bring about a reduction in the generation of wastes?

13
Hazardous Waste Management

The improper and insufficient management of hazardous waste world over has been a growing concern to industrial organizations, governments and the public at large. For various reasons, in spite of considerable interest in their impact on the environment there is no consistent definition of hazardous wastes. In addition to this, hazardous waste problem has not been put into the right perspective in comparison with other environmental concerns. As a consequence, developments in the control and minimization of hazardous waste, have been occurring very slowly.

13.1 DEFINITION OF HAZARDOUS WASTE

One of the primary difficulties encountered by the government and industries engaged in hazardous waste control programmes is to realistically define the term 'hazardous waste'. Within different jurisdictions there have been multiple levels of sophistication in identifying hazardous wastes. For instance, the Resource Conservation and Recovery Act (RCRA) of the United States of America has come out with a definition of a hazardous waste as a solid waste, or combination of solid wastes, which owing to its quantity, concentration physical, chemical or infectious characterstics may

1. cause or significantly contribute to an increase in mortality or an increase in serious, irreversible or incapacitating illness, and
2. pose a substantial present or potential hazard to human health or the environment when improperly treated, stored, transported or disposed of or otherwise managed.

While the definition refers to solids, it has been used to interpret semisolids, liquids and contained gases as well.

The US Environmental Protection Agency (EPA) has defined a waste to be hazardous under the legislation if it

1. exhibits characteristics of ignitability, corrosivity, reactivity and/or toxicity,
2. is a non-specific source waste (generic waste from industrial processes),
3. is a specific source waste from specific, industrial establishments.
4. is a specific commercial chemical protector intermediate,

5. is a mixture consisting of a listed hazardous waste,
6. is a substance that is not excluded from regulation under the Resource Conservation and Recovery Act.

13.2 THE MAGNITUDE OF THE PROBLEM

A review of information available indicates that there are four major pathways for the generation of hazardous waste and its escape to the environment.

(i) The first and most significant pathway is the contribution from the continuous discharge of hazardous wastes to land. These include
  (a) wastes which, owing to environmental restrictions have been removed from air or water effluent streams;
  (b) wastes generated by a manufacturing facility as process losses or by-products;
  (c) distrained products (products which are seized and held) which become wastes as a result of government regulations or restrictions;
  (d) wastes from government or other institutional operations;

(ii) The second source is the addition to the environment of hazardous substances that occurs during accidental spillage of higher magnitude.

(iii) The third one is the discharge of hazardous wastes in small quantities or concentrations from a variety of sources including aqueous effluent and air emission streams that are not currently subject to pollution control measures.

(iv) The fourth pathway includes situations such as mine tailing, dumps and disposal sites that are abandoned with little or no control over their long-term effects on the environment.

The problem has assumed a serious dimension as the harmful affects of hazardous wastes have been identified during the last two or three decades. Therefore, there is a dire need for stringent laws to be enacted and enforced to prevent further damage to the environment.

Table 13.1 lists some examples of toxic and hazardous chemicals which are found in the industrial effluents.

13.3 INDUSTRY AND GOVERNMENT PERSPECTIVE

Concern about the public health consequences of waster disposal, fears of long-term legal liability and adverse publicity have made generators of hazardous wastes to improve on-site control and reduce its quantity. The priorities of industry for the management of hazardous wastes are in accordance with the following ranking which favours land disposal:

1. Land disposal and storage
2. Treatment
3. Reduction of waste generation
4. Recycle and recovery
5. Incineration

Table 13.1 Examples of Toxic Chemicals Found in Industrial Effluents

Type	*Example*	*Industrial application or source*	*Probable destination of the waste (the contaminant)*
Metals and inorganics	Cyanides	Electroplating baths	Water
Pesticides	Chlordane	Manufacture of pesticides	Sediment, biota
Poly chlorinated biphenyls	PCB arochlors	Transformer coolant	Sediment, biota
Halogenated aliphatics	Dichloromethane (methylene chloride)	Solvent	Water
	Tetrachloromethane (carbon tetra chloride)	Solvent and degreaser	Water
	Chloroethane (vinyl chloride)	Manufacture of plastics	Water
Ethers	Vinyl ether	Pharmaceutical wastes	Water, sediment
Monocyclic aromatics	Ethyl benzene, toluene	Solvent	Sediment
Phenols and cresols	Phenol, pentachloro phenol	Refinery wastes, wood preservation	Water sediment, biota
Phthalate esters	Dimethyl phthalate	Manufacture of cellulose acetate	Sediment, biota
Polyacrylic aromatics	Naphthalene	Manufacture of dyes and synthetics	Sediment, biota
Nitro amines and others	Acrylonitrile	Manufacture of plastics	Water, sediment

Land disposal has been the dominant alternative to hazardous waste treatment as it is less expensive than techniques such as incineration, neutralization or other non-land based choices. But land disposal of hazardous waste has contaminated ground and surface waters, resulting in possible adverse effects on human health and destruction of aquatic life.

The law now discourages the underground injection well disposal of hazardous wastes, land treatment facilities and waste piles as they pose potential danger to groundwater supplies.

The government/pollution control boards have the following priorities for the management of hazardous wastes:

1. Waste minimization
2. Recycle and recovery
3. Treatment and incineration
4. Land disposal

It can be easily observed that the order of priorities differs from those of the industry. This is because land disposal of hazardous waste is still the lowest of the available technologies. However, the industries are motivated by giving incentives if the wastes are recycled and reused. Here too, the major stumbling block has been the lack of means to match supply and demand, such as centralized information service.

There is a need for improved long-term planning in the management of hazardous waste. While secure land disposal is the least desirable way for managing hazardous waste there will always be a need for some disposal of residual hazardous waste in the environment. The planning process must ensure that the design of short-term treatment and disposal techniques is compatible with the technologies that will be employed to minimize failure of the ultimate disposal sites.

As was in the case for solid waste management, the US, EPA has proposed a hierarchy of preferred hazardous waste management practices.

They are, in order of priority as follows:

1. Waste reduction
2. Waste separation and concentration
3. Waste exchange
4. Energy/material recovery
5. Incineration/treatment
6. Secure and disposal

Elimination or reduction of hazardous wastes through changes in the manufacturing process is the first objective of a hazardous waste management plan. After this it is essential that the remaining hazardous wastes be accounted for, from their origin to ultimate disposal on extensive documentation by a manifest or waybill system for recording waste movements.

Federal regulations under RCRA in the US, recognize that the individual states are responsible for a hazardous waste plan with certain requirements including

1. The registration, of all hazardous *small quantity generators* producing more than 100 kilograms of wastes per month
2. The collection and transport of hazardous wastes only by licensed handlers operating under a manifest system
3. Ultimate disposal of an approved hazardous waste treatment or disposal facility

The hazardous waste inventory may exclude the following:

1. Dilute waste waters treated on site and discharged to sewers or surface waters
2. Waste disposed of by deep well injection
3. Wastes from commercial operations like fuel stations, repair shops, cleaners and research institutions
4. Waste from spills and site remediation

13.4 HAZARDOUS WASTE CHARACTERIZATION

It is of significant importance to have a clear understanding of the nature of materials involved to discuss the techniques for the disposal and recovery of wastes. The understanding is limited not only to the physical state or composition, but also to a complete scheme of characterization that includes all information necessary for disposal or recycling after treatment.

A means of characterizing the waste is needed by the manufacturer, disposer and the recycle processor. Also some of the responsibility for the safe disposal of waste rests with the manufacturer. He should, therefore, characterize his waste to comply with legislation and give sufficient information to the disposal contractor to enable him to handle it safely. Experience suggests that the producer does not have sufficient information, motivation or incentive to carry out this task effectively. Therefore, a greater responsibility is placed on the shoulders of the disposal contractor, whose characterization requirements are probably much more detailed; he accepts a wide variety of wastes, some of which may be poorly defined. Waste recycle processor should also be well informed about the characterization since he is the one who actually deals with the waste management.

With an idea of helping the producer, disposer and recycle processor in the identification of hazardous waste, EPA selected four characteristics as inherently hazardous in any substances. These are:

1. Ignitability
2. Corrosivity
3. Reactivity
4. Extraction Procedure (EP) toxicity

Environmental Protection Agency (EPA) used two criteria in selecting these characteristics. The first criterion was that the properties defining characteristic be measurable by standardized and available testing protocols. The second criterion was adopted because the primary responsibility for determining whether a solid waste exhibits any of the characteristic rests with the waste generators. EPA believed that unless generators were provided with widely available and uncomplicated methods for determining whether their wastes exhibited the characteristics, the identification system would be unworkable.

Mainly because of the second criterion, EPA did not include carcinogenicity, mutagenicity bioaccumulation potential and phototoxicity to the characteristics. The available testing protocols for the aforementioned characteristics, are considered by the EPA, to be either insufficiently developed, too complex or too highly dependent on the use of skilled personnel or professional equipment. Also, in view of the current state of knowledge concerning such characteristics, EPA did not feel, it could define with any confidence the numerical threshold at which wastes exhibiting these characteristics would present a substantial hazard.

13.4.1 Characteristic of Ignitability

Ignitability is the characteristic used to define as *hazardous*, those wastes that could cause a fire during transport, storage or disposal. Examples of ignitable wastes include waste oils and used solvents.

A waste exhibiting the characteristic of ignitability has any one of the following properties:

1. It is a liquid, other than an aqueous solution containing less than 24 per cent alcohol by volume and as a flash point less than 60° C (140° F) as determined using to Pensky-Mortens closed cup tester or by a Seta-Flash closed cup tester.
2. It is not a liquid and is capable, under standard temperature and pressure; of causing fire through friction; absorption of moisture; spontaneous chemical changes, and when ignited burns-off so vigorously and persistently that it creates a hazard.
3. It is an ignitable, compressible gas.
4. It is an oxidizer.

13.4.2 Characteristic of Corrosivity

Corrosivity, mostly measured by pH, is an important characteristic in identifying the hazardous wastes, as waste with high or low pH value is likely to react dangerously with other wastes or entail the migration of toxic contaminants from certain wastes. Examples of corrosive wastes would include acidic wastes and used pickle liquor from steel manufacture.

A waste exhibiting the characteristic of corrosivity has any one of the following properties:

1. It is aqueous and has a pH value of less than or equal to 2 or greater than or equal to 12.5 as estimated by the pH meter employing an EPA test method.
2. It is a liquid and it corrodes steel (SAE, 1020) at a rate greater than 6.35 mm (0.250 inch) per year at a test temperature of 55° C (130° F).

13.4.3 Characteristic of Reactivity

Reactivity was selected as an identifying characteristic of hazardous waste because unstable wastes can pose an explosive problem at any stage of the waste management cycle. Examples include water from Trinitrotoluene (TNT) operations and used cyanide solvents.

A waste exhibiting the characteristic of reactivity has one or more of the following properties:

1. It is normally unstable and it readily undergoes violent change without detonating.
2. It reacts violently with water.
3. It forms potentially explosive mixtures with water.
4. When mixed with water, it gives rise to toxic gases, vapours or fumes in a quantity, adequate to present a danger to human health or environment.
5. It is a cyanide or sulphide bearing waste which when exposed to pH between 2.0 and 12.5 can give rise to toxic gases, vapours of fumes, in an amount adequate to pose a danger to human health or the environment.
6. It is capable of detonation or explosive decomposition reaction at standard temperature and pressure.
7. It is capable of detonation or explosure reaction if subjected to a strong initiating source or if heated under confinement.
8. It is a forbidden explosure.

13.4.4 Characteristic of Extraction Procedure Toxicity

Extraction Procedure (EP) toxicity is designed to identify wastes that are likely to attain hazardous concentrations of particular toxic constituents of a groundwater as a consequence of improper management. During the procedure the wastes are leached in a manner to simulate the leaching actions that occur in landfills. The extracts are analyzed to determine if they possess certain toxic contaminants. If the concentration goes beyond the regulatory levels, then the waste is classified as hazardous.

Some of the regulatory levels of the contaminants in the reached extract are presented in Table 13.2.

Table 13.2 Regulatory Levels of the Contaminants

Contaminant	*Maximum concentration* (mg/l)
Arsenic	5.0
Barium	100.0
Cadmium	1.0
Chromium	5.0
Lead	5.0
Mercury	0.2
Selenium	1.0
Silver	5.0

13.5 CATEGORIZATION OF HAZARDOUS WASTES

EPA of the USA has promulgated three lists of hazardous wastes. They are as follows:

13.5.1 Non-specific Source Waste (Appendix A)

These are generic wastes, commonly produced by manufacturing and industrial processes. Examples from this list include spent halogenated solvents employed in waste water treatment sludge from electroplating processes.

13.5.2 Specific Source Wastes (Appendix B)

This consists of wastes from specifically identified industries such as wood preserving, petroleum refining and organic chemical manufacturing plants. These wastes typically include sludges, still bottoms, waste waters, spent catalyst and residues, e.g. waste water treatment sludge from production of pigments.

13.5.3 Commercial Chemical Products (Appendix C)

The third list comprises specific commercial chemical products or manufacturing chemical intermediates. These include chemicals; such as chloroform and creosote, acids such as sulphuric acid and hydrochloric acid and pesticides; such as DDT and kepone (Wentz, C.A.).

13.6 HAZARDOUS WASTE MANAGEMENT

13.6.1 Waste Minimization

Benefits of hazardous waste reduction. As is well known reduction is always better than management. A major shift in the policy of industries is required to bring life to this principle.

The reason behind this is that while pollution control operations constitute a familiar operation, waste reduction impels innovation. More often than not waste reduction is overshadowed by other measures of materials productivity like labour productivity and energy consumption.

Waste reduction is a way of improving profitability and competitiveness. It minimizes the necessity of abiding by pollution control regulations that forces the spending of more and more money for increasingly smaller increments of environmental protection.

Waste reduction poses a threat to product quality. This is a major factor causing concern in industrial circles as it involves tinkering with or changing an existing industrial process solely for the purpose of reducing waste.

The conservative method of pollution control does not help in finding solution to problems. It only brings about alteration in its shifting from one form to another. The form of water may be changed, but it does not disappear. Conventional methods are likely to cause more pollution than the removal and consume resources out of proportion to the benefits derived. What emerges is an environmental paradox.

It takes resources to remove pollution. Pollution removal generates residue. It takes more resources to dispose of this residue and in the process what results is more pollution.

In order to make waste reduction successful, it must result in an environmental benefit through the prevention of pollution. For such an industry, this will result in an economic improvement. Once this is done, the aforementioned paradox is resolved.

13.6.2 Approaches to Hazardous Waste Reduction

There is a clear distinction between waste treatment and waste reduction or minimization. While the former is essentially an addition to the end of the individual processes, the latter is intricately involved in all aspects of the production process. Waste reduction succeeds only if it is art of everyday consciousness of all workers and managers involved with production process rather than only those responsible for complying with environmental regulations.

The approach to waste reduction has been lackadaisical. It is seldom seen as a criterion to measure job performance or performance in meeting government environmental requirements, developing source of inter-industrial competition, and so on. Most generators view these as a long-term ideal development rather than one that is immediately feasible. Waste reduction has been a by-product rather than a focus of altered industrial processes.

There are many approaches to waste reduction

1. Recycling potential waste or portion of it to the generator site
2. Altering primary source of waste generator by improving process technology and equipment
3. Improving plant operations such as better housekeeping, improved materials handling and equipment maintenance, automating process equipment, better monitoring and improved waste tracking
4. Introducing substitute raw materials in the process concerned which have a lesser potential of generating hazardous waste
5. Redesigning or reformulating end products

13.6.3 Recycling

It is usually the step before pollution control, but the economic limitations have to be taken into consideration.

Improving process technology: This is difficult to implement in many industries, but is important in the sense that often an entire waste stream can be eliminated.

Improving plant operations: This can be done with minimal capital investment through selected raw material changes.

It is important to define and understand the technical means to implement waste reduction and to evaluate all alternatives from an economic point of view. A fundamental antipathy exists in the industry against government involvement in front end production. Stringent government and industry product specifications wherever feasible should be made more flexible to allow for product substitution and process modification.

13.7 PRIORITIES IN HAZARDOUS WASTE MANAGEMENT

Proper management implies more than just careful disposal. There is a range of options to be considered which in turn depend on such factors as characteristic, volume and location of the waste. In order of priority the desired options for managing hazardous wastes are:

1. Minimizing the quantities generated by modifying the industrial processes involved
2. Transferring the waste to another industry that can utilize it
3. Reprocessing the waste to recover energy or materials
4. Separating hazardous from non-hazardous waste at the source and concentrating it, which reduces handling, transportation and disposal costs
5. Incinerating the waste or treating it to reduce the degree of hazard
6. Disposing the waste in a secure landfill by secure landfill is meant that, it is located, designed, operated and monitored in a manner that safeguards life and environment.

One of the more feasible options is the transference of hazardous waste to another industry. This is based on the principle that one organization's waste is another's raw material. This option may take two forms:

1. Materials exchange—equipped to handle treat and physically exchange waste; information exchange—acts as a cleaning house only.
2. A related option is recovery of energy or materials. This is economical in the sense that it requires less energy than that spent in mining and processing of virgin materials.

A method of completely detoxifying wastes is incineration. This method is environment-friendly and hence, suitable for adaptation.

Reduction of the toxicity of hazardous waste can be effected by one of the three basic methods—physical processes, chemical processes and biological processes. These entail a reduction in the quantity of hazardous waste that can be disposed of on land.

However, the least expensive, environmentally sound method of disposal is a secure landfill. Waste reduction is a near-term practical option even though it is not possible to accurately estimate how much is technically and economically feasible.

We have to recognize the necessity to reduce the quantity of hazardous waste in the bargain of recovering valuable resources. This will result in laudable balance between the need to conserve the nation's supply of raw materials and the dire necessity of controlling hazardous contaminants that are otherwise discharged into the environment.

13.7.1 Selection of a Waste Minimization Process

The selection of an actual minimization process should take into account all factors that fit into the overall programme goals (Figure 13.1). The most pragmatic procedure to evaluate solid waste management technique is to record the specific characteristics of a municipality (or a corporation) and determine which method(s) of solid waste disposal or resource recovery are properly suited to alleviate its solid waste management problems. Municipal characteristics of significance are population density, land costs, climate and power availability. Other characteristics include topography, geology, hydrology and socio-economic conditions. These constitute the primary factors that need to be considered.

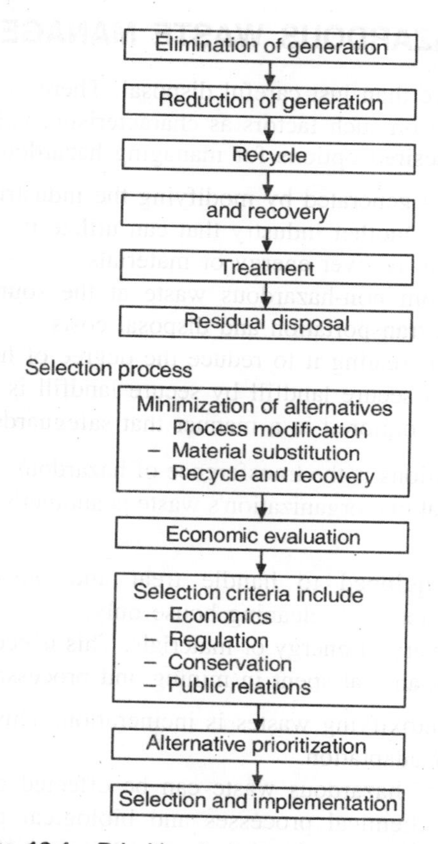

Figure 13.1 Priorities in hazardous waste management.

A second major consideration is the assessment of process parameters that determine process applicability and desirabilty, ease and convenience of procurement, conservation of materials and energy.

The compositional characteristic of the refuse in the generation area must also be taken into consideration. If the material recovery percentage is large, then it is compatible with a materials recovery system. On the other hand, high temperature incineration cannot be employed successfully when moisture content of refuse is high.

Economics, naturally, turns out to be most important process parameter. This provides the impetus for an industry to choose between recovery of desirable materials from waste and their disposal. The cost of system per ton of refuse process would contribute heavily to preliminary system selection. Budgetary constraints on process technology have brought about hindrance on the development of this environmentally desirable objective.

Dealing with the selection of process brings up the question of characterization of solid waste management systems as short-term and long-term objectives. Short-term solid waste disposal methods include

1. Landfilling
2. Incineration
3. Composting

Long-term methods include

1. Refuse as a supplementary fuel
2. Waste materials recovery
3. Pyrolysis
4. Methane production by anaerobic fermentation

Any short-term process selected must be compatible in a phased manner with long-term methods. For instance, a short-term conventional incineration process would not phase in well with long-term use of refuse as a supplementary fuel. On the other hand, a short-term conventional landfill would be very compatible with long-term pyrolysis process.

In order to make waste reduction and recycling work in an effective manner, support has to be extended by the corporate management. Everyone in the organization must devote time and energy to the selection process, then, only innovative solutions are possible and will be forthcoming.

13.8 TREATMENT OF HAZARDOUS WASTE—CHEMICAL, PHYSICAL AND BIOLOGICAL TREATMENT METHODS

All the waste products from the manufacturing operations must be subjected to some treatment or the other to render them harmless to the environment. The various treatment procedures can be classified as:

13.8.1 Chemical Treatment

The chemical treatment procedure involves the use of chemical reactions to convert hazardous waste treatment into less hazardous substances. The chemical treatment produces useful by-products and residual effluents that are environmentally acceptable. Chemical reactions either reduce the volume of the wastes or convert the wastes to a less hazardous form.

Solubility

Hazardous wastes may either be organic or inorganic substances which have to be dissolved in a suitable solvent. Water is the chief solvent which will dissolve many of the ingredients of hazardous waste. The solubility of a substance is very important in any chemical treatment process.

Neutralization

Wastes which are acid or alkaline are treated to get them neutralized. The neutralization lowers the corrosive nature of the wastes and renders them less hazardous.

Neutralization of a waste that is acid or base involves the addition of a chemical substance to change the pH to a more neutral level in the range of 6.0 to 8.0. It is necessary to neutralize acid wastes with a base and to neutralize high pH wastes with an acid according to the equation

$$\text{Acid} + \text{Basic} \rightarrow \text{Salt} + \text{Water}$$

Acid waste waters are neutralized with slaked lime $Ca(OH)_2$, caustic soda (NaOH) or soda ash (Na_2CO_3). The slaked lime is added to acidic waste water in an agitator vessel that has a pH sensor to control the slaked lime feed rate.

Alkaline waste waters may be neutralized with strong mineral acid such as H_2SO_4 or HCl or with CO_2. Agitator vessels are employed with the sensors that control the acid feed rate.

Precipitation

Undesirable heavy metals are always present in liquid waste streams. If the concentrations of the heavy metals are sufficiently high, the heavy metals have to be removed. The usual method for the removal of inorganic heavy metals is chemical precipitation. The metals are precipitated at varying pH levels, depending on the metal ion resulting in the formation of an insoluble salt. Neutralization of an acid waste stream can bring about precipitation of heavy metals and the heavy metals can be removed as a sludge residue by clarification, sedimentation or filtration.

The choice of the reactant is the first consideration in the precipitaiton of heavy metals. The second consideration is solubility. Since solubility is affected by temperature it is another important factor in precipitation reactions.

Precipitation of heavy metals can also be achieved by the addition of sulphide chemicals, such as sodium sulphide (Na_2S) or sodium bisulphide (NaH). The addition of these soluble sulphide compounds must be carefully controlled to minimize odour and potential toxicity problem.

Care must be taken, so that hydrogen sulphide which is a potential health hazard, is not generated during sulphide precipitation. This can be accomplished by carrying out the precipitation at an alkaline pH (Figure 13.2).

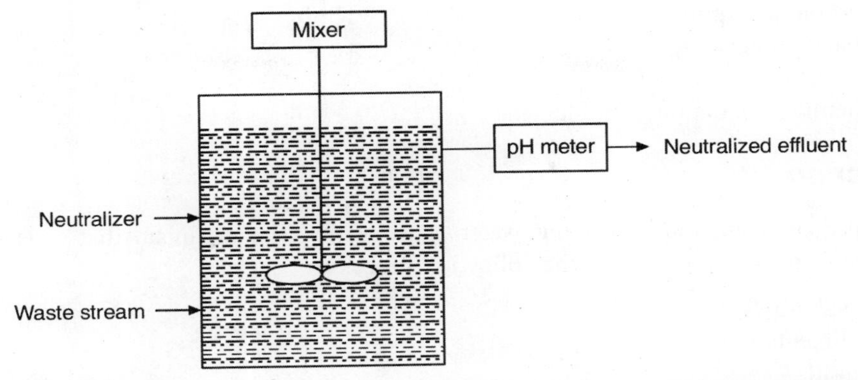

Figure 13.2 Chemical neutralization treatment system for waste management.

Coagulation and flocculation

The effectiveness of the precipitation process can be enhanced through the addition of various water-soluble chemicals and polymers that promote coagulation and flocculation. These operations bring about the separation of suspended solids from liquids when their normal sedimentation rates are too slow for filtration.

Coagulation is the addition and rapid mixing of a coagulant to neutralize charges and collapse the colloidal particles so that they can combine and settle down. The colloidal species in waste water are clay, silica, heavy metals and organics.

Alum [$Al_2(SO_4)_3$], ferric chloride ($FeCl_3$) and ferric sulphate ($Fe_2(SO_4)_2$) are common coagulants used in treating aqueous waste streams. Water soluble organic polymers are often more effective than alum or iron salts in promoting coagulation.

Flocculation is the agglomeration of the colloidal particles that have been subjected to the coagulation treatment. Flocculation requires gentle addition to allow bridging of the flocculent chemical between colloidal particles to form large flows. Slow mixing promotes flocculation and high-mixing speeds up the tearing part.

Oxidation and reduction

Oxidation and reduction can be utilized to convert toxic contaminants to either harmless or less toxic substances. For instance, hexavalent chromium is toxic and injurious to the environment. Hexavalent chromium is reduced to trivalent chromium and then it can be precipitated as chromic hydroxide.

$$SO_2 + H_2O \rightarrow H_2SO_3$$

$$2CrO_3 + 3H_2SO_3 \rightarrow Cr_2(SO_4) + 3H_2O$$

$$Cr_2(SO_4)_3 + 3Ca(OH)_2 \rightarrow 2Cr(OH)_3 + 3CaSO_4$$

Colour removal

The chemicals in the waste effluent are responsible for the colour. The colour can be reduced by modifying upstream processing conditions. The colour removal processes are:

- Carbon adsorption
- Coagulation
- Flocculation
- Chemical oxidation with chlorine

Disinfection

The purpose of disinfecting drinking water is to destroy the organism that causes diseases. Micro-organisms are removed by the following processes:

1. Coagulation
2. Sedimentation
3. Filtration

13.8.2 Physical Treatment

Physical treatment procedures of hazardous wastes include a number of separation processes commonly employed in industry. For a waste containing liquids and solids, physical separation is of first importance, as it is simple and very economical to carry out. The physical processes for separation of liquids and solids are screening, sedimentation, clarification, centrifugation, floatation, filtration, sorption, adsorption, absorption, evaporation and distillation, stripping, and reverse osmosis.

The tolerance levels for residual solids in the treated effluent is an important criterion in the selection of particular treatment process.

Screening

By the use of bar racks, strainers and screens, large solids such as plastics, wood and paper are removed.

Sedimentation

Sedimentation is the removal of suspended solids from liquids by gravitational settling. The velocity of the effluent stream must be such that enough time is available for the suspended solids to settle down in the sedimentation tank. The rate of settling of the solids is influenced mainly by the size, shape, density of the solid as well as by the density of the liquid phase.

The liquid displaced by the particles moves upwards and the particles settle down with no apparent flocculation between the particles.

Clarification

Clarifiers are employed to achieve faster sedimentation and removal of solids from liquid wastes. The objective of clarification is to produce a clear liquid effluent rather than a dense sludge.

Large settling ponds are employed to clarify waste water if adequate land sources are available. The liquid overflow remaining after the removal of solids can either be sent for discharge or for further treatment. A typical clarifier is shown in Figure 13.3.

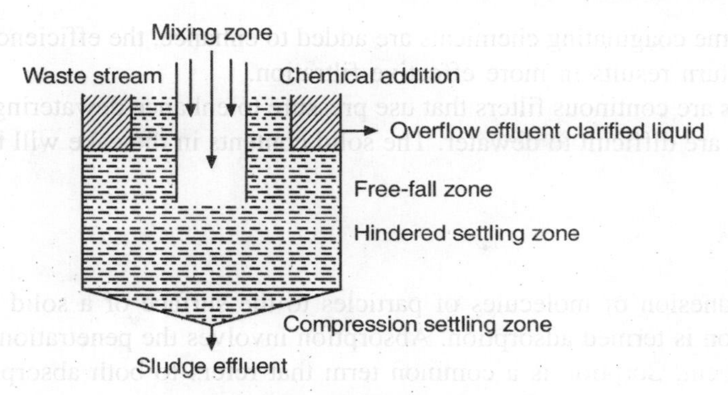

Figure 13.3 Centre-feed clarifiers.

Centrifugation

Centrifuges are employed in concentrating the waste sludges from 10 per cent to 40 per cent solids. The process results in the production of a solid cake, which can be conveniently disposed to final disposal site.

These are usually employed in dewatering applications owing to the fact that they are compact and simple to operate. The time required to settle the solids in centrifuge is very short when compared to a gravity setting tank.

The important design variables for the centrifuge are:

- Bowl type
- Bowl rotational speed and
- Scroll speed

Floatation

Low density solids and hydrocarbon solids can be separated from liquids by air floatation. The air which is introduced into the waste liquid in the form of finely divided bubbles attach to the particles to be removed. The particles that rise to the surface of the floatation cell can be removed by skimming.

The microbubbles of air become attached to the particles by contact or by actual formation at the solid-liquid interface. The air contacted particles are less dense than water and therefore, rise to the surface. The particles are then removed by skimming.

Filtration

In filtration the liquid effluent is made to pass through a porous medium to remove the suspended solids. The solids deposited increases the thickness of the medium.

Multimedia filters offer greater advantages compared to single-medium filters. The media should have different grain sizes and specific gravities. Ground anthracite solids and silica sand are used in multimedia filter. Filtration efficiencies can be increased by the addition of a third medium such as garnet.

Vaccum filters, belt presses and filter presses are often used for dewatering sludges and they concentrate the solids to about 50 per cent.

Usually, some coagulating chemicals are added to enhance, the efficiency of the rotary drum filter which in turn results in more effective filtration.

Belt presses are continous filters that use pressure to enhance dewatering. They are desirable for sludges that are difficult to dewater. The solid contents in the cake will be from 10 to 40 per cent.

Sorption

The physical adhesion of molecules of particles to the surface of a solid adsorbent without a chemical reaction is termed adsorption. Absorption involves the penetration of the particles into the solid absorbent. Sorption is a common term that refers to both absorption and absorption.

Adsorption

The removal of organic and inorganic substances from aqueous waste with activated carbon is effected through the adsorption of the chemical substances onto a carbon matrix. In waste water treatment activated carbon is the most widely employed adsorbent.

When activated carbon comes in contact with the waste effluent containing organics, adsorption of the organic solute occurs.

Absorption

Gas absorption takes place when soluble components of a gas mixture are dissolved in a liquid. The absorption may be purely a physical one or it may involve a chemical reaction. Examples of absorption include the water absorption of ammonia or hydrogen chloride from air. The solute is usually recovered by stripping or distillation and the absorbing liquid is recycled back to the absorber.

Evaporation and distillation

Liquids with difference in their relative volatility can be separated by distillation. Soluble salts and other waste impurities in the liquid will decrease the vapour pressure and elevate the boiling point.

Physical separation techniques usually precede evaporation processes, since separation reduces solid formation which enhances the heat transfer efficiencies.

Stripping

This separation technique is also based on the variation in the relative volatility of the liquids concerned. The principle on which a stripper functions is the same as that of a distillation column.

Reverse osmosis

By the process of osmosis, solvent flows through a semipermeable membrane from a dilute to a more concentrated solution.

Osmotic pressure of the solution is that pressure which when applied to the solution will prevent the passage of the solvent through semipermeable membranes. In reverse osmosis, a differential pressure that exceeds the osmotic pressure is applied to the membrane, causing the solvent to move from the concentrated portion to the dilute region.

13.8.3 Biological Treatment

Biological treatment can be an efficient and a cost-effective way to remove hazardous substances from effluents.

The microbes for biological treatment may be categorized as heterotrophic or autotropic depending upon their source of nutrients. Heterotrophs use organic matter as the source of nutrition while autotrophs use inorganic matter.

Mirobes are further, classified according to their utilization of oxygen. In aerobic processes, oxygen molecules are required to decompose organic matter. Anaerobic microbes use oxygen that is present in chemical combination with other elements such on nitrates, carbonates or sulphates.

Aerobic organisms are commonly employed to treat waste effluent streams. Anaerobic systems are confined to the treatment of strong organic wastes or organic sludges.

The ability of bacteria to consume organics is measured by Biochemical Oxygen Demand (BOD) which is the quantity of oxygen utilized by microorganisms in the aerobic oxidation of organics at 20° C.

Hazardous waste materials are toxic to some of the microorganisms. But a substance that is hazardous to one group of organisms may turn out to be a valuable food source for another group. They can be treated biologically provided the proper organism distribution can be established.

Biological systems can lower the cost of downstream processes by reducing organic load if they are supplemented by other physical or chemical treatment steps.

Most of the organisms capable of treating hazardous substances grow well in the pH range of 6.0 to 8.0. Many hazardous organics are readily degraded aerobically.

Temperature is a significant factor in the biological treatment of hazardous wastes. The rate and extent of removal of organics varies appreciably with the temperature of the system.

The common technologies for suspended-growth biological systems are:

1. Continuous flow system
2. Sequencing batch reactor

In a continuous flow system pre-treated waste stream enters a stirred bioreactor for organism growth and accompanying substrate control. The suspension passes from the bioreactor to a clarifier. The biomass is separated from the treated liquid effluent, which may undergo further treatment. The bottoms of the clarifier are returned to the bioreactor to maintain the desired biomass concentration.

In the sequencing batch reactor (SBR) the process takes place in distinct steps. The steps involved in the SBR are: 1. filling, 2. settling, 3. reacting, 4. decanting or drawal, and 5. idling.

During the filling stage, pre-treated waste enters the vessels containing acclimated biomass. Aerobic and anaerobic reactions proceed in the next state viz., the reacting stage.

The biomass is allowed to settle for a predetermined period of time by shutting down the mixing and aeration equipment. The treated and clarified effluent is removed to the drawal or decanting stage. During idling the liquid flow is switched into the other reactor, beginning a new cycle.

13.8.4 Thermal Processes

Most hazardous wastes consist of carbon, hydrogen, oxygen, halogens, sulphur, nitrogen and heavy metals. These should be the major considerations in applying incineration technology to the thermal destruction of hazardous wastes.

If the waste can be destroyed or reduced to carbon dioxide, water and other inorganic substances the organics should be rendered harmless.

Incineration is particularly useful disposal technology when dealing with large quantities of organic hazardous wastes.

13.9 TRANSPORTATION OF HAZARDOUS WASTES

13.9.1 Containers for Disposal of Solid Wastes

The discharge of hazardous waste is defined as the accidental or intentional spilling, leaking, pumping, pouring, emitting, emptying or dumping of hazardous waste into air or any land or water. Ensuring the safe transportation of hazardous matter is a complex activity. The accidental release of hazardous materials poses a serious threat to human safety, property and to the environment.

Containers employed for shipping hazardous waste materials have to comply with certain basic requirements. If there is a spillage taking place inadvertently under normal transportation conditions, there should be no significant release of hazardous material to the environment and effectiveness of packaging should not be reduced during transport.

Hazardous products are transported in bulk by vessels, tank cars, etc. in containers such as cylinders, drums, barrels, cans, boxes, bottles and casks. The specification of the packaging depends on the nature of the hazardous material and the strength of the containers.

13.9.2 Storage

Storage of hazardous waste is a short-term proposal for later collection of material and reprocessing. The objectives in storage are summarized in the chart given hereinafter.

Chart

Waste disposal	*Storage*	
For future use	*Short-term;* accumulation of sufficient material to make recovery viable	e.g. solvents
	Long-term; intentional	e.g. metal hydroxide sludge
For safety		e.g. radioactive waste

Dumping for future recovery is employed in short-term for liquid wastes such as solvents, which are accumulated until recovery is practical and viable, but long-term storage has only been seriously proposed for solid wastes except for radioactive liquid wastes which is a special case.

The safety measures to be taken during storage include

1. Good housekeeping
2. Compliance with limits, set for stocks of potentially hazardous chemicals
3. Storage, segregation, handling of gas cylinder and supporting of cylinders
4. Segregation of incompatible materials (e.g. nitrates, chlorates from carbonaceous materials, oxidizing acid from organic material)

Underground storage tanks constitute one of the important methods of storage. Much information is available on the parameters on which the tank owner should focus his resources to minimize overall risk. For instance, an older unprotected tank located in the recharge zone of a major public drinking water supply will warrant expedient replacement with a double walled system. On the other hand, a new tank in a less environmentally sensitive areas may be economically retrofitted at a later time to meet minimum requirements. The important parameters to be considered while handling storage tanks are potential release characteristics, tankage, corrosion protection controls, leak detection controls, tank design and soil corrosivity.

The other face of the coin is site vulnerable data, such as soil hydraulic conductivity, surrounding population, aquifer use and proximity to surface water.

13.9.3 Reconditioned Drums as Hazardous Waste Containers

The normal practice of waste disposal is through accumulation in drums and containers. It is therefore, essential that these containers are maintained in good condition to prevent corrosion and therefore, pose a threat to human life and the environment. Frequent inspection of the containers is, therefore, an absolute necessity.

The drum reconditioning industry employs hazardous waste generating unit as a potential source of drums. But there is an inherent long-term risk in the supply of used drums to reconditioners.

13.9.4 Bulk Transport—Highway Transport

Cargo tanks are the main carriers of bulk hazardous material over roads. These are usually made of steel and aluminium alloys. Titanium, nickel or stainless steel may also be used as materials of construction. Cargo tanks usually have a life time of 8–10 years. Those carrying corrosives have much shorter life spans. Between loads, cleaning has to be done. This brings about a further reduction in the life span.

13.9.5 Rail Transport

The two major classifications of rail tank cars are pressure and non-pressure for transporting both gases and liquids. These two differ in the types of discharge valves, pressure relief systems and type of thermal shielding. The commodities most commonly transported by rail are flammable liquids and corrosive materials. The most common material of construction is steel. Aluminium is the second most widely used metal.

Precautions: There is a possibility of puncturing of flammable pressurized cars. This may result in excessive heating and consequent expansion of the contents, resulting in explosion. Safety relief devices have to be provided. In addition, top and bottom shelf couplers which are less risky than relief devices may be employed. For flammable gas, ethylene oxide and ammonia gas tank cars, installation of head shields is mandatory as further protection against coupler damage.

13.9.6 Water Transport

The largest bulk containers for water transport are ships, tankers and tank barges. The most common materials that are transported are petroleum products and crude oil. Chemicals such as sulphuric acid, sodium hydroxide, alcohols, benzene and toluene constitute the rest.

The quantity of material transported in this manner is large and vessels travel slowly. Sufficient safety measures are, therefore, taken and statistically this mode is the safest.

All shipments are subjected to strict regulation. The tankers' captains and operators should demonstrate thorough knowledge of pollution containment and cleanup, methods for disposal of sludge and waste materials. Regular inspection of the tankers is essential and this may be the reason for safety record of water transportation.

13.9.7 Non-bulk Transport

Materials that constitute non-bulk transport are fiberboard, plastic, wood, glass, fiberglass, and metal. Packages within packages are often employed for transportation of hazardous waste materials. For compressed gases, independent units such as steel drums and cylinders are used.

Regulations: Regulations are suggested to monitor and obtain information from shippers and carriers regarding the nature, mode of transportation and disposal of hazardous waste material. Licensing, registration and permit requirement enable state and local bodies to achieve the above. These, in turn, are used to target enforcement activities, plan and develop emergency response programmes and optimize routing. The hazardous materials' guidelines include procedures for analyzing routes within a jurisdiction. The risk assessment is based on the chance of an accident and the effect or impact that might be felt in the affected zone.

Hazardous waste management as a field has grown at such a fast rate that appropriate technology cannot keep pace with its ambitious programmes. For any hazardous waste management programme to be effective, there has to be close coordination between governmental agencies and the industries concerned. Successful waste reduction efforts have always been a consequence of attempts to increase the efficiency of industrial operations. The realization that waste reduction is long-time ideal is the starting point of efficient waste management.

Typical treatment methods for waste containing hazardous contaminants are listed in Table 13.3.

Table 13.3 Typical Treatment Methods for Hazardous Wastes

Method	*Typical processes*	*Description*
Biological methods	1. Suspended growth processes (aerobic, anoxic and anaerobic) 2. Attached growth processes (aerobic, anoxic and anaerobic) 3. Combined suspended and attached growth processes (aerobic, anoxic and anaerobic)	Biological processes are employed to treat 1. Liquids contaminated groundwater, industrial process effluent and landfill leachate 2. Slurries (sludges and contaminated soils with clean or contaminated water) 3. Solids (contaminated soils) and 4. Vapours from other treatment processes
Physico-chemical processes	1. Carbon adsorption 2. Chemical oxidation 3. Gas stripping 4. Steam stripping 5. Membrane separation 6. Super critical fluids extraction and super critical water extraction	Granular activated carbon adsorption is employed for the sorption of organic compounds from liquids, powdered activated carbon is typically used in conjunction with the activated sludge treatment process. Chemical oxidation is employed to detoxify a wide array of organic compounds. Stripping processes are used to remove volatile and semivolatile organics from industrial process waters. Membrane separation is employed to remove contaminants from a variety of process waters and liquids. Super critical fluids extraction and super critical water extraction are employed to remove organics from water sediments and soil.
Stabilization and solidification	1. Cement-based solidification 2. Pozzalan based aggregate 3. Thermoplastic 4. Organic polymers	Stabilization and solidification processes are employed for the treatment of 1. industrial wastes 2. a variety of wastes including incinerator bottom and fly ash before placement in a secure landfill and 3. large quantities of contaminated soil. These processes are all employed to immobilize hazard waste contaminants
Thermal processes	1. Vapour, liquid and solid combustion 2. Catalytic volatile organic chemical (VOC) combustion 3. Fluidized-bed incinerators 4. Pyrolysis reactors	Thermal processes are employed to destroy organic fraction of hazardous waste contaminants found in all types of waste streams including gases and vapours, liquids slurries and solids
Land disposal	1. Municipal landfills 2. Monofill landfills 3. Land farming 4. Impoundment and storage facilities 5. Deep-well injection	The objective of land disposal is to ensure that wastes placed in such facilities do not migrate off-site. In land farming the objective is to bring about natural and biological decay of hazardous waste materials. The problem with deep-well injection is that the final location of the injected wastes is unknown.

EXERCISES

13.1 Give the definition of a hazardous waste.

13.2 How are hazardous wastes characterized. Give an account of their characteristics.

13.3 Describe the various approaches to hazardous waste reduction.

13.4 Outline the physical, chemical and biological methods of treatment of hazardous wastes.

13.5 Discuss the various aspects of transportation of hazardous wastes.

CASE STUDIES

Pollution Prevention at Source

Liquid, solid and gaseous effluents are invariably generated during the manufacture of any product. Besides posing environmental pollution problems, these effluents represent losses of valuable materials and energy from the production processes and a significant investment in pollution control. As is well known, traditionally pollution control focuses on *end-of-pipe* and *out the back door* points of view. The control of pollution in this way needs manpower, energy, materials and capital investment. Such an approach removes the pollution from one source, such as liquid effluent treatment or air pollution abatement, but places the pollutants elsewhere such as landfill.

The factors such as more stringent regulations, higher disposal expenses and increased liability costs have made the policy-makers to commence critical examinations of the end-of-pipe pollution control measures. The value and advantage of waste reduction at source have become so apparent to the industrial establishments that they have started looking at broader environmental management objectives rather than focusing solely on pollution control.

Pollution prevention, on the other hand, is the process of stopping pollution at source through changes in production, operation and materials use. It has the attendant advantages such as reductions in material usage, pollution control and liability costs. In addition, it can also help protect the environmental conditions and reduce risks to the workers' health and safety.

CASE STUDY 1

Pollution Prevention and Waste Minimization of a Petroleum Refining Unit

This case study is aimed at perusing how a reduction in pollution in the petroleum refining industry can be effected. It describes the general base from which one would want to improve the options available for mitigating pollution from these effluent streams(71). Some of the important aspects of this case study are given here.

In the following presentation, the first section describes the major effluent streams associated with the process of petroleum refining and the methods commonly employed to treat the effluents. It may be mentioned in this context that the description presented here gives a baseline from which one would like to improve. The second section describes three categories of possibilities for reducing pollution load. They are

1. Pollution prevention possibilities
2. Material substitution possibilities
3. Recycling possibilities
4. Effluent treatment possibilities

Some of the specific possibilities within each of those categories have been identified.

POLLUTION AND MANAGEMENT OF EFFLUENTS

The contaminants that are associated with a petroleum refining facility typically include Volatile Organic Compounds (VOCs), carbon monoxide (CO), oxides of sulphur (SO_x), oxides of nitrogen (NO_x), particulates, ammonia (NH_3), hydrogen sulphide (H_2S), metals, spent acids and a number of organic compounds. The impurities in crude oil give rise to sulphur and metals. The other wastes represent losses of input and final product.

These pollutants may emanate as air emissions, liquid effluents or solid wastes. All of these wastes can be subjected to treatment. However, air emissions are more difficult to capture than liquid effluents or solid wastes. Therefore, it turns out that air emissions are the largest source of untreated effluents released to the environment.

Air Emissions and Effluent Treatment

The sources that contribute to air emissions include point and non-point sources. Point sources are emissions that emanate as stack gases and therefore, can be monitored and treated. Non-point sources fall under the category of fugitive emissions which are difficult to locate and capture. Fugitive emissions take place throughout refineries and originate from the numerous valves, pumps, tanks, pressure relief valves, flanges, etc. It can be mentioned here that while individual leaks are typically small, the sum of all the fugitive leaks at a refinery facility can turn out to be one of its largest emission sources.

The large number of process heaters employed in refineries to heat process streams or to generate steam (boilers) for the purpose of heating or steam stripping, can be possible sources of the contaminants viz. SO_x, NO_x, CO, particulates and hydrocarbons.

Under the right operating conditions and while burning cleaner fuels, such as refinery fuel gas, fuel oil or natural gas, these emissions are relatively low. However, if combustion does not go to completion or heaters are fired with refinery fuel pitch or residuals, emissions can turn out to be appreciable.

The majority of gas streams coming out of each refinery process consists of varying quantities of refinery fuel gas, H_2S and NH_3. These streams are collected and sent to the gas treatment and sulphur recovery units to recover the refinery fuel gas and sulphur. Gaseous emissions coming out of the sulphur recovery unit typically consist of some quantity of H_2S, SO_x and NO_x. Other emission sources from refinery process emanate from regeneration of catalysts which is done periodically. These processes give rise to streams that are likely to contain relatively high levels of carbon monoxide, particulates and volatile organic compounds. Such off-gas streams may be treated first through a CO boiler to burn CO and VOCs, and subsequently be made to pass through an electrostatic precipitator or cyclone separator to get rid of particulates before being discharged to the atmosphere.

Sulphur is removed from various process off-gas streams in order to meet the SO_x emission limits and also to recover soluble elemental sulphur. Process off-gas steams from the coker, catalytic cracking unit, hydrotreating units and hydroprocessing units can contain high concentrations of H_2S mixed with light refinery fuel gases. Before the recovery of elemental sulphur, the fuel gases (which basically consist of methane and ethane) need to be separated from the H_2S. This is typically carried out by dissolving the H_2S in a chemical solvent. Solvents most widely employed for this purpose are amines such as diethanolamine. Dry adsorbents, such as molecular sieves, activated carbon, iron sponge and zinc oxide are also employed. In the amine solvent process, diethanolamine solution or some other amine solvent is pumped to an absorption tower where they come into contact with the gases. The hydrogen sulphide gets dissolved in the solution. The fuel gases are removed for use as fuel in the process furnaces in other refinery operations. The amine H_2S solution is then heated and subjected to stream stripping to remove the H_2S gas.

The methods currently used for removing sulphur from the H_2S gas streams are typically a combination of two processes: the Claus process followed by the Beaven, SCOT, or Wellman-Land process. The Claus process consists of partial combustion of the H_2S-rich gas stream (with the quantity of air being one third the stoichiometric value) and then reacting the resulting SO_2 and unburned H_2S in the presence of a bauxite catalyst to produce elemental sulphur.

Owing to the fact that the Claus process by itself removes only about 90 per cent of the H_2S in the gas stream, the Beaven, SCOT or Wellman-Lord processes are often employed to bring about further recovery of sulphur. In the Beaven process, the H_2S in the relatively low concentration gas stream from the Claus process can almost be completely removed by absorption in a quinine solution. The dissolved H_2S is oxidized to form a mixture of elemental sulphur and hydroquinone. The solution is injected with air or oxygen to oxidize the hydroquinone back to quinine. The solution is then subjected to filtration or centrifuging to remove the sulphur and the quinone is then reused. The Beaven process is also quite effective in removing small quantities of SO_2, carbonyl sulphide that are not affected by the Claus process. These compounds are first converted to H_2S in elevated temperature conditions (cobalt molybdate is employed as the catalyst) prior to being fed to the Beaven unit. Air emissions from sulphur recovery units will comprise H_2S, SO_x and NO_x in the form of process tail gas as well as fugitive emissions and also as releases from vents.

The SCOT process is also used to a large extent for removing sulphur from the Claus tail gas. The sulphur compounds in the Claus tail gas are converted to H_2S by heating and passing tail gas through a cobalt molybdenum catalyst with the addition of a reducing gas. The gas is then cooled and contacted with a solution of di-isopropanolamine (DIPA) which enables the removal of all, but trace quantities of H_2S. The sulphide rich DIPA is sent to a stripper where H_2S gas is removed and sent to the Claus plant. The DIPA is then sent back to the absorption column.

Many of refinery process units and equipments are manifolded into a collection unit termed the blowdown system. Blowdown systems provide for the safe handling and disposal of liquids and gases that are either automatically vented from the process units through pressure relief valves or that are manually drawn from units. Recirculated process streams and cooling water streams are frequently manually drawn from units and are manually purged to prevent the occurrence of continued build-up of contaminants in the stream. Part or all of the contents of equipment can also be purged to the blowdown system prior to shutdown before normal or emergency shutdowns. Blowdown systems employ a series of flash drums and condensers to separate the blowdown into its vapour and liquid components. The liquid typically comprises mixtures of water and hydrocarbons containing sulphides, ammonia and other contaminants which are sent to the effluent treatment plant. The gaseous component typically consists of hydrocarbons, H_2S, ammonia, mercaptans, solvents and other constituents and is either discharged directly to the atmosphere or is subjected to combustion in a flare. A major part of air emissions from blowdown systems are hydrocarbons in the case of direct discharge to the atmosphere and sulphur oxides when flared.

Liquid Effluent Generation and Treatment

Liquid effluents from a petroleum refinery facility comprise cooling water, process water, storm water and sanitary sewage water. A large portion of water used in a petroleum refining unit is used for cooling. Most cooling water is recycled repeatedly. Cooling water typically does not come into contact with process oil streams and naturally contains less contaminants than process waste water. However, there is a possibility of its containing some oil contamination due to leaks in the process equipment. Water employed in processing operations contributes to a significant portion of the total waste water. Process effluents originate from desalting crude oil, stream stripping operations, pump gland cooling, product fractionator reflux drum drains and boiler blowdown. Owing to the process water coming into contact with oil often, it is usually contaminated to a large extent. Storm water that is surface water run-off is intermittent and will consist of constituents from spills to the surface, leaks in the equipment and any materials that may have got collected in drains. Run-off surface water also includes water coming from crude and product storage tank roof drains.

Liquid effluents are treated in on-site effluent treatment facilities and then discharged to community treatment facility. Petroleum refineries mostly utilize primary and secondary effluent treatment. Primary liquid effluent treatment is composed of the separation of oil, water and solids in two stages. During the first stage of treatment, an API separator, a corrugated plate interceptor or other separator design is employed. Waste water moves very slowly through the separator letting free oil to float to the surface and be skimmed-off and solids to settle to the bottom and be scraped-off to a sludge collecting hopper. The second stage of effluent treatment makes use of physical or chemical methods to separate emulsified oils from the liquid effluent. Physical methods may involve the use of a series of settling ponds with a long retention time or the use of Dissolved Air Floatation (DAF). In dissolved air floatation, air is bubbled through the liquid effluent and both oil and suspended solids are skimmed-off the top. Chemicals such as ferric hydroxide or aluminium hydroxide can be employed to coagulate impurities into a froth or sludge which can be more easily skimmed-off the top. Some wastes which are associated with the primary treatment of liquid effluents at petroleum refineries may be considered to be hazardous and these wastes include API separator sludge, primary treatment sludge, sludges coming out from other gravitational separation techniques float from DAF units and waste originating from settling ponds.

After the primary treatment is carried out, the liquid effluent can be discharged to a community treatment works or made to undergo secondary treatment before it is discharged to surface water. In secondary treatment, dissolved oil and other organic contaminants are likely to be consumed biologically by microorganisms. In biological treatment the addition of oxygen through a number of different techniques including activated sludge units, trickling filters and rotating biological contactors may be required. Secondary treatment generates bio-mass waste which is usually treated anaerobically and then dewatered.

In some refinery units, an additional stage of effluent treatment termed polishing is adopted in order to meet the discharge limits. In the polishing step, the use of activated carbon, anthracite coal or sand is made to filter out any remaining impurities, such as biomass, silt, trace metals and other inorganic chemicals as well as any remaining organic chemicals. Certain refinery effluent streams are treated separately prior to their being sent to effluent treatment plant, to remove contaminants that would not be amenable for treatment after being mixed with other liquid effluents. The sour water drained from the distillation reflux drums is one such effluent stream. Sour water contains dissolved hydrogen sulphide and other organic sulphur compounds and ammonia which are stripped in a stripping column with gas or steam before being discharged to the effluent treatment plant.

Liquid effluent treatment plants invariably are a significant source of refinery air emissions and solid waste. Air releases originate from fugitive emissions from the numerous tanks, ponds, and sewer system drains. Solid wastes are generated in the form of sludges from many treatment units.

Solid Wastes and Their Treatment

Solid wastes are generated from a number of refining processes, petroleum handling operations besides waste water treatment. Both hazardous and non-hazardous wastes are produced, treated and disposed. Refinery wastes are usually in the form of sludges (including sludges from effluent treatment), spent process catalyst, filter clay and incinerator ash. Treatment techniques followed for these wastes include techniques such as incineration, land treating off-site, landfilling on-site, landfilling off-site, chemical fixation, neutralization and others.

An appreciable portion of the non-petroleum product outputs of refinery units is transported off-site and sold as by-products. These outputs include sulphur, acetic acid, phosphoric acid, and recovered metals. Metals from catalysts and from the crude oil that have deposited on the catalyst during the process of manufacture often are recovered by third party recovery facilities. Storage tanks are employed throughout the refining process to store crude oil and intermediate process feeds for cooling and further processing. It is a common practice to keep finished petroleum products in storage tanks before transport off-site. Storage tank bottoms are mixtures of iron rust from corrosion, sand, water and emulsified oil and wax which tend to accumulate at the bottom of tanks. Tank bottom liquids comprising primarily water and oil emulsions, are periodically drawn-off to prevent their continued build-up. Tank bottom liquids and sludges are also removed during periodic cleaning of tanks for the purpose of carrying out an inspection. It is likely that tank bottoms contain some amounts of tetraethyl or tetramethyl lead (although this is increasingly rare owing to the phase out of leaded products), other metals and phenols. Solids generated from leaded gasoline storage tank bottoms are listed as RCRA hazardous waste.

MITIGATION OF POLLUTION

A number of options are available for reducing pollution from refineries. These include pollution prevention options, recycling options and effluent treatment options. In the following section, these options are listed out.

Pollution Prevention Option

Good housekeeping practices

Good housekeeping practices prevent waste by better handling of both inputs and waste without incorporating significant modifications to the current production technology. If inputs are handled in a better way, they are less likely to turn out to be wastes inadvertently through spills or outdating. If effluents are handled in a better way, they can be managed in the most cost-effective manner. Some effective housekeeping options that have been identified in the petroleum industry are the following:

Segregation of process waste streams: A significant portion of refinery waste originates from oily sludges found in combined process/storm sewers. Segregation of the relatively clean rain water run-off from the process streams can effect a reduction in the quantity of oily sludges generated. Furthermore, there is a much higher possibility of recovery of oils from smaller, more concentrated process streams.

Controlling solids entering sewers: Solids released to the effluent sewer system can account for a large portion of a refinery's oil sludges. Solids that enter the sewer system (primarily soil particles) become coated with oil and are deposited as oily sludges in the API oil/water separator. Because a typical sludge has a solid content of 5 to 30% by weight, preventing a kilogram of solid from entering the sewer system can eliminate 2 to 10 kg of oily sludge. Techniques employed to control solids include: using a street sweeper on paved areas, paving unpaved areas, planting ground cover on unpaved areas, relining sewers, cleaning solids from ditches and catch basins besides reducing heat exchanger bundle cleaning solids by employing antifoulants in cooling water.

Identification of benzene sources and installation of upstream water treatment: Benzene in effluent stream can often be treated more easily and effectively at the point of its generation rather than at the effluent treatment plant after it gets mixed with other waste water.

Training personnel to reduce solids in sewer: A training programme conducted for the personnel which lays emphasis on the importance of keeping solids out of the sewer systems will help in reducing that portion of waste water treatment plant sludge originating from the daily activities of refinery personnel.

Training personnel to prevent soil contamination: Contaminated soil can be reduced by educating personnel on the available methods to avoid leaks and spills.

Modifications in the manufacturing process

Manufacturing process modifications involve capital equipment, layout and process changes that help in reducing the quantity or toxicity of waste that are generated. Some options for process modifications in the petroleum refining industry are hereinafter.

Installation of the vapour recovery for barge loading: Even though barge loading is not a factor to be considered for all refineries, it is an important emissions source for many petroleum refining facilities. One of the largest sources of VOC emissions identified during a study conducted is fugitive emissions from loading of tanker barges. It has been observed that these emissions could be reduced by 98 per cent through installing a marine vapour loss control system. Such systems could comprise a vapour recovery or VOC destruction in a flare.

Replacement of old boilers: Older refinery boilers can turn out to be significant source of gaseous pollutants viz. SO_x and NO_x and also particulate emissions. It is possible and worthwhile to replace a large number of old boilers with a single new cogeneration plant with emission controls.

Reduction on the use of drums: Replacement of drums with bulk storage can minimize the chance of leaks and spills.

Installation of high pressure power washer: Chlorinated solvent vapour degreasers can be replaced by high pressure power washers as these do not generate spent solvent hazardous wastes.

Refurbishing or eliminating underground piping: Underground piping can turn out to be a source of undetected releases to the soil and groundwater. Inspecting, repairing, or replacing underground piping with surface piping can reduce or eliminate these potential sources.

Eliminating the use of open ponds: Open ponds used to cool, settle out solids and store process water can be a significant source of volatile organic compound (VOC) emissions. Waste water from coke cooling and coke VOC removal is occasionally cooled in open ponds where VOCs easily escape to the atmosphere. In many instances, open ponds can be replaced by closed storage tanks.

Placing secondary seals on storage tanks: One of the largest sources of fugitive emissions from petroleum refineries is storage tanks containing gasoline and other volatile products. These losses can be significantly brought down by installing secondary seals on storage tanks. It has been observed that VOC losses from storage tanks could be reduced by 75 to 90 per cent.

Minimizing solids leaving the desalter: There is a probability of the solids entering the crude distillation units to eventually attract more oil and produce additional emulsions and sludges. The quantity of solids removed from the desalting unit should, therefore, be maximized. The techniques that can be employed include using, low shear mixing devices to mix desalter wash water and crude oil; using lower pressure water in the desalter to avoid turbulence; and replacing the water jets used in some petroleum refineries with mud rakes which add less turbulence when removing settled solids.

Minimizing cooling tower blowdown: The dissolved solids' concentration in the recirculating cooling water is controlled by purging or blowing down a portion of the cooling water stream to the effluent treatment system. Solids in the blowdown eventually give rise to additional sludge in the effluent treatment plant. However, the quantity of cooling tower blowdown can be lowered by minimizing the dissolved solids' content of the cooling water. An appreciable portion of the total dissolved solids in the cooling water can originate from the cooling water makeup stream in the form of naturally occurring calcium carbonates. Such solids can be controlled either by choosing a source of cooling water, makeup water with less dissolved solids, or by removing the dissolved solids from the makeup water stream. Common treatment methods are cold lime softening, reverse osmosis or electro-dialysis.

Control of heat exchanger cleaning solids: In many refinery plants, using high-pressure water to clean heat exchanger bundles generates and releases water and entrained solids to the refinery effluent treatment system. Exchanger solids may then attract oil as they move through the sewer system and may also produce finer solids and stabilized emulsions that are more difficult to remove. Solids can be got rid of at the heat exchanger cleaning pad by installing concrete overflow weirs around the surface drains or by covering drains with a screen. Other ways to effect reduction in solids' generation are by employing anti-foulants on the heat exchanger bundles to prevent the formation of scales and by cleaning with reusable cleaning chemicals that also enable easy removal of oil.

Control of surfactants in effluents: Surfactants entering the petroleum refinery effluent streams will increase the quantity of emulsions and sludges generated. Surfactants can enter the system from a variety of sources including; washing unit pads with detergents, treating gasolines with an end point over 200 degrees celsius thereby producing spent caustics; cleaning tank interiors; and using soaps and cleaners for miscellaneous tasks. Besides this, the over use and mixing of the organic polymers used to separate oil, water and solids in the effluent treatment plant can actually stabilize emulsions. The operators can be advised to minimize the use of surfactants. Their use can be minimized also by routing surfactant sources to a point downstream of the dissolved are floatation unit and by using drycleaning, high pressure water or steam to clean oil surfaces of oil and dirt.

Installation of ruptured discs and plugs: Fugitive emissions can be reduced by installing rupture discs on pressure relief valves and plugs in open ended values.

Options in Material Substitutions

Using non-hazardous degreasers: The quantity of spent conventional degreaser solvents can be brought down or eliminated through substitution with less toxic and/or biodegradable products.

Eliminating chromates as an anti-corrosive: Chromate containing wastes can be reduced or eliminated in cooling tower and heat exchanger sludges by replacing chromates with less toxic alternatives such as phosphates.

Using high quality catalysts: By employing catalysts of a higher quality, process efficiencies can be significantly increased while the required frequency of catalyst replacement can be reduced.

Replacing ceramic catalyst support with activated alumina supports: Activated alumina supports can replace ceramic catalyst support. Activated alumina supports can be recycled with spent alumina catalysts.

Options on Recycling

Even though pollution is reduced very much if wastes are prevented in the first place, the second best option for bringing down pollution is to treat wastes so that they can be transformed into useful products. A few recycling possibilites applicable to the petroleum refinery are given here.

Recycling and regenerating spent caustics: Caustics employed to absorb and remove hydrogen sulphide and phenol contaminants from intermediate and final product streams can be recycled. Spent caustics may be sold to chemical recovery companies if concentrations of phenol or hydrogen sulphide are high enough. Process modifications in the refinery may be necessary to increase the concentration of phenols in the caustic to make recovery of the contaminants economical. Caustic containing phenols can also be recycled on-site by reducing the pH of the caustic until the phenols become insoluble thereby enabling physical separation. Caustic can then be treated in the refinery effluent system.

Using oily sludges as feedstock: Many oily sludges can be sent to a coking unit or the crude distillation unit where it becomes part of the refinery products. Sludges sent to the coker can be injected into the coke drum with quench water, injected directly into the delayed coker, or sent to the coker blowdown contactor employed in separating the quenching products. Application of sludge as a feedstock has increased appreciably in recent past and is presently carried out by most refineries. The amount of sludge that can be sent to the coker is restricted by coke quality specifications which may impose limitation on the quantity of sludge solids in the coke. Coking operations can be upgraded however, to increase the quantity of sludge that they can handle.

Controlling and reusing fluidized bed catalytic cracking unit (FCCU) and coke fines: Significant amounts of catalyst fines are frequently present around the FCCU catalyst hoppers and reactor and regeneration vessels. Coke fines are often present around the coker unit and coke storage areas. The fines can be collected and recycled before being washed to the sewers or migrating off-site through the wind. Collection techniques normally used are dry sweeping the catalyst and coke fines and sending the solids to be recycled or disposed of as non-hazardous wastes. Coke fines can also be recycled for fuel use. Another collection technique involves the application of vacuum ducts in dusty areas (and vacuum hoses for manual collection) which run to a small bag house for collection.

Option on Effluent Treatment

When pollution prevention and recycling options are not economically viable, pollution can still be brought down by subjecting the wastes to treatment, so that they are transformed into less environmentally harmful wastes or can be disposed of in a less environmentally harmful media. The following are a few treatment options available for the petroleum refineries:

Thermal treatment of applicable sludges: The toxicity and volume of some deoiled and dewatered sludges can be further brought down through thermal treatment. Thermal sludge treatment units use heat to vapourize the water and volatile components in the feed and leave behind a dry solid residue. The vapours are condensed for separation into the hydrocarbon and water components. Non-condensible vapours can be either flared or sent to the refinery amine unit for treatment and subsequent use as refinery fuel gas.

Improving recovery of oils from oily sludges: In view of the fact that the oily sludges make up a large portion of the refinery solid waste, any improvement in the recovery of oil from the sludges can appreciably reduce the volume of waste. Presently, a number of technologies are used to mechanically separate oil, water, and solids which include belt filter presses, recessed chamber pressure filters, rotary vacuum filters, scroll centrifuges, disc centrifuges, shakers, thermal dryers and centrifuge dryer combinations.

Reducing the generation of tank bottoms

Tank bottoms from crude oil storage tanks constitute a large percentage of refinery solid waste and pose a particularly difficult disposal problem owing to the presence of heavy metals. Tank bottoms consist of heavy hydrocarbons, solids, water, rust, and scaled deposits. Minimization of tank bottoms is carried out most beneficially through careful separation of the oil and water, remaining in the tank bottom. Filters and centrifuges are also useful for recovering the oil for the purpose of recycling.

Regeneration or elimination of filtration clay

Clay from refinery filters needs to be replaced periodically. Spent clay often contains significant quantities of entrained hydrocarbons and therefore should be designated or termed as hazardous waste. Back washing spent clay with water or steam can result in the reduction of the hydrocarbon content to levels at which it can be reused or handled as a non-hazardous waste. Another method employed to regenerate clay is to wash the clay with naphtha, dry it by steam heating and then feed it to a burning kiln for regeneration. In some cases, clay filtration can be substituted entirely by hydrotreating.

Minimizing fluidized catalytic cracking unit (FCCU) decant oil sludge: Decant oil sludge from the FCCU can contain significant concentrations of catalyst fines. These fine particles often prevent the use of decant oil as a feedstock or require treatment which generates oily catalyst sludge. Catalyst in the decant oil can be minimized by using a decant oil catalyst removal system. One such catalyst removal system incorporates high voltage electric fields to polarize and capture catalyst particles in the oil. The quantity of catalyst fines reaching the decant oil can be minimized by installing high efficiency cyclone separator in the reactor to shift catalyst fines losses from the decant oil to the regenerator where they can be collected in the electrostatic precipitator.

CASE STUDY 2

Environmental and Process Safety Guidelines for the Manufacture of Polymers of Petroleum Origin

The guidelines presented in this section are applicable to petroleum based polymer manufacturing in which the polymerization of monomers takes place and the finished products are in the form of granules and pellets. These pellets and granules can subsequently be used in industries.

Potential environmental problems that are confronted in the polymer manufacturing processes include

- Air emissions
- Liquid effluents
- Hazardous materials
- Wastes

AIR EMISSIONS

Volatile Organic Compounds (VOCs) from the Drying and Finishing Operations

The most common air emissions from polymer plants are volatile organic compound (VOC) emissions from the drying, finishing and purging operations. Effective measures to control VOC in the aforementioned operations include the following: (World Bank, 2007)

1. Separation and purification of the polymer downstream to the reactor
2. Flash separation of solvents and monomers
3. Stripping with steam or hot nitrogen
4. Degassing stages in extruders, preferably under vacuum conditions.
5. Condensing VOCs at low temperatures or in adsorption beds, before the venting of exhaust air occurs. Drying should recycle exhaust air or nitrogen with volatile organic chemical condensation

6. Using closed-loop nitrogen purge systems and degassing extruders and collecting off-gases from the extrusion process in polyolefin plants owing to the fire-hazard related to the flammability of the hydrocarbons and to the high temperatures involved
7. Vent gases emitted from the equipments viz. reactors, blow-down tanks and stripping columns consisting of appreciable levels of Vinyl Chloride Monomer (VCM) should be collected and purified before they are let into the atmosphere. Water has appreciable levels of VCM. For instance, water used in applications such as the cleaning of reactors containing VCM, transfer lines and suspension or latex stock tanks should be passed through a stripping column to get rid of VCM in polyvinyl chloride (PVC) manufacturing, employing the suspension process
8. Use of stripping columns specifically designed to strip suspensions in PVC manufacturing employing the suspension process
9. Production of stable latexes and use of appropriate technologies in emulsion PVC plants which combine the processes viz. emulsion polymerization and open cycle spray drying
10. Multistage vacuum devolatilization of molten polymer to reduce the residual monomer at low levels in polystyrene and generally in styrenic polymers production
11. Prevention of leak and spill in acrylic monomer emulsion polymerization, owing to the very strong, pungent, low-threshold odour of all acrylic monomers
12. Treatment of gaseous effluents by catalytic oxidation or equivalent techniques in polyethylene terephthalate production
13. Wet scrubbing of vent gases in polyamide production
14. Catalytic or thermal treatment of gaseous and liquid effluents in all thermoset polymer production
15. Installation of closed systems with vapour condensation and vent purification, in phenol formaldehyde resins production, owing to the high toxicity of both the main monomers; and
16. VOCs from the finishing sections and vents of the reactors should be subjected to treatment through thermal and catalytic incineration techniques before they are let into the atmosphere.

Volatile Organic Compounds from Process Purges

Process purges are associated with purification of raw materials, filling and emptying of reactors and other equipment, removal of the by-products of the reaction in polycondensation, vacuum pumps and depressurization of vessels. Effective pollution prevention and control steps are listed out hereinafter.

1. Process vapours purges should be recovered by compression or refrigeration and condensation of components which can be liquefied. Else, they should be sent to a high efficiency flare system that can ensure efficient destruction.
2. The gases which are incondensable should be fed to a waste-gas burning system specifically designed to ensure a complete combustion with low emissions and prevention of dioxins and furan formation.

3. In polyvinyl chloride plants, VCM-polluted gases (air and nitrogen) coming from VCM recovery section should be collected and subjected to treatment by VCM absorption or adsorption, by incineration techniques or by catalytic oxidation before they are let into the atmosphere.
4. In high impact polystyrene sheets production, air emissions from polybutadiene dissolution systems should be minimized by employing continuous systems, vapour balance lines and vent treatment.
5. In the case of unsaturated polyesters and alkyl resin units, gaseous effluents generated from process equipment should be subjected to treatment by thermal oxidation or if emissions concentrations permit, by the method of adsorption by activated carbon.
6. Using glycol scrubbing columns or sublimation boxes for the recovery of anhydride vapours from unsaturated polyester and alkyl resins storage tank vents.
7. In phenolic resins manufacture, VOC contaminated process emissions, especially the ones from reactor vents should be recovered or incinerated.
8. In aliphatic polyamide manufacturing, wet scrubbers, condensers, activated carbon adsorbers can be employed together with thermal oxidation.

Volatile Organic Compounds from Fugitive Emissions

In polymer manufacturing facilities, fugitive emissions are mainly associated with the release of VOCs from leaking piping, valves, connections, flanges, packings, open-ended lines, floating roofs, storage tanks and seals, gas conveyance systems, compressor seals (e.g. ethylene and propylene compressors) and pressure relief valves. They are also associated with loading and unloading operations of raw materials and chemicals (e.g. cone roof tanks), preparing and blending of chemicals (e.g. preparations of solutions of polymerization aids and polymer additives) and liquid effluent treatment units. The design of the process should be done keeping in view the need for minimizing fugitive emissions of toxic and hydrocarbon gases.

The following are the effective prevention measures:

1. In the production of polyethylene monomer leakages from reciprocating compressors employed in high-pressure polyethylene plants, should be recovered and recycled to the low-pressure suction stage.
2. In the manufacturing process of polyvinyl chloride, opening of reactors for maintenance should be kept to the minimum and automatic cleaning system should be adopted.

Particulate Matter

Emissions of particulate matter (i.e. polymer fines and/or additives and antistick agents, etc.) go along with polymer drying and packaging operations. Other sources from which particulate matter emission occurs include pellet conveyance, transfer and dedusting. The following measures will help very much in the management of particulate matter.

1. Optimization of dryer design
2. Use of gas closed loop
3. Reduction of particulates at source and capture through elutriation facilities
4. Installation of equipment viz. electrostatic precipitators, bag filters or wet scrubbing tower
5. Installation of automatic bagging systems and efficient ventilation in packaging operations
6. Good housekeeping

Venting and Flaring

Venting and flaring are important safety measures applied in the facilities for the manufacture of polymers, to make sure that all process gases, coming from storage as well as from process units are safely disposed of in the event of a safety disc or valve opening, emergency power or equipment failure or other plant upset conditions. Emergency discharges from reactors and other crucial process equipment should be conveyed to blow-down tanks where the reactants could be recovered (e.g. by steam or vacuum stripping) before the treated effluents are discharged, or through scrubbing and high-efficiency flaring. Some of the measures to contain pollution are given here.

1. Ethylene vented from high pressure, low density polyethylene (LDPE) plants cannot be conveyed to the flare due to opening of the reactor safety discs at high pressures, but should be vented to the atmosphere through a stack, after having been diluted with steam and cooled by water scrubbing to minimize risks of explosive clouds.
2. Pressure safety valves should be used in polymerization plants to effect a reduction on the quantity of chemicals released from an overpressure/relief device activation where release goes directly to the atmosphere.
3. In the production of polyvinyl chloride the occurrence of emergency venting from the polymerization reactors to atmosphere owing to runaway reaction should be minimized by one or more of the following techniques:
   (a) Specific control instrumentation for reactor feed and operational conditions
   (b) Chemical inhibitor system to stop the reaction
   (c) Emergency reactor cooling capacity
   (d) Emergency power for reactor stirring, and
   (e) Controlled emergency venting to VCM recovery system
4. In case foaming takes place during emergency venting, it should be be reduced by antifoam addition, to avoid plugging of venting system.
5. During emergency venting, the content of the reactor should be discharged to a blow-down tank and steam stripped before disposal.
6. In the production of acrylic-latexes, emergency venting to flare system from reactors due to runaway polymerization should be prevented by one more of the following:

(a) Continuous computer controlled addition of reactants to the reactors, based on actual polymerization kinetics
(b) Chemical inhibitor system to arrest the reaction
(c) Emergency reactor cooling capacity
(d) Emergency power for reactor agitation, and
(e) Discharge of reactor content to a blow-down tank.

Acid Gases

Hydrogen chloride traces, originated from the hydrolysis of chlorinated organic compounds by the catalyst, can be found in exhaust air from drying of polymers produced by ionic catalysis. Although acid is normally present at low level, gas stream testing is recommended. If concentration levels become appreciable, wet scrubbing can be done to prevent contamination of the environment.

Dioxins and Furans

Gaseous, liquid and solid waste incineration plants are typically present as one of the auxiliary facilities in polymer manufacturing plants. The incineration of chlorinated organic compounds (e.g. chlorophenols) could give rise to dioxins and furans. Also the formation of furans and dioxins can be facilitated, by certain catalysts in the form of transition metal compounds like copper.

The following prevention and control strategies have been found to be useful.

1. Operation of incineration facilities according to approved technical standards
2. Maintaining proper operating conditions such as adequately high incineration and flue gas temperatures, to prevent the formation of dioxins and furans

LIQUID EFFLUENTS

Industrial Process Liquid Effluents

Process liquid effluents from plant units, may contain hydrocarbons, monomers and other chemicals, polymers and other solids (either suspended or emulsified), surfactants and emulsifiers, oxygenated compounds, acids, inorganic salts and heavy metals.

The liquid effluent management strategies which have been observed to be quite useful are listed here.

1. Liquid effluent containing volatile monomers, e.g. VCM, styrene, acrylonitrile, acrylic esters, vinyl acetate, caprolactum and/or polymerization solvents (e.g. condensate from steam stripping of suspensions or latexes, condensate from solvent elimination or liquid effluents from equipment maintenance) should be recycled to the processes where possible or otherwise subjected to treatment by flash distillation or equivalent separation to get rid of VOC, before letting it to the facility's liquid effluent treatment system.

2. Organics need to be separated and recycled to the process, when possible or subjected to incineration.
3. Non-recyclable contaminated streams, such as liquid effluents originated from polyester or from thermoset polymer plant, should be catalytically or thermally incinerated.
4. To the extent possible biodegradable emulsions and suspension polymerization aids should be chosen, as they enter the effluent stream during the process of recovery of polymer.
5. Spent reactant solutions should be sent to specialized treatment for disposal.
6. Acidic or caustic effluents from dimineralized water preparation should be treated by neutralization before being let into the unit's effluent treatment system.
7. Oily effluents, such as process leakages need to be collected in closed drains, decanted and discharged to the unit's effluent treatment system.

Process Effluent Treatment

Source segregation and pre-treatment of concentrated effluent streams are the techniques for treating industrial process effluents in the petroleum-based polymer manufacturing sector. Effective effluent treatment steps include, grease traps, skimmers, dissolved air floatation or oil-water separators for separation of oils and floatable solids, filtration for solids which can be filtered, sedimentation for suspended solids reduction employing clarifiers, biological treatment such as aerobic treatment for reduction of soluble organic matter (BOD), chlorination of effluent incase disinfection is needed, dewatering and disposal of residuals in designated hazardous effluent landfills.

There is a need for additional engineering controls for (1) containment and treatment of volatile organics stripped from various unit operations in the liquid effluent treatment system, (2) advanced metals removal processes employing membrane filtration or other physical/ chemical treatment technologies, (3) removal of non-biodegradable COD and recalcitrant organics using activated carbon or advanced chemical oxidation, and (4) reduction in effluent toxicity employing appropriate technology like reverse osmosis, ion-exchange, activated carbon, etc.

Solid Polymer Wastes

Polymer wastes are generated during normal plant operation (e.g. latex filtering and sieving, screening and granule grinding), start-up and maintenance and emergency shutdowns of polymer processing equipment.

The measures for pollution prevention control are given here.

1. Waste streams can be recycled or reused where possible in lieu of disposal. One recycling option is sale of wax to wax industry.
2. Some treatment technique to get rid of and separately recover VOCs (e.g. steam stripping).

3. Segregation and storage in a safe location. Polymer wastes like oxidized polymer recovered during dryer maintenance, process plant crusts without anti-oxidants might be unstable and susceptible to self-heating and self-ignition. These wastes can be stored in a safe manner and disposed of by incineration as early as possible.

Process Safety

Process safety management would involve the following actions:

1. Physical hazard testing of materials and reactions.
2. Conducting hazard analysis studies to review the process chemistry and engineering practice including thermodynamics and kinetics.
3. Scrutinization of preventive maintenance and mechanical integrity of the process equipment and utilities.
4. Imparting training to operators, and
5. Development of operating instructions and emergency response procedures.

In the following section, process safety measures applicable to specific production processes in the polymer industry are presented.

Polyethylene production

In the production of polyethylene, a specific process hazard has a bearing on the possible release of large quantities of hot ethylene to the atmosphere and subsequent cloud explosion. Leaks from gaskets or the maintenance operations result in accidents. In the case of LDPE production plants, accidental events can include opening of the safety disc of the reactor and explosion of the high pressure separator. The following are a few process safety measures:

1. Ethylene vented due to opening of the reactor safety discs at high pressure cannot be conveyed to the flare, but should be vented to the atmosphere by a short stack, after diluting it with steam and cooling with water scrubbing to avoid risks of explosion clouds.
2. Product decomposition in tubular reactors should be prevented from occurring through heat transfer, temperature profile control, high-speed flow and good pressure control.
3. Explosion of high pressure separators should be prevented by vessel reactor design measures, addition of peroxides in correct proportion, careful control of polymerization temperature, quick detection of uncontrolled exothermic reactions and rapid isolation/depressurizing and good maintenance of reactors and separators.

In the case of high density polyethylene (HDPE) production process fire hazards originate from high-pressure and high-temperature conditions in the polymerization reactor and desolventizer operating at a temperature close to the self-ignition temperature of the solvent, together with the high flow rates of hydrocarbon solvent.

In HDPE slurry process a spill from the reactor can give rise to an explosive cloud owing to flash evaporation of isobutene and propylene. Adoption of recognized engineering standards for equipment and piping design, good maintenance, plant layout and location/frequency of emergency shut-off valves will help in the prevention of spills and explosive clouds.

PVC production

The opening of pressure safety valves of a reactor due to runaway polymerization can bring about accidental venting to the atmosphere of vinyl chloride monomer (VCM) with the subsequent formation of an explosive and toxic cloud. Preventive steps include degassing and steam flushing of reactor before opening.

VCM is easily oxidized by air to polyperoxides during recovery operations after polymerization. After recovery, VCM is held in a pressurized or refrigerated holding tank. A chemical inhibitor, such as a hindered phenol is sometimes added to prevent polyperoxide formation. Under normal circumstances, any polyperoxide formed is kept dissolved in VCM, where it reacts slowly and safely to form PVC. However, if liquid VCM containing polyperoxides is evaporated, there is a possibility of polyperoxides precipitating and decomposing exothermally with the attendant risk of explosion and resultant toxic cloud.

Batch polymerization process

Batch polymerization can give rise to a hazard of runaway polymerization and reactor explosion in the event of incorrect dosing of reactants or failure in the agitation or heat exchange systems. These can be prevented by process safety measures such as limiting the practice of batch polymerization and the application of process control methods including the provision of back-up emergency power, cooling, inhibitor addition systems and blow-down tanks.

Compounding, finishing and packaging processes

Compounding, finishing and packaging operations pose risks of fire in blenders and in extruders (in case the polymer is overheated) and in equipment involving mixtures of polymer powders and air, such as dryers, pneumatic conveyers and grinding equipment. Use of recognized electric installation standards including grounding of all equipment and installation of specific fire fighting systems can be adopted.

CASE STUDY 3

An Environmental Assessment of the Process of Manufacture of Polyvinyl Chloride

Minimization of pollution can be realized in practice by adequate process design. Studies have proved that proper design and operation can cut down the emissions of pollutants by as much as 50 per cent in majority of the processes. There are no fixed rules for minimizing pollution. Each case has to be treated on its own merit. However, stepwise approach coupled with innovation will definitely come up with a good result. A six-step approach suggested is as follows:

1. Study the basic process critically and identify the waste streams. Find out whether there is any alternate process with less waste. If both the processes emit pollutants within the prescribed limits, compare their economic aspects when the cost of treating waste is considered.

   A process involving a highly toxic substance is not to be preferred even if this substance is not a part of the waste stream.
2. Choose the design which will keep the materials in the best way inside the equipment. Though this is very apparent, yet according to various studies, 50 per cent or more of material losses and consequent pollution are due to leaks and spills. Containment is the key to minimizing pollution control cost and a good design makes a lot of difference.
3. The third important point is to keep the quantity of wastes to a minimum. Also, unwarranted pollution should be avoided. It is obvious that large volumes incur more expenditure and low concentrations make removal more different.
4. In any process, accidental losses are bound to happen at some stage or the other; such accidental losses should be anticipated and adequate precautions should be taken to minimize pollution. This is important for both safety and environmental control and the project should be reviewed from both perspectives at the same time.
5. All effluents from a process must be analyzed. This is done to ensure the success of the control system which would otherwise be a failure if designed based on insufficient or inaccurate data.
6. The remaining problems should be dealt with by the pollution control experts.

The environmental evaluation of the processes of the manufacture of polyvinyl chloride is presented in the following section:

In this era of increasing use of plastics and simultaneous increase in health and environmental concerns, the Vinyl Chloride (VC) industry is of particular interest. The principal use of vinyl chloride is in the production of Polyvinyl Chloride (PVC). PVC is a versatile polymer which is widely used in the production of plastic materials, such as floor tiles, pipes, electrical insulation, etc. A great majority of PVC resins used to manufacture PVC plastics are homopolymers of vinyl chloride.

The plastics industry may be looked at from the environmental perspective and the sources of pollution can be identified. Once the sources are identified, some remedial measures to minimize ecological hazards can be suggested. It may be mentioned in this context that the carcinogenic nature of vinyl chloride may curtail its use.

Polyvinyl Chloride Manufacture

PVC is manufactured by four types of processes. Three of them are batch processes and the fourth one is continuous process. The batch processes which account for the bulk of the production are suspension polymerization, dispersion or emulsion polymerization and bulk polymerization. The solvent polymerization is a continuous process. A brief outline of the various commercial processes with the sources of pollution indicated, is given here.

Suspension polymerization

Figure C3.1 represents the simplified flow diagram for the suspension process. The general overall process involves the suspension of liquid vinyl chloride over a continuous water phase.

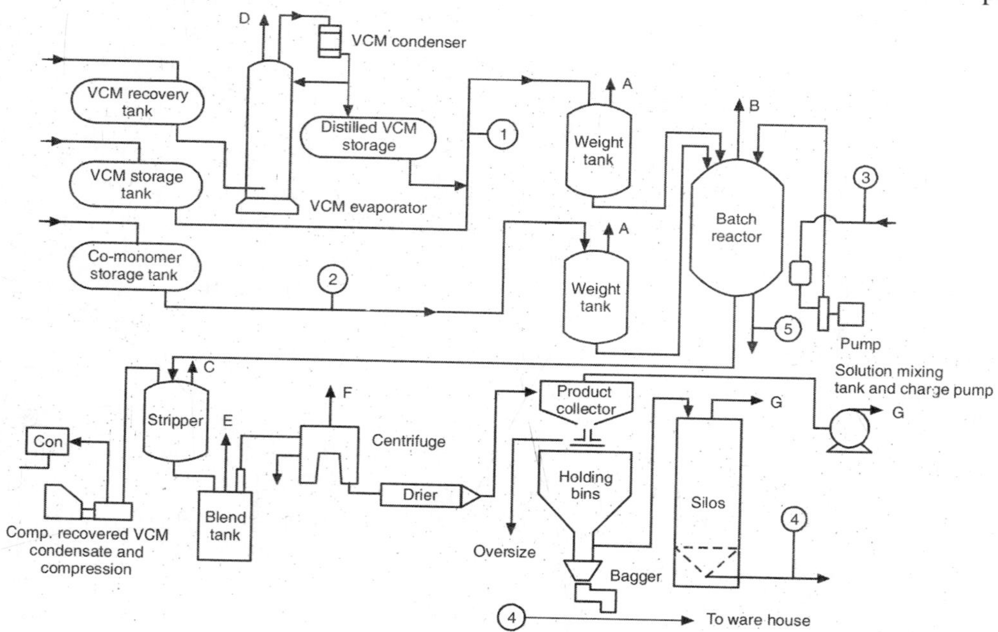

Figure C3.1 Suspension process.

A free radical catalyst dissolved in the Vinyl Chloride Monomer (VCM) feed is used to initiate the polymerization reaction. VCM droplets are kept under suspension with continuous agitation. After the polymerization reaction has been completed (85–90 per cent), the monomer is stripped from the slurry. The slurry is then transferred to the blend tank, passes through centrifuge to remove as much water as possible and then into a rotary drier, the discharge of which is classified and stored.

Excellent heat transfer rates are attainable. The major disadvantage is that water must be separated from the polymer.

Emulsion (dispersion) polymerization

The emulsion process (Figure C3.2) is similar to the suspension process. PVC produced by the emulsion process tends to have a higher molecular weight. Since the particle sizes obtained are much smaller than those obtained by suspension process spray driers are generally used.

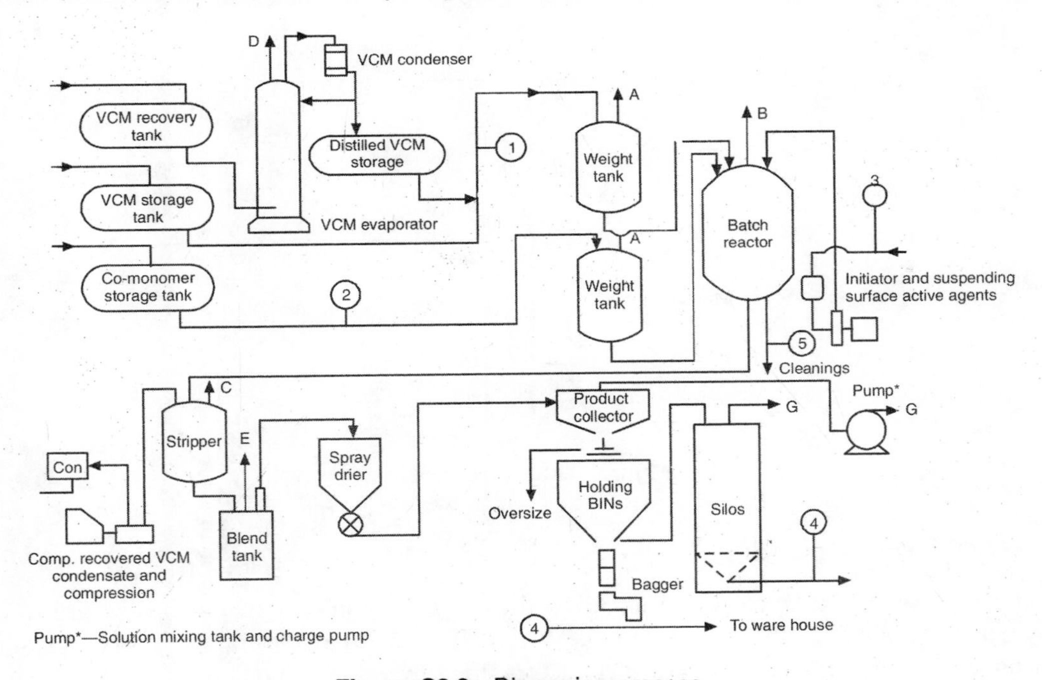

Figure C3.2 Dispersion process.

Bulk polymerization

The Pickiney-Saint Gobain process consists of making seed PVC from liquid VCM in an autoclave using active initiators at 40° to 70°C and pressures of 70 to 170 psi. Polymerization is carried out to completion (85 to 90 per cent) in a large autoclave. Reflux of part of the VC provides better temperature control. The residual monomer is vacuum stripped and recovered. The final classification and storage operations are similar to those employed in the proceeding two processes (Figure C3.3).

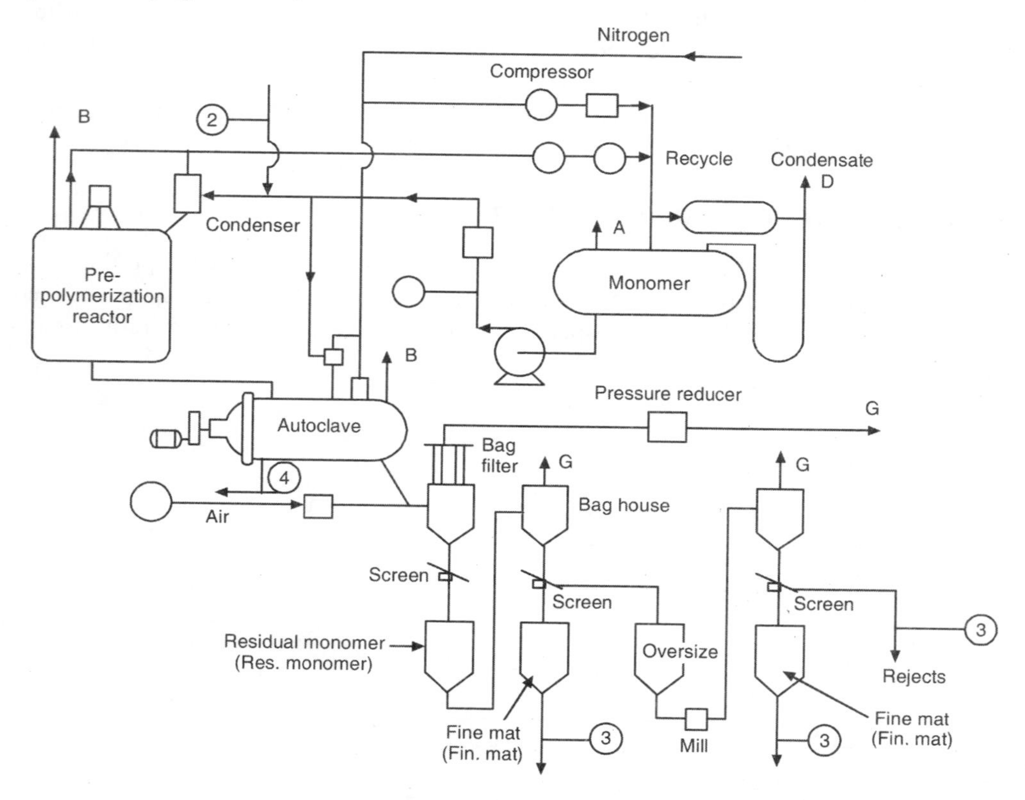

Figure C3.3 Bulk process.

Solvent polymerization

In this process, a mixture of a solvent and vinyl chloride monomer is continuously charged into the autoclave along with the initiator, at 40°C. PVC is filtered from the slurry drawn of continuously from the reactor. The filter cake is dried by flash evaporation.

A volatile solvent (like n-butane) if used, aids in the easy removal of unreacted VCM. The absence of water simplifies the recovery of the final product and is a major advantage (Figure C3.4).

Environmental Problems of the Industry

The occupational exposure problems faced by the workers, in plants handling VCM and PVC came to light few decades ago. The US Environmental Protection Agency sought to identify the environmental problems resulting from the manufacture and use of vinyl chloride (VC) and

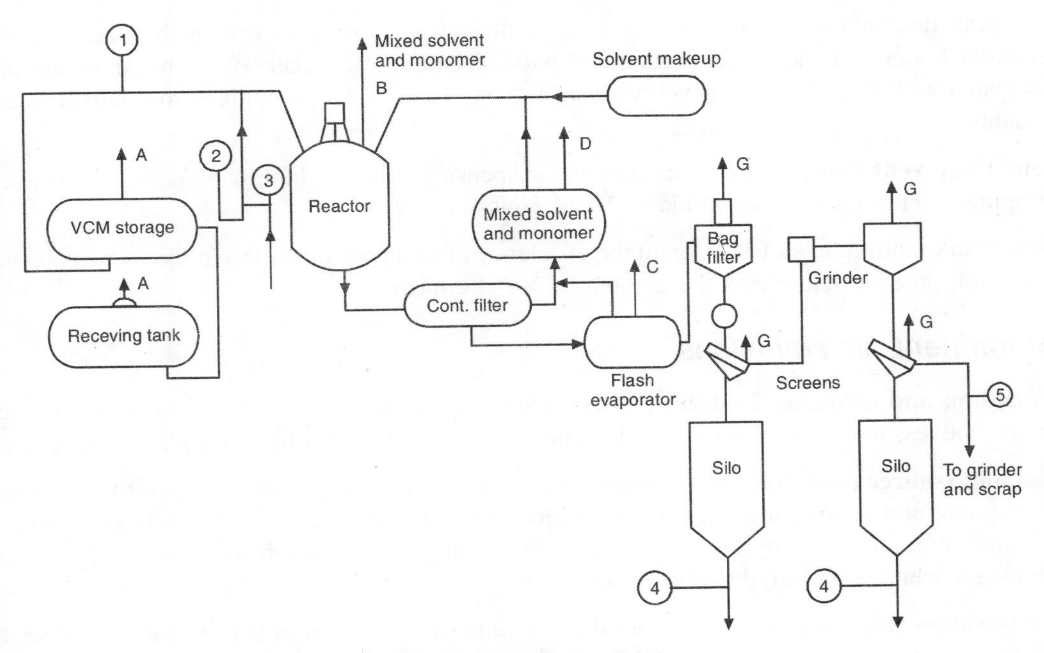

Figure C3.4 Solvent process.

PVC. While air, water and solid waste disposal are all possible routes for entry of VC into the environment, the agency concluded that the air route poses the most significant environmental problem.

Toxicology studies have implicated VC as a human chemical carcinogen. Incomplete combustion of PVC products can also result in the emission of VC as entrapped monomer. PVC products not being easily biodegradable are harmful to the ecology. In the US it is believed that 90 per cent of all VC atmospheric emissions emanate from the PVC plants. With present-day technology, it is believed that these emissions can be reduced by more than 75 per cent.

Exposure standards in the work place have been set by the US Occupational Safety and Health Administration (OSHA) for VC.

Environmental control of air pollution—Continuous air emission

VCM recovery system inerts (source area D): This results from the necessity to purge inerts from the monomer recovery system. High concentrations of VC are present in this major emission stream. Emissions can be reduced by condensing under low pressure and low temperature cooling (10–30° F). It can be most effectively reduced by using a vent scrubber or carbon absorber.

Drier stacks and other solid handling vents (source area G): Here VCM is present in low concentrations (< 0.1 per cent) and in large volumes. The amount of VCM depends on the effectiveness of the stripping operation upstream.

Spray drier exhaust gas flow is similar to that of a rotary drier except that the stream is generally hotter and more nearly saturated with moisture. Particulate PVC emissions are also associated with these vent streams, cyclones and bag filters can reduce them to relatively small amounts.

Centrifuge vent (source area F): For the suspension process the effectiveness of upstream stripping determines the magnitude of VCM emission.

Blend tank (source area E): Again the efficiency of stripping operation in the suspension and dispersion process determines the amount of VCM emitted.

Intermittent air emissions

Unloading and charging facilities (source area A): Losses of VCM at unloading facilities, pumps, valves, meters, etc. is inevitable. These can be minimized by good plant practices.

Reactors (source area B): If a runaway reaction occurs, it is generally, stopped by releasing the pressure and venting to stacks. Periodic cleaning of reactors (every 2–6 batches) also result in some emission. New improved methods of cleaning reactors such as water jet or solvent cleaning systems greatly reduce these emissions.

Safety valves (includes in source area B): A runaway condition in the reactor can cause the relief values to open resulting in VCM emissions.

Fugitive emissions

Source areas may be leaks at pumps, flanges, filters, seals, strainers, etc. Only good maintenance can keep these fugitive emissions low at all times. Some of these emissions are particulate PVC.

It has been observed that the continuous solution polymerization process is environmentally safer than the other three. Also VCM emissions for the bulk process are seen to be lower than that for suspension process. Since the bulk process uses no water, there are no drier or waste water emissions of VC. However, emissions equivalent to the emissions from the drier in the suspension and dispersion process take place in all operations downstream of the reactors. Emissions are also lower in the solvent process since it is a continuous process.

Control Technology for Air Emissions

The control techniques that can be applied to reduce VC emissions, fall into the following general categories:

1. Add-on type control systems such as sorbers, refrigeration systems or incinerators which reduce emissions from point sources.
2. Installation of effective seals on rotating or reciprocating shafts or by enclosing the equipment.
3. Reducing emissions by altering the manufacturing process such as improving the stripper performance to reduce emissions from slurry blend tanks, centrifuges and driers

PVC particulate emissions can be controlled by installation of centrifugal separators and fabric filters.

In a process waste water entering the treatment ponds, there can be as much as 1.33 milligrams of VC per gram of water. Reactors, centrifuges, water seals in the compressors and the recovery system are all sources of in-process waste water. All the VC in the waste water is released to the atmosphere in the treatment ponds. This emission source can be controlled by stripping out the VC in stripping column prior to the treatment ponds. Scrap polymer must be recycled or destroyed. However, HCl can be recovered by pyrolysis of this scrap polymer.

The pollution from the PVC plants may be minimized by adopting following measures:

1. It is economical to recover most of the unreacted VCM as the polymerization goes to 95 per cent completion. About 99 per cent of VCM is recoverable from the latex at 5 inches of mercury vacuum absolute and also if the vapours are compressed and chilled (–30°F, 70 Psig).
2. The vents to remove inerts from the VCM recovery system is a major source of emission. A good scrubbing system that would enable easy recovery of VCM is necessary.
3. A good solvent cleaning system for cleaning the reactors is essential. Although dimethyl formamide and tetrahydrofuran work well, the problem has been to separate the solvent from the dissolved PVC economically.
4. The attendant problem of foaming, during the stripping operation requires the development of efficient chemical foam or mechanical foam breaking devices, if emissions are to be reduced.

CASE STUDY 4

Waste Minimization and Pollution Prevention in Sugar Industry

Some of the sugar mills are based on sugar beet and others use sugar cane as the raw materials. Most of the sugar mills are located in agricultural areas, close to sugar cane fields from where the raw material is procured. A number of chemicals including lime, sulphur, phosphoric acid, bleaching powder, poly electrolytes, floatation aid and decolouring agents find use in sugar making process. The quantity of these chemicals is likely to vary from mill to mill due to the variations in the type of process.

THE ENVIRONMENTAL POLLUTION PROBLEMS

The environmental challenge for the sugar mills is associated with liquid effluents, gaseous effluents and solid wastes. There are three major departments in a sugar manufacturing facility. They are mill house, process house and boiler house. Table C4.1 presents the main input of each stage and the important effluent streams and by-products that emerge from it. For the production of industrial alcohol, the distillery stage is added.

Solid Waste

Two types of solid wastes are generated during the manufacture of sugar. Bagasse is produced in the mill house in a quantity of about 30 per cent of the crushed cane. The bagasse consists of 50 per cent moisture. Press mud or filter cake is produced in vacuum filters and press filters. The mud is produced in a range of 3–8 per cent of the crushed cane, depending on the nature of the process of manufacture of sugar. The mud contains nearly three quarters of moisture. On dry basis mud normally contains about 70 per cent organic matter and about 30 per cent minerals.

Press mud is normally used by nearby farmers as manure although in some situations its application as manure is preceded by bio-compostings. Bagasse is used as fuel for boilers. It is estimated that 70 per cent of the power requirements of sugar mills is fulfilled by this way. Bagasse also finds applications in for chip-board and paper manufacture.

Table C4.1 Wastes Generated from During Sugar Production

Process stage	*Main inputs*	*Wastes and by-products*
Mill house	Sugar cane	• Waste water containing suspended solids and oil content • Water employed for cleaning the floor containing sugar spills • Bagasse
Process house	Sugar juice	• Filter cake • Washing of different equipments, such as evaporator, juice heater, vacuum pan, clarifiers, etc.; generate effluents with high BOD_5, COD and TDS concentrations • Molasses
Boiler house	Bagasse and furnace oil	• Fly ash • Smoke • Flue gases • Liquid effluent from scrubbers
Cooling pond	Water and chemicals	• Liquid effluent
Distillery	Molasses	• Waste water (stillage) containing very high BOD_5, COD and suspended solids

Air Emissions

The major source of air pollution in sugar mills relates to boiler emissions. Boilers are operated under three different conditions; fuel oil only, bagasse only, and mixed fuel (fuel oil and bagasse). The choice of fuel source significantly changes the nature of emissions. Sulphitation process in sugar refining is also responsible for generating SO_x gases. Most of the sugar mills are equipped with dust collectors or cyclones to capture the particulate matter. However, continuous attention of the mills is needed for effective and effluent utilization of these installations. The following data highlight some important points in respect of air emissions from sugar mills:

1. Hydrogen sulphide, sulphur dioxide, oxides of nitrogen and trace metals could be well maintained under stipulated standards.
2. Although carbon monoxide levels are generally under the stipulated standard, significant scope for improvement remains and the mills should carefully monitor their own emission levels.
3. The variation in the data for particulate matter could be very high. Individual mills should have their particulate matter emissions profile studied in order to comply with the stipulated standards.
4. In boilers that use fuel oil, smoke levels are at or above the set limits.
5. Bagasse combustion also generates ash and fly ash. Flue gas contains about 4500 mg/cu.m of fly ash on an average. This is a visual nuisance as well as a health concern. These problems are particularly high in boilers which are not equipped with scrubbing or cyclone systems.

Liquid Effluent

The liquid effluent has normally very high pollution levels. They are particularly severe in the case of effluents from distilleries. For instance, the COD level is over 1000 mg/l and the observed level for the distillery is much higher. The situation is equally bad for pH value, BOD_5 and TSS.

Other Health Hazards

Bagasse dust, fly ash, and high noise levels (reaching upto 110 dB) are particular health hazards in sugar mills. Excessive exposure to fly ash and bagasse dust may cause irritation to eyes, asthma and other respiratory disorders. Dust prone areas, including cane preparatiion and the boiler house, need to be carefully monitored. Nuisance dust levels should be controlled properly. Dermatitis or skin disease is a major health complaint resulting from chemical burns and contact with lime and sugar. Toxic gases including sulphur dioxide, caustic fumes are emanated at various stages of the process and are health hazards.

WASTE MINIMIZATION AND POLLUTION PREVENTION METHODS

A variety of steps can be taken to address the environmental pollution problems faced by sugar mills and distilleries. The most immediate of these relate to waste reduction at source which are presented here.

Waste reduction at source; In-house improvement steps

1. Flow measurement through (flow meters) and monitoring at inlet and outlet of each consumer unit at the mill for effective water management practices.
2. Use of optimum inhibition rate to save energy in terms of steam consumption and to reduce organic and hydraulic load from the process house.
3. Dry cleaning of mill floors with bagasse.
4. Efficient functioning of evaporators will reduce waste disposal problems and enhance sugar recovery. Overloading of evaporators and vacuum pans, boiling at excessive rates, operating them at incorrect liquid levels and variation of vacuum lead to a loss through condenser water. Improper design of these units, particularly the entrainment separator may result in irregular boiling and splashing.
5. Recycling of cooling and condenser water.
6. The simple measure of controlling spillovers of molasses can very significantly lessen the organic pollution content of the effluent stream.
7. Segregation of oil from other effluent will allow for the recovery and reuse of lubricating oil and reduce soil contamination when liquid effluent is applied for irrigation.
8. Controlling the mixing of filter mud with effluent can very significantly reduce the organic and inorganic pollution content of waste water stream.
9. Routine inspection of units particularly pumps, conveyors, pipes and other vessels will help much in reducing waste at source.

10. Effecting a reduction in the amount of water used for floor sweeping and washing by recovering water from various mill processes. It may be reused for cleaning purposes.
11. Detaining filter cloth washing in a holding tank for a short time before being allowed to mix with other effluents from the mill, will reduce the contamination in the effluent stream.
12. Installation of circular mist eliminators and demisters made of stainless steel in the multiple effect evaporators can eliminate sugar entry in the condensate water.
13. Bagasse management is of utmost importance in establishing overall energy efficiency in the mill. Steam and power generation and reduction in fuel oil consumption are largely dependent on an adequate supply and efficient utilization of bagasse. Benefits will also result from ensuring that maximum moisture has been removed before bagasse is used in the boiler.
14. As far as air emissions are concerned, the first step should be to set up a system of regular monitoring of stack emissions with periodic boiler tune-ups. This can considerably increase boiler efficiency and bring about minimization of emissions.

The technologies and methods for air and noise pollution control in sugar mills are given here.

Air pollution control for particulate matter including fly ash involves the use of

1. Settling chamber
2. Cyclones
3. Wet collectors
4. Electrostatic precipitators, and
5. Gas scrubbers.

Noise pollution control methods include

1. Sound reduction at source, e.g. silencers, design of fans, etc.
2. Interrupting the path of the sound, e.g. sound barrier.
3. Protecting the recipient ear muffs, plugs.

The most important environmental challenge faced by sugar mills and distilleries is in respect of liquid effluents.

Table C4.2 presents details about the major technologies that could be employed to treat liquid effluents from sugar mills and distilleries.

Lagoons, owing to a large land area requirement, odour problem, possible ground water contamination may not be a preferable solution liquid effluent treatment. Trickling filter owing to low efficiency is also not recommended for sugar mills in general. However, with low pollution load this option can also be explored. Activated sludge system needs a very large amount of energy for aeration and therefore, operation and maintenance cost is very high for this option. Upward aerobic sluge blanket (UASB) system for treatment of highly concentrated liquid effluent from agricultural industries is increasingly popular and liquid effluent from sugar mills and distilleries can be treated for significant reduction in contamination levels. The attendant advantage is the methane gas produced in UASB system can be used as an energy source.

Table C4.2 Brief Review of Liquid Effluent Treatment Technologies

Solutions	*Technical characteristics of the treatment system*	*Operational characteristics of the treatment system*
Lagoons	• Anaerobic lagoons are deep earthern basins employed for high strengths organic wastewater with high solid concentration • Facultative lagoons are earthern basins filled with screened or primary effluents in which stabilization of effluent is brought about by a combination of aerobic, anaerobic and facultative bacteria • Anaerobic lagoons are large, shallow earthern basins employed for treatment of liquid effluent by natural processes involving both algae and bacteria. • Maturation ponds are low-rate stabilization ponds usually designed to provide for secondary effluent polishing and seasonal nitrification	• BOD_5 loading (kg/cu.m/d) is least efficient • BOD_5 removal efficiency in the range 85–90 per cent • Energy requirement for aeration (in kWh/kg) BOD treated is moderately efficient • BOD_5 treated is moderately efficient • Hydraulic detention time is quite high • Mechanical complexity is low • Reactor resilience for power failure and shock loads are moderate to high • No by-product • On-site environmental impacts are soil infiltration and aerosols dispersion • Large land requirement • Skilled human resource is required • Moderate frequency of repair and maintenance
Trickling filters	• Effluent flows from top to bottom, dispersed over filter material (stones, lava or plastic) during which soluble compounds are removed and to lesser extent, solids are taken up into the bio-film adhered to the carrier material	• BOD_5 Loading kg/cu.m per day is least efficient • BOD_5 removal efficiency is in the range of 85–90 per cent • Energy requirement for aeration in kWh/kg BOD treated is most efficient (natural ventilation) • Hydraulic detention time is most efficient (recirculation is required) • Low mechanical complexity • Reactor resilience for power failure and shock loads is moderate • No by-product • On-site environmental impact is the presence of insects • Small land requirement • Skilled manpower requirement is needed • Low frequency of repair and maintenance
Upflow Anaerobic Sludge Blanket (UASB) reactor	• The basic idea of this system is that the flocs of anaerobic bacteria will have a tendency to settle under gravity, when applying a moderate upflow velocity of water. In this way no separate sedimentation tank is needed. • The effluent passes the reactor from the bottom to top. To ensure adequate contact between the incoming effluent and the bacteria in the sludge layer, the effluent is fed evenly over the bottom of the reactor. Further mixing is brought about by the generation of the gas. • The organic compounds are consumed by anaerobic bacteria during passage of effluent through the sludge layer and produces bio-gas.	• BOD_5 loading (kg/cu.m./d) is very efficient • BOD_5 removal efficiency is in the range 80–90 per cent • Energy requirement for aeration (kWh/kg) BOD treated, is most efficient only for pumping. • Most efficient hydraulic detention time • Low mechanical complexity • Moderate reactor resilience for power failure and shock loads • Biogas by-product • No on-site environmental impact • Small land requirement • Requirement of highly skilled human resources • Low frequency of repair and maintenance

(*Contd.*)

Table C4.2 Brief Review of Liquid Effluent Treatment Technologies (*Contd.*)

Solutions	*Technical characteristics of the treatment system*	*Operational characteristics of the treatment system*
Activated sludge treatment (sequential batch reactor)	Many variations of activated sludge treatment exist depending on load characteristics. Sequential batch reactor is most appropriate for high organic pollution loads. It is most successfully applied if hourly flow is low. • System comprises only aeration tank (operated as fill and draw system) mechanical surface aerators. Aeration and sedimentation occur in the same reactor on the cyclical principle given here. Feeding and the aeration of the reactor during a certain period, switching-off of aeration, followed by settling of sludge and discharge of the effluent	• BOD_5 loading kg/cu.m/d is quite efficient • BOD_5 removal efficiency is in the range 85–95 per cent • Energy requirement for aeration (kWh/kg) BOD treated is least efficient • Moderately efficient hydraulic detention time • High mechanical complexity • Moderate reactor resilience for power failure • No by-product • On-site environmental impacts are aerosol dispersion and noise • Moderate land requirement • Requirement of highly skilled human resources • Very high frequency of repair and maintenance

Effective Methods for Sugar Sector

Liquid effluent is the most important environmental challenge confronting sugar mills. This problem requires major focus and calls for a lot of investment and attention. It is desirable to make in-house improvements towards minimization of wastes at source. There is so much of potential in this, to reduce the pollutant as well as the hydraulic load to a level at which end-of-pipe treatment technologies may then be able to bring it down further to levels near or below the set standards. It is obvious that such measures are also to result in net savings for the sugar mills in the long run.

A strategy for liquid effluent treatment is presented here.

Reduction

1. Improvement of filter press to prevent juice leakages. This will bring down the sugar content in the effluent. The juice vapours may also be prevented from seeping into the condenser cooling water by improving the performance of evaporators and pans. This will, in turn, improve the function of the condenser.
2. Setting up of a pit for collecting juice contaminated water to bring about a reduction in the sugar content in washing and cleaning water.
3. Collected water can subsequently be used as maceration of sugar cane milling process.
4. The quality and quantity of pollution sources need to be monitored and followed up in order to prevent additional contaminating effluent.

Recycling

The clean warm water needs to be separated from the contaminated stream for reuse. After cooling down by cooling towers about 90 per cent tail water can be recovered, the remaining 10 per cent can be discharged and replaced by fresh water. After removal of grease and oil, cooled mill water can be recycled for use.

Regeneration

Precipitation and filtration of flue gas washing water will regenerate it into a colourless and transparent state. This regenerated clean water can be put to reuse in the flue gas washing system.

The end-of-pipe treatment in sugar mills is given here.

Combination of lagoons in sequence (anaerobic, facultative and maturation)

For a sugar mill generating 6000–7000 cu.m/day of liquid effluent containing 2000–3000 mg/l BOD_5, it is suggested that to achieve 90 per cent removal efficiency, the lagoons system would need a total surface area of 120 hectares, a total volume of 1.5 million cu.m. and a total retention time of 216 days. The excessive level of retention time, land requirement and high cost of lining make this option inadvisable for sugar mills. However, unlined lagoons, though present serious threat to ground water quality, can also be taken into consideration after making a careful analysis of environmental condition around the mill.

Combined treatment system comprising a UASB reactor and an activated sludge system of processing

When employed alone, the activated sludge system has a high operational cost owing to the energy requirement for aeration. Combination of UASB reactor with this reduces the cost appreciably. The UASB reactor can remove around 80–90 per cent BOD_5; besides this, there is a by-product of this process viz.; methane which can be used as a source of energy for the boiler. After passing through the UASB reactor, the effluent can be made to pass through the activated sludge system for the treatment the remaining BOD_5. The combined system will accomplish the same contaminant removal efficiency as the activated sludge system, but at lower cost. This combined system will bring the present BOD_5 and suspended solids level near or below the set standards.

The end-of-pipe treatment for distillery's effluent

There are two options for the end-of pipe treatment of effluent from distilleries attached to sugar mills. As in the case of effluent from sugar mills, the first option is to place a combination of lagoons in the same sequence as in the treatment of sugar mills' effluents–anaerobic, facultative and maturation. This option is totally outside the realm of those feasible for distilleries. A set of lagoons designed for a hydraulic load of 200–300 cu.m/day, containing 60,000 mg/l BOD_5

and 110,000 mg/l COD, would require 60 hectares of land, 700,000 cu.m of volume and a very long retention time to attain 90 per cent removal efficiency.

Taking into consideration the higher organic load in the effluents from distilleries, the second option is a combined treatment system that consists of a UASB reactor and a sequential batch reactor (activated sludge system). Such a combined system can achieve about 98 per cent overall removal efficiency with a reduced cost of operations.

However, it is felt that although such a system would bring about a large reduction in the organic load in the effluent, it will be considerably higher than the set limits for COD and BOD_5. Notwithstanding this limitation, the combined system is recommended for effluent treatment in the industrial alcohol distilleries, in view of the fact that such a system would drastically reduce the environmental stress being placed by distillery effluent.

Composting of filter mud cake with additional moistening with stillage could be an attractive alternative solution to handle both the waste streams and to produce valuable by-product (fertilizer). Various alternatives can be applied, they are: (1) window composting with natural aeration, (2) static pile composting with forced aeration, and (3) UASB treatment of stillage and static pile composting with forced aeration.

APPENDICES

APPENDIX A

Priority Pollutants

Priority Pollutant	*Type of Chemical Substance*
Acenophthene	Aromatic
Accnapthylene	Aromatic
Acrolein	Organic
Acrylonitrile	Organic
Aldrin	Pesticide
Anthracene	Aromatic
Antimony	Metal
Arsenic	Metal
Asbestos	Mineral
Beryllium	Metal
Benzene	Aromatic
Bensidine	Substitute aromatic
Benzo (a) anthracene	Aromatic
3, 4-Bensofluranthane	Aromatic
Benzo (to) fluranthane	Aromatic
Benzo (ghi) perylene	Aromatic
Benzo (e) pyrene	Aromatic
e-BHC-α	Pesticide
b-BHC-β	Pesticide
r-BHC-α	Pesticide
g-GHC-β	Pesticide
bis 2-chloroethyxy methane	Chlorinated ether
bis 2-chloromethyl ether	Chlorinated ether
bis-chloromethye ether	Chlorinated ether
bis (2-chloropropyl phthalate)	Phtolate ester
Bromoform	Chlorinated alkane
4-Bromophenyl phenyl ether	Chlorinated ether
Butyl benzyle phthalate	Phthalate ester
Cadmium	Metal
Carbontetrachloride	Chlorinated alkane

Priority pollutants	*Type of Chemical Substance*
Chlordane	Pesticide
Chlorobezene	Chlorirnated alkene
Chloroethane	Chlorinated alkane
2-chloroethyl vinyl ether	Chlorinated ether
Chloroform	Chlorinated alkane
2-chlorophenol	Phenol
4-chlorophenyl ether	Chlorinated ester
2-chlorophythalene	Chlorinated aromatic
Chromium	Metal
Chrysene	Aromatic
4,4 DDD	Pesticide
4,4 DDE	Pesticide
4,4 DDT	Pesticide
Dibenzo (a,b) anthracene	Chlorinated aromatic
1,3 Dichlorobenzene	Chlorinated aromatic
3,3 Dichlorobenzidene	Substituted aromatic
1,4 Dichlorobenzene	Substituted aromatic
Dichlorodifluromethane	Chlorinated alkane
1,1 Dichloroethane	Chlorinated alkane
1,2 Dichloroethane	Chlorinated alkane
1,1 Dichloroethylene	
2,4 Dichloro phenol	Chlorinated alkane phenol
1,2 Dichloro propane	Chlorianted alkane
Dieldrin	Pesticide
Diethyl phtalate	Phthalate ester
2,4 Dimethyl phenol	Phenol
Dimethyl phthalate	Phthalate ester
Di n-butyl phthalate	Phthalate ester
4,6 Dinitro-o-cresol	Phenol
2,4 Dinitiro phenol	Phenol
2,4 Dinitro toluene	Substituted aromatic
2,6, Dinitro toluene	Substituted aromatic
Di-n octyl phthalate	Phthlate ester
1,2 Diphenyl hydrozine	Substituted aromatic
A-Endosulfan-α	Pesticide
B-Endosulfan-β	Pesticide
Endosulfan sulphate	Pesticide
Endrin	Pesticide
Endrin aldehyde	Pesticide
Ethyl benzene	Aromatic
Fluoranthene	Aromatic
Fluorene	Aromatic
Hapthalene	Aromatic
Heptachlor	Pesticide

Priority pollutant	*Type of chemical substance*
Heptachlor epoxide	Pesticide
Hexachlorobenzene	Chlorinated aromatic
Hexachlorobenzene	Chlorinated alkane
Hexachlorocyclopentadiene	Chlorinated alkane
Hexachlorethane	Chlorinated akane
Isophorene	Organic
Mercury	Metals
Methyl bromide	Chlorinated alkane
Methyl chloride	Chlorinated alkane
Methylene chloride	Chlorinated alkane
Nickel	Metal
Nitrobenzene	Substituted aromatic
2-Nitrophenol	Phenol
4-Nitrophenol	Phenol
n-Nitrosodimethylamine	Organic
n-Nitrosodi-N-propylamine	Organic
n-Nitrosodiphynylamine	Organic
Para-chlor-meta cresol	Phenol
PCB-1016	Chlorinates biphenyl
PCB-1221	Chlorinated biphenyl
PCB-1232	Chlorinated biphenyl
PCB-1242	Chlorinated biphenyl
PCB-1254	Chlorinated biphenyl
PCB-1260	Chlorinated biphenyl
Pentach, orophenol	Phenol
Phenathane	Aromatic
Phenol	Phenol
Pyrene	Metal
Selenium	Metal
Silver	Metal
2,3,7,8 Tetrachloridibenzo p-dioxin	Chlorinated organic
1,1,2,2, Tetrachloroethane	Chlorinated alkane
Tetrachloroethylene	Chlorinated alkane
Thallium	Metal
Toluene	Aromatic
1,2 Trans Dichloroethylene	Chlorinated alkane
1,2,4, Trichlorobenzene	Chlorinated aromatic
1,1 Trichloroethane	Chlorinated alkane
Trichlorofluromethane	Chlorinated alkane
2,4,6 Trichloro phenol	Chlorinated alkane
Vinyl chloride	Chlorinated phenol
Zinc	Metal

Appendix B

Hazardous Wastes from Non-specific Sources

Industry and EPA hazardous wastes number from non-specific source	*Hazardous wastes*	*Hazard code**
Generic F 001	The following spent halogenated solvents are used in degreasing: tetrachloroethylene, trichloroethylene, methylene chloride, 1,1,1, tricholoroethane, carbontetrachloride and chlorinated fluorocarbons; all spent solvent mixture blends used in degreasing contained before use a total of 10 per cent more of (by volume) one or more of the above halogenated solvents or those solvents listed in F 002, F 003, and F 004 and still bottoms from the recovery of these solvents and spent solvent mixtures.	(T)
F 002	The following spent halogenated solvents are tetrachloroethylene, methylene chloride, trichloroethylene, 1,1,1, trichloroethane, chlorobenzene 1,1,2 trichloroethane, 1,2,2, trifluoroethane orthodichloro benzene, trichlorofluoromethane and 1,1,2 trichloroethane all spent solvent mixture/blends containing before use a total of 10 per cent or more (by volume) of one or more the above halogenated solvent or those listed in F 001, F 004 or F 005 and still bottoms from the recovery of these spent solvents and spent solvent mixtures.	(T)
F 003	The following spent non-halogenated solvents; ethylbenzene, ethyl ether, methyl isobutyl ketone, butyl alcohol, cyclohexonone and methanol, all spent solvent mixtures/blends containing before use one or more solvents listed in F 001, F 002, F 004 and F 005 and still bottoms from these spent solvents and spent solvent mixtures.	(I)

Industry and EPA hazardous wastes number from non-specific source	*Hazardous wastes*	*Hazard code**
F 004	The solvent spent non-halogenated solvents: cresols and cresylic acid and nitrobenzene all spent solvent mixtures/blends containing before use a total of 10 per cent or more (by volume) of one or more of the above non-halogenated solvents of those solvents listed in F 001, F 002 and F 005; and still bottoms from the recovery of these spent solvents and spent solvent mixtures.	(T)
F 005	The following spent non-halogenated solvents: toluene, methyl ethyl ketone (MEK) carbon disulphide, isobutanol, pyridine, benzene, 2-ethoxyethanol and 2-nitropropane: all spent solvent mixtures/blends containing before use, a total of 10 per cent or more (by volume) of one or more of the aforementioned non-halogenated solvents or those solvents listed in F 001, F002 or F 004 and still bottoms from the recovery of these spent solvents and spent solvent mixtures.	(I,T)
F 006	Effluent treatment sludges from electroplating operations except from the following processes: (1) sulphuric acid anodizing or aluminium, (2) tin plating or carbon steel, (3) zinc plating (segregated basis) on carbon steel, (4) aluminium of zinc-and-aluminium plating on carbon steel, (5) cleaning/striping associated with tin, zinc and aluminium plating on carbon steel; and (6) chemical etching and milling of aluminium.	(T)
F 019	Waste treatment sludges from the chemical conversion coating of aluminium	(T)
F 007	Spent cyanide plating bath solutions from electroplating operations	(R,T)
F 008	Plating bath residues from the bottom of plating baths from electroplating operations in which cyanides are employed in the process	(R,T)
F 009	Spent stripping and cleaning bath solutions coming out of electroplating operations in which cyanides are employed in the process	(R,T)
F 010	Quenching bath residues coming out of oil baths from metal heat treating operations where cyanides are employed in the process	(R,T)
F 011	Spent cyanides coming out of salt bath pot cleaning from metal heat-treating techniques (operations)	(R,T)
F 012	Quenching waste water treatment sludges coming out of metal heat treating operations in which cyanides are employed in the process	(R,T)

Industry and EPA hazardous wastes number from non-specific source	*Hazardous wastes*	*Hazard code**
F 024	Wastes, including but not limited to distillation residues, heavy ends tars and reactors clean outs wastes coming out of the process of production of chlorinated aliphatic hydrocarbons, having carbon content from one to five, utilizing free radical catalyzed processes (it may be noted that, this listing does not include light ends, spent filters and filter aids, spent dessicants, effluent treatment sludges, spent catalysts and wastes listed in Appendix C)	(R,T)
F 020	Wastes (excluding waste water and spent carbon from hydrogen chloride purification) from the production or manufacturing use (as a reactant chemical intermediate, or component in a formulating process) or trichlorophenol or tetrachlorophenol or of intermediate employed to make their pesticide derivatives. (It may be mentioned here that the listing given here excludes wastes from the production of hexachlorophene from highly purified 2, 4, 5 trichlorophenol)	(H)
F 021	Wastes (excluding waste water and spent carbon from hydrogen chloride purification) from the production use (as a reactant, chemical intermediate or a component in a formulating process) of pentachlorophenol or of intermediates employed to make its derivatives	(H)
F 022	Wastes (excluding waste water and spent carbon from hydrogen chloride purification) from the manufacturing use (as a reactant, chemical intermediate or component in a formulating process) of tetra, penta or hexachlorobenzenes under alkaline conditions	(H)
F 023	Wastes (excluding waste water and spent carbon coming out of hydrogen chloride purification process) from the production of materials on equipment previously employed for the manufacturing application (as a reactant, component or chemical intermediate in a formulating technique) of tri and tetrachlorophenols (it may be mentioned here that the listing given does not include waste from equipment employed only for the production of hexachlorophene from highly purified 2,4,5-trichlorophenol)	(H)
F 026	Waste (excluding waste water and spent carbon from hydrogen chloride purification) coming out of the production process of materials on equipment previously employed for the manufacturing use (as a reactant, component or chemical intermediate in a formulating process) or tetra, penta- or hexachlorobenzene under alkaline conditions	(H)

Industry and EPA hazardous wastes number from non-specific source	*Hazardous wastes*	*Hazard code**
F 027	Discarded unused formulations comprising tri, tetra or pentachlorophenol or discarded unused formulations consisting of compounds derived from these chlorophenols (it may be noted that in the listing given here, formulations consisting of hexachlorophene syntho-sized from prepurified 2,4,5-trichlorophenol as the sole component are not included)	(H)
F 028	Residues generated from the incineration or thermal treatment of soil contaminated with EPA hazardous waste nos. F 020, F 021, F 022, F 023, F 026 and F 027	(H)

* Definitions of the hazard codes

C denotes corrosivity

E denotes extraction procedure toxicity

H denotes acute hazardous waste

I denotes ignitability

R denotes reactivity

T denotes toxic waste

APPENDIX C

Hazardous Wastes from Specific Sources

Industry and EPA hazardous wastes number	*Hazardous waste*	*Hazard code*
Wood preservation		
K 001	Bottom sediment sludge from the treatment of effluents from wood preserving processes that employ creosote and or pentachlorophenol	(T)
Inorganic pigments		
K 002	Effluent treatment sludge coming out of the manufacturing facility of chrome yellow and organic pigments	(T)
K 003	Effluent treatment sludge coming out of the production unit of molybdate orange pigments	(T)
K 004	Effluent treatment sludge emanating from the manufacturing plant of zinc yellow pigments	(T)
K 005	Effluent treatment sludge coming out of the production unit of chrome green pigments	(T)
K 006	Effluent treatment sludges emanating from the production plant of chrome oxide green pigments (anhydrous and hydrated)	(T)
K 007	Effluent treatment sludge coming out of the production unit of iron blue pigments	(T)
K 008	Over residue from the reduction of chrome oxide green pigments	(T)
Organic chemicals		
K 009	Distillation bottoms coming out of the production facility of acetaldehyde from ethylene	(T)
K 010	Distillation side cuts from the production plant of acetaldehyde from ethylene	(T)
K 011	Bottom streams from the waste water stripper in the manufacturing plant of acrylonitrile	(R,T)

Industry and EPA hazardous wastes number	*Hazardous waste*	*Hazard code*
K 013	Bottom stream coming out of acetonitrile column in the production facility of acrylonitrile	(R,T)
K 014	Bottoms from the acetonitrile purification column in the production unit of acrylonitrile	(T)
K 015	Still bottoms coming out of the distillation column of the benzyl chloride plant	(T)
K 016	Heavy ends or distillation residues coming out of the production plant of carbontetrachloride	(T)
K 017	Heavy ends (still bottoms) coming out of the purification column of the production unit of epichlorohydrin	(T)
K 018	Heavy ends from the fractionation column in the ethyl chloride production facility	(T)
K 019	Heavy ends from the distillation of ethylene dichloride in the ethylene dichloride production unit	(T)
K 020	Heavy ends from the distrllation of vinyl chloride in the vinyl chloride monomer production unit	(T)
K 021	Aqueous spent antimony catalyst effluent from the production unit of fluoronethane	(T)
K 022	Distillation bottom tars coming out of the production plant of phenol acetone from cumene	(T)
K 023	Distillation light ends from the production plant of phthalic anhydride (from naphthalene)	(T)
K 024	Distillation bottoms from the production unit of phthalic anhydride (from naphthalene)	(T)
K 093	Distillation light ends coming out of the production plan for phthalic anhydride (from orthoxylene)	(T)
K 094	Distillation bottoms from the production unit of phthalic anhydride (from orthoxylene)	(T)
K 025	Distillation bottoms from the production plant of nitrobenzene (by the nitration of benzene)	(T)
K 026	Stripping still tails from the production plant of methyl ethyl pyridines	(T)
K 027	Centrifuge and distillation residues from the toluene disocyanate production plant	(T)

Industry and EPA hazardous wastes number	*Hazardous waste*	*Hazard code*
K 028	Spent catalyst from the hydrochlorinator reactor of the production unit of 1,1,1,1 trichloroethane	(T)
K 029	Waste from the product steam stripper in the production unit of 1,1,1-trichloroethane	(T)
K 095	Distillation bottoms from the production unit of 1,1,1-trichloroethane	(T)
K 096	Heavy ends coming out of the heavy end column of the production unit of 1,1,1 trichloroethane	(T)
K 030	Column bottoms of the heavy ends from the combined production facility of trichloroethylene and perchloroethylene	(T)
K 083	Distillation bottoms from the production unit of aniline	(T)
K 103	Process residues from the aniline extraction process in the production of aniline	(T)
K 104	Combined waste water streams generated in the process of manufacture of nitrobenzene/aniline	(T)
K 085	Distillation or fractionation column bottoms from the production plant of chlorobenzenes	(T)
K 105	Separated aqueous streams from the reactor product washing step in the manufacture of chlorobenzenes	(T)
K 111	Product wash waters from the production of dinitrotoluene by the process of nitration of toluene	(C,T)
K 112	Reaction by-product water coming out of the drying column in the production of toluenediamine via hydrogenation of dinitrotoluene	(T)
K 113	Condensed liquid light ends coming out of the purification process of toluenediamine in the production of toluenediamine by the method of hydrogenation of dinitrotoluene	(T)
K 114	Vicinals from the purification of toluenediamine in the production of toluenediamine by the process of hydrogenation of dinitroluene	(T)
K 115	Heavy ends from the purification process of toluenediamine in the manufacture of toluenediamine by the method of hydrogenation of dinitrotoluene	(T)
K 116	Organic condensate coming out of the solvent recovery column in the production of toluene diisocyanide via phosgenation of toluenediamine	(T)
K 117	Effluent from the reactor vent gas scrubber in the production of ethylene dibromide by the process of bromination of ethane	(T)

Industry and EPA hazardous wastes number	*Hazardous waste*	*Hazard code*
K 118	Spent adsorbent solids from the purification process of ethylene dibromide via bromination of ethane	(T)
K 136	Still bottom from the purification process of ethylene dibromide in the production of ethylene dibromide by the technique of bromination of ethane	(T)
Inorganic chemicals		
K 071	Brine purification muds from the mercury cell process in the production of chlorine in which separately prepurified brine is not used	(T)
K 073	Chlorinated hydrocarbon waste from the purification step of the diaphragm cell process employing graphite anodes in chlorinc production	(T)
K 106	Effluent treatment sludge from the mercury cell process in the production of chlorine	(T)
Pesticides		
K 031	By-product salts generated in the production of MSMA and cacodylic acid	(T)
K 032	Effluent treatment sludge from the production of chlordane	(T)
K 033	Waste water and scrub water from the chlorination of cyclopentadiene in the production plant of chlordane	(T)
K 034	Filter solids from the filtration of hexachloro-cyclopentadiene in the production unit of chlordane	(T)
K 097	Vacuum stripper discharge from the chlordane chlorinator in the production plant of chlordane	(T)
K 035	Effluent treatment sludges generated in the manufacturing plant of creosote	(T)
K 036	Still bottoms from toluene reclamation distillation in the production unit of disulfoton	(T)
K 037	Effluent treatment sludges from the production of disulfoton	(T)
K 038	Effluent from the washing and stripping of phorate production	(T)
K 039	Filter cake from the filtration of diethylphosphor odithioc acid in the production of phorate	(T)
K 040	Effluent treatment sludge from the production unit of phorate	(T)
K 041	Effluent treatment sludge from the production plant of toxaphene	(T)

Industry and EPA hazardous wastes number	*Hazardous waste*	*Hazard code*
K 098	Untreated process effluent from the production unit of toxaphene	(T)
K 042	Heavy ends of distrillation residues from the distillation of tetrachlorobenzene in the process of manufacture of 2,4,5-T	(T)
K 043	2,6-Dichlorophenol waste from the production of 2,4, D	(T)
K 099	Untreated effluent from the production of 2,4,D	(T)
K 123	Process effluent from the production of ethylene-bisditicarbomic acid and its salt. The process effluent includes suppermates, filtrates and wash waters	(T)
K 124	Reactor vent scrubber water from the production of ethylene-bisdithiocarbomic acid and its salts	(C, T)
K 125	Filtration, evaporation and centrifugation solids from the production plant of ethylenedithiocarbomic acid and its salts	(T)
K 126	Baghouse dust and floor sweepings in milling and packaging operations from the production and formulation of ethylenedithio-carbomic acids and its salts	(T)
Explosives		
K 044	Effluent treatment sludges from the manufacturing and processing of explosives	(R)
K 045	Spent carbon from the treatment of effluent containing explosives	(R)
K 046	Effluent treatment sludges from manufacturing, formulation and loading of lead-based initrating	(T)
K 047	Pink/red water from trinitrotoluene operations	(R)
Petroleum refining		
K 048	Dissolved air floatation (DAF) float from the petroleum refinining unit	(T)
K 049	Slop oil emulsions from the petroleum refining unit	(T)
K 050	Heat exchanger bundle cleaning sludge from the petroleum refining unit	(T)
K 051	API separator sludge from the petroleum refining unit	(T)
K 052	Tank bottoms (loaded) from the petroleum refining unit	(T)
Iron and steel		
K 061	Emission control dust/sludge from the primary production of steel in electric furnaces	(T)
K 062	Spent pickle liquor generated by steel finishing operations of the facilities within the iron and steel industry (510 codos 331 and 332)	(T)

Industry and EPA hazardous wastes number	*Hazardous waste*	*Hazard code*
Secondary lead		
K 069	Emission control dust/sludge from secondary lead smelting	(T)
K 100	Waste leaching solution from acid leaching of emission control dust/sludge from secondary lead smelting	(T)
Veterinary pharmaceuticals		
K 084	Effluent treatment sludges generated during the production of veterinary pharmaceuticals from arsenic or organoarsenic compounds	(T)
K 101	Distillation tar residues from the distillation of aniline-based compounds in the manufacture of veterinary pharmaceuticals from arsenic or organoarsenic compounds	(T)
Ink formulation		
K 086	Solvent washes and sludges, caustic washes and sludges or water washes and sludges from cleaning tubs and equipment employed in the formulation of ink, from pigments, droers, soaps and stabilizers containing chromium and lead	(T)
Coking		
K 060	Ammonia still lime sludge from coking operations	(T)
K 087	Decanter tank tar sludge from coking operations	(T)

Hazard code definitions

C denotes corrosivity

E denotes extraction procedure toxicity

H denotes acute hazardous waste

I denotes ignitability

R denotes reactivity

T denotes toxic waste

References

Alan, P. Rossister, Dennes Spriggs, H. and Howard Klee Jr., *Chemical Engineering Progress*, January 1993.

Arthur, C. Stern (Ed.), *Engineering Control of Air Pollution*, Vol. IV, 3rd ed., Academic Press, New York, 1977.

Azad, Hardom Singh, Editor-in-Chief, *Industrial Wastewater Management Handbook*, McGraw Hill, New York, 1976.

Berglund, R.L. and Snyder, G.E., *Hydrocarbon Processing*, April 1990.

Bill, T. Ray, *Environmental Engineering*, PWS Publishing Company, New Delhi, 1995.

Blackman Jr., W.L., *Basic Hazardous Waste Management*, Lewis Publishers, Boca Raton Flo, 1993.

Bridgewater, A.V. and Mumford, C.J., *Waste Recycling and Pollution Control Handbook*, Von Nostrand Reinhold Environmental Engineering Series, 1974.

Butner, R.S., Pollution Prevention in Process Development and Design, *Industrial Pollution Prevention Handbook,* H.M. Freeman, Ed., McGraw Hill, New York, 1995.

Calvert, Seymour and Harold, M. England (Ed.), *Handbook of Air Pollution Technology*, John Wiley & Sons, New York, 1989.

Carl, H. Tromm, Michael, S. Callahan, Harry, M. Freeman and Maruin Drabkin, *Chemical Engineering*, September 1987.

Chad Scott, P.E. and Shishir Mohan, *Chemical Engineering Progress*, August 2002.

Chadha, Nick and Charles S. Parmele, *Chemical Engineering Progress*, January 1993.

Chanlett, E.T., *Environmental Protection*, McGraw Hill, New York, 1979.

Cheremisinoff, P.N. and Richard, A. Young, *Pollution Control Practice Handbook*, McGraw Hill, New York, 1990.

Cheremisinoff, P.N., *Encyclopedia of Environmental Technology*, Vol. 4., Hazardous Waste Contaminant and Treatment, McGraw Hill, New York, 1990.

Cheremisinoff, Paul, N. and Richard A. Young, *Pollution Engineering Practice Handbook*, McGraw Hill, New York, 1990.

Chester, W. Spicer, et al., *Hazardous Air Pollutant Hand Book—Measurement Properties and Fate in Ambient Air*, Lewis Publishers, Michigan, 1995.

Cooper, C.D. and Alley, F.C., *Air Pollution Control: A Design Approach*, Waveland Press, Prospect Heights, Illinois, 1996.

Culp, Gordon, George Wasner, Robert Williams and Mark, V. Hughes Jr., *Wastewater Reuse and Recycling Technology—Pollution Technology Review-72*, Noyes Data Corporation, New Jersey, 1980.

Curl, Kriton (Ed.), *Treatment and Disposal of Liquid and Solid Industrial Wastes*, Pergamon Press, New York, 1980.

Davis, M.L. and Cornwell, D.A., *Introduction to Environmental Engineering*, McGraw Hill, New York, 1991.

De Renzo, D.J. (Ed.), 'Pollution Control Technology for Industrial Wastewater', *Pollution Technology Review*, No. 80, Noyes Data Corporation, New Jersey, 1981.

Down, R.D., *Pollution Prevention through Process Design in Industrial Pollution Prevention Handbook*, H.M. Freeman (Ed.), McGraw Hill, New York, 1995.

Edmund, B. Besselieve P.E., *The Treatment of Industrial Wastes*, McGraw Hill, New York, 1993.

Edwin, R. Bennett, 'Air and Water Pollution', *Proceedings of the Summer Workshop*, University of Colorado, Adam Hilger, London, August 1970.

Freeman, H.M. (Ed.), *Standard Handbook of Hazardous Waste Treatment and Disposal*, McGraw Hill, New York, 1989.

Fresenius, W., Schneider, W., Bobnhe, B., and Poppinghaus, K., *Waste Water Technology, Origin, Collection, Treatment and Analysis of Wastewater*, Springer-Verlay, Berlin, 1991.

Gavaskar, L., *Process Equipment for Cleaning and Degreasing in Industrial Pollution Prevention Handbook*, H.M. Freeman (Ed.), McGraw Hill, New York, 1995.

Gray, N.F., *Water Technology—An Introduction for Environmental Scientists and Engineers*, Viva Books, New Delhi, 1999.

Gregory, G. Bond, *Chemical Engineering Progress*, September 1999.

Halverson, R. and Ruby, M.G., *Benefit-cost Analysis of Air Pollution Control*, Lexington Books, Lexington, Mass, 1982.

Hammer, Mark, J., *Water and Wastewater Technology*, John Wiley & Sons Inc., 1998.

Henry, J.G. and Gary, W. Heinke, *Environmental Science and Engineering*, Prentice-Hall Inc., 1996.

Hesketh, H.E., *Understanding and Controlling Air Pollution*, Ann Arbor Science, Ann Arbor Michigan, 1974.

Higgins, T., *Hazardous Waste Minimization Handbook*, Nt Lewis Publishers, Chelsea, 1989.

Homer, W. Parker, *Air Pollution*, Prentice-Hall Inc., New Jersey, 1977.

Hunt, G.E., 'Overview of waste reduction techniques leading to pollution prevention,' *Industrial Pollution Prevention Handbook*, H.M. Freeman (Ed.), McGraw Hill, New York, 1995.

International Finance Corporation, Environmental, Health and Safety Guidelines Petroleum-based Polymers Manufacturing, World Bank Group, April 2007.

Irvin, Gerard Kiely, *Environmental Engineering*, McGraw Hill International Editions, Irwin, McGraw Hill, 1998.

Irvin, Gerard Kiely, *Environmental Engineering*, McGraw Hill, New York, 1985.

James, G. Mann and Liu, Y.A., *Industrial Water Reuse and Wastewater Minimization*, McGraw Hill, New York, 1985.

Karoll, Maro, *Chemical Engineering Progress*, November 2003.

Kemmer, F.N., *The Nalco Water Handbook*, McGraw Hill, New York, 1979.

Kenneth, L. Mulholland and James, A. Dyer, *Chemical Engineering Progress*, May 1999 and January 2000.

Kenneth, L. Mulholland, *Chemical Engineering Progress*, June 2003.

Kenneth, Noll, *Industrial Air Pollution Control*, Ann Arbor Science, Michigan, 1996.

Krieth, Frank, 'Air and Water Pollution', *Proceedings of the Summer Workshop*, University of Colorado, Adam Hilger, London, August 1970.

Krofchak, Davin and Neil Stone, J., *Science and Engineering for Pollution-Free Systems*, Ann Arbor Science Publishers Inc., Michigan, 1975.

Kuhre, W.L., *Practical Management of Chemical and Hazardous Wastes*, Prentice-Hall, New Jersey, 1995.

Lagrega, M.D., Buckingham, P.L. and Evans J.C., *Hazardous Waste Management*, McGraw Hill, New York, 1994.

Lee, C.C. and Shun Dar Lin, *Handbook of Environmental Engineering Calculations*, McGraw Hill, New York, 2000.

Long, R.B., *Separation Process in Waste Minimization*, Marcel Dekker, New York, 1995.

Masters, G.M., *Introduction to Environmental Engineering and Science*, Prentice-Hall, New Jersey, 1991.

Metcalf and Eddy Inc. Revised by George Tchobanoglous and Franklin L. Burton, *Wastewater Engineering—Treatment, Disposal and Reuse*, 3rd ed., McGraw Hill Inc., 1991.

Metclaff and Eddy Inc. Revised by George Tchobanoglous, *Wastewater Engineering, Treatment, Disposal, Reuse*, 2nd ed., Tata McGraw Hill, 1979.

Moores, C.W., *Before Environmental Assessment—Checklist*, Hydrocarbon Processing, March 1977.

Nancy, J. Sell, *Industrial Pollution Control—Issues and Techniques*, Van Nostrand Reinhold Co, New York, 1981.

Nelson, K.E., *Hydrocarbon Processing*, Houston, TX, March 1990.

Nemerrow, N.L., *Liquid Wastes of Industries—Theories, Practices and Treatment*, Addison-Wesley Inc., New York, 1971.

Noyes Development Corporation, *Air Pollution Control Processes and Equipment*, London, 1968.

Osantowski, R.I., Liello Jc. and Applegate C.S., 'Generic Pollution Prevention,' *Industrial Pollution Prevention Handbook*, H.M. Freeman (Ed.), McGraw Hill, New York, 1995.

Pandey, G.N. and Corney, G.C., *Environmental Engineering*, Tata McGraw Hill, New Delhi, 1993.

Patterson, James W., *Industrial Wastewater Treatment Technology*, 2nd ed., Butterworths, New York, 1985.

Pavoni, J.L., Heerjr, J.E. and Hagerty, D.J., *Handbook of Solid Waste Disposal—Materials and Energy Recovery*, Van Nostrand Reinhold, Environmental Engineering Series, New York, 1975.

Peavey, H.S., Rowe, D.R., and Tchobanoglous, G., *Environmental Engineering*, McGraw Hill, New York, 1985.

Peirce, Jeffrey J., Rutb F. Woiner, Aarne Visilind, A., *Environmental Pollution and Control*, Butterworth-Heinemann, LA, 1998.

Pertins, H.C., *Air Pollution*, McGraw Hill, New York, 1974.

Quano, E.A.R., Lohani, B.N. and Thanh, N.C., *Water Pollution Control in Developing Countries*, Proceedings of the International Conferences, Bangkok, Thailand, 1990.

Rao, J.J., *Chemical Engineering Progress*, November 2001.

Reynolds, T.D., *Unit Operations and Processes in Environmental Engineering*, PWS Publishers, Boston, 1982.

Richard, A. Jacobs, *Chemical Engineering Progress*, June 1991.

Robert, A. Kabrn, *Chemical Engineering Progress*, July 1991.

Robert, W. Rittmeyer, *Chemical Engineering Progress*, May 1991.

Romalho, R.S., *Introduction to Wastewater Treatment Processes*, Academic Press, New York, 1977.

Ronald, E. Bartleft, *Wastewater Treatment*, Applied Science Publishers, London, 1971.

Roy, M. Harrison (Ed.), *Pollution—Causes, Effects and Control*, 2nd ed., Royal Society of Chemistry, London, 1989.

Sawyer, C.N. and McCarty, P.L., *Chemistry for Environmental Engineering*, McCraw Hill, New York, 1978.

Seinfeld, J., *Air Pollution*, McGraw Hill, New York, 1975.

Shen, T.T., *Industrial Pollution Prevention Handbook*, Springer-Verlag, Berlin, 1995.

Sincero, Arcadio, P. and Sincero, Gregoria, A., *Environmental Engineering—A Design Approach*, Prentice-Hall of India, New Delhi, 2002.

Singh, Vijay P. and Ram Narayan Yadav, *Environmental Pollution*, Allied Publishers, New Delhi, 2003.

Smith, R., *Wastewater Minimization in Waste Water Minimization through Process Design*, A.P. Rossiter (Ed.), McGraw Hill, New York, 1995.

Stephen, Adler F., *Chemical Engineering Progress*, April 2001.

Stumm, W. and Morgan J.J., *Aquatic Chemistry, An Introduction Emphasizing Chemical Equlibrium in National Waters*, John Wiley & Sons, New York, 1981.

Tchobanoglous, G., Theisen, H. and Eliassen, R., *Solid Waste Engineering Principles and Management Issues*, McGraw Hill, New York, 1977.

Terry, L. Collet, *Chemical Engineering Progress*, December 1991.

Thom, J. and Higgins, T., *Solvents Used for Cleaning, Refrigeration, Firefighting and Other Uses in Pollution Prevention*, T. Higgins (Ed.), CRC Press, Boec Roton, Florida, 1995.

Thomas, S.T., *Facility Manager's Guide to Pollution Prevention and Waste Minimization, Bureau of National Affairs*, Washington DC, 1945.

Van Raven Swaay, E.C., Adapted from US, EPA Profile of Petroleum Refining Industry, EPA, 310-R-95-013, North Carolina, September 1995.

Verschuereron, K.A., *Handbook on Environmental Data on Organic Chemicals*, Van Nostrand, New York, 1983.

Vesilind, P.A., Peirce, J.J. and Wooner, R.F., *Environmental Engineering*, Butterworth, Boston, 1988.

Waste, R.C., *CRC Handbook of Chemistry and Physics*, Chemical Rubber Co., Cleveland, Ohio, 1975.

Wastewater Engineering, Treatment, Disposals and Reuse, 2nd ed., URBAIR, 1993.

Wentz, C.A., *Hazardous Waste Management*, McGraw Hill, New York, 1989.

Wesley Eckenfelder, W., Jr., *Industrial Water Pollution Control*, 2nd ed., McGraw Hill, New York, 1989.

Williamson, H., *Fundamentals of Air Pollution*, Addison-Wesley, New York, 1973.

Young, Murray Moo and Grahame, J. Farquhar (Ed.), *Waste Treatment and Utilization Theory and Practice of Waste Management*, Pergamon Press, 1979.

Index